Biologically-Inspired Systems

Volume 17

Series Editor

Stanislav N. Gorb, Department of Functional Morphology and Biomechanics,
Zoological Institute, Kiel University, Kiel, Germany

Motto: Structure and function of biological systems as inspiration for technical developments.

Throughout evolution, nature has constantly been called upon to act as an engineer in solving technical problems. Organisms have evolved an immense variety of shapes and structures from macro down to the nanoscale. Zoologists and botanists have collected a huge amount of information about the structure and functions of biological materials and systems. This information can be also utilized to mimic biological solutions in further technical developments. The most important feature of the evolution of biological systems is multiple origins of similar solutions in different lineages of living organisms. These examples should be the best candidates for biomimetics. This book series will deal with topics related to structure and function in biological systems and show how knowledge from biology can be used for technical developments in engineering and materials science. It is intended to accelerate interdisciplinary research on biological functional systems and to promote technical developments. Documenting of the advances in the field will be important for fellow scientists, students, public officials, and for the public in general. Each of the books in this series is expected to provide a comprehensive, authoritative synthesis of the topic.

More information about this series at http://www.springer.com/series/8430

Adam Gadomski

Editor

Water in Biomechanical and Related Systems

 Springer

Editor
Adam Gadomski
Department of Physics
UTP University of Science and Technology
Bydgoszcz, Poland

ISSN 2211-0593   ISSN 2211-0607   (electronic)
Biologically-Inspired Systems
ISBN 978-3-030-67226-3   ISBN 978-3-030-67227-0   (eBook)
https://doi.org/10.1007/978-3-030-67227-0

This Springer imprint is published by the registered company Springer Nature Switzerland AG
The registered company address is: Gewerbestrasse 11, 6330 Cham, Switzerland

Preface: Various Faces of Water

This book contains a survey of versatile studies on water as it is seen in different contexts pertinent to selected biomechanical and related systems. The contexts basically involve three types of distance (also force/pressure and energy) scales, ranging from a microscopic scale over a mesoscopic (or intermediate) scale to a macroscopic scale.

Water itself is a vivid subject of permanent debate on its molecular, aggregation and phase-transition involving properties. It is also a medium that expresses many colligative as well as synergistic properties, such as those ranging from contribution to facilitated lubrication of articulating joints (especially on supramolecular level of biomatter organization) to adhesion of suction cups.

The contributions collected by the book have included:

(i) General discussions about sense and nonsense about water (a critical and knowledge-refreshing study), and water nanoclusters in cosmology, astrobiology, the RNA world, and biomedicine – a future-prospect study.

(ii) Microscopic scale involving considerations as regarding: solvent-induced effects on protein folding, protein intramolecular and solvent bonding in a major synovial fluid component, water behavior near the lipid bilayer, and water molecules organization surrounding ions, amphiphilic protein (viz hyaluronan) residues.

(iii) Mesoscopic scale addressing studies on: pathological water science (a critical and on-experiment-based study), a powdery mildew fungus turning oak leaf surface to the highly hydrophobic state (toward superhydrophobicity), physics and/or biomimetics of suction cups in air and in water, water transport through synthetic membranes as a biomimetic inspiration coming from biological membranes, and travelling waves connected to blood (having an appreciable content of water) flow and motion of arterial walls.

(iv) Macroscopic scale unravelling case studies on: fractal properties of flocs, filtration cakes and biofilms in water, and wastewater treatment processes; soil
hydrology; pollution and patterns of water stock; and water on livestock. The
two latter studies seem to suit partly for decision- and policy-makers; of course,
ecologists can be invited for reading it too.

The collection of (i)–(iv) determines a real space for looking at scaling properties
of the water containing biosystems. The present monograph has found it useful and
practical, as it has been illustrated as well as argued by a current overview
contained in it. This overview (Chap. 1) gives in a concise way an insight into the
subject matter of the book. The main thought is that "water is both readily percolating and predictably interacting agent." At a molecular level it is a dipole, and it takes
preferentially on a tetrahedral form. It interacts with biosurfaces, and accommodates its "adhesive layer" to be structured viz ordered. It flows readily and quite
predictably, being accepted as a Newtonian fluid. Water molecules and their clusters
as well as volumetric contributions (at the macroscale) undergo principal dynamics'
laws: such as that of Newton (at macroscale), the one by Coulomb, contributing
preferentially to the electrostatic interactions, as well as that of (attractive and quantum-mechanics invoking) London-Van der Waals interaction set.

It is worth unveiling that the book as a whole is intended to help find physical
scale (or distance vs. force/pressure and energy) dependent solutions that will be
capable of transferring the detailed knowledge accumulated about the subject of the
role of water for biosystems' mechanics and the related (percolation addressing)
issues into biotechnologically efficient and biomimetically well-posed solutions for
the future.

Bydgoszcz, Poland   Adam Gadomski

Contents

Chapter 1
Current Overview on the Role of Water in Biomechanical and Related Systems

Adam Gadomski

Abstract The monograph puts emphasis on a superior role of water in biosystems exposed to a mechanical stimulus of versatile nature. It is well known that water plays an extraordinary role in our life. It feeds our (mammalian) organism after distributing over its essentially whole volume to support certain physiological and cognitive processes to mention but two of them, both of extreme relevance.

The water content, not only in the mammalian organism but also in other biosystems such as whether those of soil which is equipped with microbiome or the ones pertinent to plants or livestock, having their own natural network of water vessels or routes, is always subjected to a force field.

The decisive force field applied to the biosystems makes them biomechanically agitated irrespective of whether they are subjected to external or internal force-field (or, pressure) conditions. It ought to be noted that the decisive mechanical factor shows up in a close relation with the space-and-time scale in which it causes certain specific phenomena to occur.

The scale problem, emphasizing the range of action of gravitational force, thus the millimeter or bigger force vs. distance scale, is supposed to enter the so-called macroscale approach to water transportation through soil or plant root system, or when feeding the livestock subject to farmer's control. It is related to a percolation problem, which assumes to properly inspect the random network architecture assigned to the biosystems invoked. The capillarity conditions turn out to be of prior importance, and the porous-medium and/or membrane/filter effect has to be treated, and solved in a fairly approximate way.

The deeper the scale is penetrated by a force-exerting and hydration-causing agent the more non-gravitational force fields manifest. This can be envisaged in terms of the corresponding thermodynamic forces, and the phenomena of interest are mostly attributed to suitable changes of the osmotic pressure. In low Reynolds

A. Gadomski (✉)
Group of Modeling of Physicochemical Processes, Department of Physics,
UTP University of Science and Technology, Bydgoszcz, Poland
e-mail: agad@utp.edu.pl

A. Gadomski (ed.), *Water in Biomechanical and Related Systems*,
Biologically-Inspired Systems 17, https://doi.org/10.1007/978-3-030-67227-0_1

number conditions, thus mostly in the (sub)micrometer distance-scale zone, they are related with the corresponding viscosity changes of the aqueous, e.g. cytoplasmatic solutions, of semi-diluted and concentrated (but also electrolytic) characteristics. For example, they can be observed in articulating systems of mammals, in their skin or eyes, and to some extent, in other living beings, such as lizards, geckos or even certain invertebrates. Through their articulating (pre)devices an external mechanical (or, piezoelectric) stimulus is transmitted from macro- to nanoscale.

The contents of the monograph can be distributed twofold. First, the biomechanical mammalian-type (or, similar) systems with extraordinary relevance of water for their functioning, and/or responding to some mechanical sometimes very subtle, or intimate action, will be presented, also including a presentation of water itself as a key physicochemical or biomechanical "nanoclustering agent". Second, the suitably chosen related systems, mainly of soil/natural biofilms vs. plant or livestock, also water-stock addressing origins, will be examined thoroughly. As a common denominator of all of them, it is proposed to look at their hydrophobic and/or (de) hydration effects, and how do they impact on their basic mechanical (and related, such as chemo-mechanical, etc.) properties. An additional tacit assumption employed throughout the monograph concerns, albeit put as an open problem of statistical scalability of the presented nano- via meso- toward macroscale embracing biosystems (which is equivalent to be taken for granted), a certain similarity between local and global system's properties, mostly those of biomechanical nature. Moreover, for enhancing and broadening a comprehension of the material collected, an artistic view, emphasizing the role of water in the systems under study, has been proposed.

Keywords Water distribution · Water scale dependent properties · Water phases · Water faces · Pathological science on water · Water containing interface · Force-pressure conditions · Macroscopic scale · Mesoscopic scale · Microscopic scale · Transmission of water contacts over space-time scales

1   Introduction

Water is a physicochemical component of formidable significance for functioning of the living creatures. This *ad hoc* definition promises much about involvement of water in constituting many essential properties of the systems to be named complex biophysical and/or biochemical systems.

One of the unifying features of the complex systems, except of their flow vs. aggregation as well as nonlinear propensities, appears to be that they are scalable in a useful albeit statistical way. This way calls upon the notion of fractality or statistical self-similarity; indeed this is a main concern of one of the chapters included in the monograph (Gorczyca this volume), unveiling fractal properties of flocs, filtration cakes and biofilms, especially in wastewater treatment processes. Of course, the so-called scaling concept, thus, the one bearing the fractality, and/or including the fractality notion, is implicitly but sometimes also explicitly (Strzelewicz et al. this volume) mentioned in many other chapters of the monograph, ranging from macroscale (M-scale), passing the (intermediate) mesoscale (mo-scale), and ultimately finishing at the molecular microscale (m-scale), with a special emphasis placed on the nanoscale (nm-scale).

Based on the length- and/or distance-addressing division into three scales mentioned, thus together M-mo-m scales, one should not expect to much from such fairly unsharp division but rather ought to accept that at least for the force fields (or equivalently, pressure fields) applied to the systems under consideration (the biomechanical and some related) the entities' interaction distances may play a certain distinguishable roles.

In this unique-in-spread/reach monograph, it is proposed that it shall commence with a small treatise on water itself, presented in a chapter by (Yuvan and Bier this volume), dealing with the sense and nonsense of water as a physicochemical component. After this chapter, it is proposed to take a deep breath, thus to explore the scales' survey by means of a chapter by (Johnson this volume) who offered a treatment of nanoclusters in cosmology (M-scale), astrobiology (mo-scale), while ultimately landing on the RNA level (m-scale), provided that the universe can be treated as a biosystem. For an artistic depiction of this part of the monograph, see Fig. 1.1.

What applies then for the remaining chapters can be juxtaposed as follows. First, let us start from the m-scale, giving the floor to an Israeli researcher Arieh Ben-Naim (this volume) with his study on solvent/water induced effects on protein folding. Second, let us go on with the monograph contents, as expressed in terms of the m-scale, and invoke an analysis of intramolecular and solvent bonding in sonovital fluid components by a chapter (Bełdowski et al. this volume). Third, when trying to be more specific, let us recommend the next chapter by (Kruszewska et al. this volume) on water behavior near the lipid bilayer. Next, it is appropriate to introduce another m-scale involving chapter by (Siódmiak this volume) on water molecules organization when surrounding ions, parts of amphiphiles, and biomolecules. For an artistic view of this part of the monograph, see Fig. 1.2.

The mo-scale, thus the intermediate one, is represented by the following studies. (By the way, do inspect an artistic depiction of this part of the monograph, Fig. 1.3.) Namely, one can read the book here by comprehending a chapter by (Elton and Spencer this volume) on several examples of pathological water science as exemplified, e.g. by water exclusion zone (EZ) or water under magnetic field, then going over three biomimetics' oriented and nature solutions' addressing studies by (Gorb and Gorb this volume), then by (Tiwari and Persson this volume), and also by (Strzelewicz et al. this volume), not avoiding to complete the list by (Dimitrova and Vitanov this volume) on blood flows. In (Gorb and Gorb this volume) one deals with how a powdery mildew fungus turns oak leaves to a highly hydrophobic state. In the chapter by (Tiwari and Persson this volume) one enters the proposed principia of suction cusps in air, and contrastively, in water. In the chapter by (Strzelewicz et al. this volume) one can have a look at water transport through synthetic membranes as

Fig. 1.1 Artistic depiction of a general omnipresence of dynamic states of water in the universe (blue colors) and the water molecules when grouped in clusters or staying alone, and ranging from a microscopic (atomic/molecular) scale to a galaxy scale of water involvement. Here, an artistic imaginative presentation of the Andromeda galaxy (middle curled colorful stain), also known as Messier 31 or M31 is given; it is located about 2.4 million light-years away from Earth. (It is the largest galaxy in the Local Group, which also contains our own Milky Way Galaxy and more than 50 other galaxies.) Based on an artist's free imagination, it is supposed that it may contain nanoclusters of water, see (Johnson this volume). The water nanoclusters are purposely exaggerated to some equilibrium or honeycomb forms of the polywater (Yuvan and Bier this volume; Elton and Spencer this volume). The living creatures such as bees (with their own RNA codes), our honey-producers, are figuratively stuck to the hexagonal clusters; the latter is somehow allowed to overlap with certain (light or dark) stars from the galaxy which all-in-all completes the free-of-scale fantasy invented for unveiling in an informal way the message of the first general part of the monograph; see Acknowledgements

Fig. 1.2 Artistic view of dynamics of water adjacent to a compartment of a lipid bilayer (biomembrane), cf. (Kruszewska et al. this volume), as equipped with two oversimplified protein channels, capable of transporting (macro)ions through them. In certain neighborhoods of the channel throats two snake-like, presumably unfolded (Ben-Naim this volume), biomolecules of different sizes have been allocated (Bełdowski et al. this volume), showing virtually up their ability to either penetrate the channels or to make them closed/inactivated. Some "free" water molecules represent symbolically which is the basic medium of prior relevance here, see (Siódmiak this volume). According to an artist's imagination, different colors of water, dividing it into cell-internal (cytoplasmatic) and cell-external parts, are to be applied to the overall system, making it sensitive to differences between internal and external osmotic pressures (Siwek et al. this volume); see Acknowledgements

Fig. 1.3 Imagination of an artist is embodied in this picture that by neglecting the micro- vs. macro-world proportions, lands perfectly in the mo-scale introduced. The omnipresence of water, envisaged in blueish colors, may create some exclusion zones of it (Elton and Spencer this volume), for example, in a cell seen at the left-hand side by means of a magnifying glass; the cell itself, having a tendency more to chemotaxis expression (Elton and Spencer this volume) than to follow a free fall, ultimately "intends" to reach a point-like group of bacteria beneath, or alike. A symbolic tree expressed behind, with its leaves of light- vs. dark-green colors, refers to a mildew's problem applied to the leaves (Gorb and Gorb this volume) under some deficit of water. The lizard

inspired by their natural, biological counterparts, see Fig. 1.2. Similarly to the two preceding chapters, the biomimetic conception is also expressed here in a clear and attractive way. The next chapter within this category is presented by (Dimitrova and Vitanov this volume), and describes by means of a numerically addressed but non-linear analytic approach how the waves of blood go through the blood arteries.

The M-scale involvement, expressing the distance-largest scale, consists of the following four chapters. First, in the afore mentioned (Gorczyca this volume) a summary of major developments are presented, and knowledge gaps regarding fractal properties of particle aggregates, deposits (cakes) and biofilms formed in the water and wastewater treatment processes are considered. Moreover, similarities and differences between water treatment and wastewater treatment flocs, deposits and biofilms are emphasized. Second, in (Futo and Bodnar this volume) the water uptake mechanism through the root system in a plant field is discussed. Modern irrigation systems are used to prevent water shortage (water stress) to avoid yield loss, and it is advisable to start irrigation before the onset of the visible symptoms of water deficiency. To support suitable monitoring calculation methods are given to determine the irrigation water requirement. An example on effects of an up-to-date irrigation system on maize yields is offered. Third, the chapter by (Cerqueti this volume) aims at delineating some quantitative viz. stochastic models for the analysis of the two most relevant themes pertinent to the lifecycle of the water. On the one hand, the pollution of water generated by the impact of human activities on biological occurrences is discussed. On the other hand, some arguments on the stock of available water, by including also some details on its possible exhaustion, are revealed. Fourth, in (Siwek et al. this volume) certain issues related to use of water in agriculture, especially in livestock production, are discussed. Water is detected both in every cell of the body, as well as in the intercellular spaces. Moreover, there is a difference between biological (bulk) water and cellular (nonbulk) water. Biological water is any water surrounding a biomolecule that has distinct properties compared to those in the water mass. Cellular water forms an ordered molecular structure. It is surrounded by the molecules mediating its transfer inside the cells. The amount of water in the individual body depends on such factors as the species, gender, age, and body structure.

It is known that in animal physiology, water impacts on all bodily functions, such as thermoregulation, fluid balance, and salt concentration, all of them closely related to a pressure and/or force-field distribution over the animal tissues. For an artistic view of this part of the monograph, refer to Fig. 1.4.

Fig. 1.3 (continued) that got stuck to the human type part of the three may perform its virtual walk by employing certain physical suction's conditions, cf. (Tiwari and Persson this volume), putatively, with and almost without any involvement of water. Natural membranes, such as the white cell membrane (see to the left again) or a part of "porous" soil in which the tree is built in, exemplify the subject addressed by (Strzelewicz et al. this volume). The arteries and pore type routes embodied in the human-like silhouette symbolize a feasibility of transporting both blood and water through them (Dimitrova and Vitanov this volume); see Acknowledgements

Fig. 1.4 An artistic view of the largest scale (M-scale) considered in this monograph. The presented imaginative picture demonstrates by a zooming-in, scalable graphical procedure how, for instance, one is able to pick up a geographical region from South America, then to focus on a landscape detail of the region, having involved a cow, utilizing in terms of food the presence of grass around it (Siwek et al. this volume), and around a lake (Cerqueti this volume); the grass field is naturally well irrigated (Futo and Bodnar this volume) because of its neighborhood to a lake. The cow itself consists of cells being likely well fed by water and the grass-based food. At the bottom of the lake, see white point-like objects certain material deposits and wastes are located (Gorczyca this volume). The blueish colors express an overwhelming role of water (and, its "curly" nurishing dynamics) in the life cycle; see Acknowledgements

In the remaining part of this overview-type chapter, the three scales, thus, the overall M-mo-m scale shall be considered in depth based on what the chapters' authors were able to disclose in order to fulfil the principal task of the monograph, that means, which is the basic role of water in biomechanical and related systems.

2 General and Small-to-Large Scale Considerations on Water Behavior in Biosystems

We shall start with two chapters by (Yuvan and Bier this volume) and by (Johnson this volume). In what follows, let us provide arguments speaking for ordinary and extraordinary role of water; however, the attribute of extraordinariness is given a broad and acceptable, just "quite ordinary" meaning here. Both presented chapters advocate for the fact that the notion of water evokes a huge stream of emotions. It is due to emergence of mishaps, also called "pathologies" (Elton and Spencer this volume), that are a bit sarcastically described in (Yuvan and Bier this volume), quoting imaginatively in their conclusions an analogy to a case of a Polish postal clerk from the so-called interwar times. His opportunistic and crude social behavior has been turned by the upper society, in analogy with attitudes of some well-ranked researchers of our times, to a well-posed and far-seeing attitude, as a social phenomenon, reminiscent of high level research on water. In short, new conceptions on water, such as polywater (Yuvan and Bier this volume) or memory-of-water effect, or their "unusual" magnetic properties (Elton and Spencer, this volume), ought to be so sound that they have to perform a career as if they were at a Nobel type research level, thus, of great impact on the society as a whole. Commenting thoroughly on those mishaps on water concept, or specifically a general misconception rather, they describe solely the typical properties of water, also those pertaining to water phase transitions of first order, concluding finally that "sometimes water/solvent is just the solvent" (and, nothing more).

The chapter by (Johnson this volume) assumes that the universe is a biosystem, and that it contains ubiquitous water nanoclusters, originating from in-space-time frequent collisions of interstellar pieces of matter, including a certain ice amount at their surfaces, with other material pieces of the type, equipped sometimes also with biomolecular adhesives, containing RNA molecules, the attributes of life, cf. Figs. 1.1 and 1.2. The proposed fairly hypothetical, molecular-physics and quantum-mechanics addressing conceptual framework presented by (Johnson this volume) contributes to broadening immensely our view of cosmological or bio-astrophysical alternative of testing novel phenomena, including RNA world and viruses, as seemingly equally plausible explanations on the origins of the universe, and those on first emergence of life in the universe, extended beyond its extraterrestrial existence; all of that with a predominant role of water, herein based on its solid state.

When relating the contents of both chapters presented above to the main topical motivation of the monograph, one can see the role of water in the overall m-mo-M scale, commencing the whole story with its dipolar and single-molecular properties (Yuvan and Bier this volume), then going over its bulky vs. non-bulky characteristics, while ultimately conveying the rationale toward its role in the cosmological scale, Fig. 1.1. The latter is always full of dynamic collisions, expressing themselves in a high force-field impact on the colliding matter, as well as characteristic of large magnitudes of the pressures coming out from the collisions in the large space-time scale (Johnson this volume). It is also instructive that in both chapters one can find descriptions of phenomena worth mimicking or worth making them reproducible in laboratories, in order to aiming at their biotechnological and biomedical applications. As also written above, one may inspect water addressing mishaps, firstly to disregard them from any practical point of view but virtually to learn from them toward what went wrong with their enforcement into research, and which were the basic grounds of doing so. Two figures by (Yuvan and Bier this volume) clearly demonstrate that a boom on polywater investigations is already finished more than 40 years ago (as exemplified by a logistic curve) whereas a research on "magnetic water" is still ongoing, following quite readily an exponential, thus, rapid path of occurrences of studies, lasting already over more than 30 years.

For a synthetic artistic depiction of the state-of-the-art presented by this section, see Fig. 1.1; an anticipation of having the objects' proportions as devoid of length and/or distance scale is necessary for a full comprehension thereof.

3 Biosystems' Scales of Interest

As mentioned in the preceding sections, there are three scales of utmost concern: the smallest (m), an intermediate (mo) and the biggest (M). Let us start our considerations from the m-scale.

3.1 m-scale

The m-scale is represented by four chapters. These are the following: (i) the study on protein folding in aqueous environments by (Ben-Naim this volume); (ii) a computer simulation by (Bełdowski et al. this volume), mimicking an amphiphile behavior in sonovital milieus; (iii) a computer-simulation supported analysis of interactions of water with lipids (Kruszewska et al. this volume), and (iv) complex behavior of water as unveiled in terms of radial distribution functions, characteristic of aqueous biopolymeric systems (Siódmiak this volume).

The chapter by Ben-Naim (this volume) is about how water effects on protein folding. The conveyed argumentation line relies on having a close inspection on hydrophilic contribution to the molecule conformation against the perennially alive hydrophobic effect on the intramolecular interactions. It is then rationalized that when protein falls in an unfolded state, it occupies much of the conformation space, exploring a good number of available degrees of freedom, and accumulating many of water molecules as occluding it. If, in turn, the protein becomes folded, then the water molecules are successively exuded from its interior but the overall molecule gets surrounded by water again. It means that water appears to be the main player in the game. As a consequence, it ensues in counting decisively on how water molecules will establish a robust network of bonds and will participate in especially strengthening the pairwise interactions between the hydrophilic groups at a certain distance of a small fraction of nanometers. These arguments in favor of hydrophilicity discourage almost automatically the hydrophobic effect to be dominant, see (Ben-Naim this volume).

In the chapter by (Bełdowski et al. this volume) a clear relevance of water in determining the protein structure, when using a case study with albumin, has been demonstrated, partly corroborating the arguments conveyed in (Ben-Naim this volume). The interactions of amino acids with water are the main factors that render the biomolecule takes on a specific structure. The dynamics of these interactions are dependent on amino acid types, their positions in the chain, and the strength of interactions. As revealed by a molecular dynamics study, the most stable are intramolecular hydrogen bonds, and hydrophobic contacts. Intermolecular bonding with water demonstrates a much less durable lifespan than intramolecular interactions. Amino acids in albumin behave according to their chemical characteristics. Namely, polar and charged amino acids take part in the formation of hydrogen bonds, hydrophobic amino acids, except of lysine, form hydrophobic contacts.

In (Kruszewska et al. this volume) a computer simulation of the water behavior near a lipid bilayer has been performed. It aimed at disclosing that water molecules, when shielding virtually the bilayer from the components of synovial fluid in articular cartilage, could be responsible for triggering different tribological responses of the system by changing the energy barrier for the formation of various interactions. It was observed that a faster diffusive movement of water near the bilayer lowers the energy barrier for the formation and causes breakages of hydrogen bonds between the water molecules, an effect that contributes to a nanoscale view of friction in natural articulating devices. The presence of the water molecules with long residence times at the surface of the bilayer indicates they are strongly hydrogen-bonded and are separated from the bulk water by a large free energy barrier. The interaction and motion investigations performed by the researchers (Kruszewska et al. this volume) have indicated that the water molecules undergo various types of dynamics. Therefore, they can barely be split into three groups: those performing subdiffusional (thus, confined) motions, the ones moving purely diffusionally in the bulk, and the remaining that move superdiffusively due to switchable attraction-repulsion interactions with the movable bilayer.

In (Siódmiak this volume) a molecular dynamics simulation has been performed too, unveiling an organization of water molecules surrounding ions (coming from a salt dissociation), amino acids, and proteins such as hyaluronic acid. The results of molecular dynamics simulations of bulk water have also been provided by means of an observable termed as the radial distribution function, serving to analyze the simulation results. The main interest has been to answer whether the water in the mentioned systems shows up characteristics of an ordered structure. It turns out that there are problems with the results obtained from computer simulations. If one wants to make radial distribution function exhibiting reality, the orientation radial distribution function must be employed. The reason for this is that if one selects the atom or chemical group around which one wants to check the distribution of solvent molecules, the remaining atoms of the molecule obscure the solvent molecules on the opposite side of the biomolecule. It remains then advisable to cope with structural fluctuations and molecular excitations of hydrating water molecules, exploring a broad range of space and time domains.

For an artistic cumulative view of the m-scale systems, described in this subsection in a unified way, uncovering a constructive and detective role of water as applied to the bionanomechanical entities, that is to say, the invoked amphiphiles and their aggregates, see Fig. 1.2. Detailed analyses, mainly of the computer experiments discussed, demonstrate a huge potential to help develop biotechnological or bioinformatic experiments, indispensable to serve for biomedicine, and especially, biomaterials technological development.

3.2   mo-scale

The presentation of mo-scale involving chapters includes five chapters. It commences with the critical (i.e., uncovering pathologies on water examination) chapter by (Elton and Spencer this volume), then contains two hydrophobicity and adhesion discussing chapters by (Gorb and Gorb this volume) and (Tiwari and Persson this volume), and eventually completes the mo-scale approach with two remaining chapters, exploiting (bio)matter transportation over channeling systems and performed by (Strzelewicz et al. this volume), and (Dimitrova and Vitanov this volume).

When reading the chapter by (Elton and Spencer this volume), one finds out that four areas of pathological water science – polywater, the Mpemba effect, Pollack's "fourth phase" of water, and the effects of static magnetic fields on water have been discussed. Some common water-specific issues emerge such as the contamination and confounding of experiments with dissolved solutes and nanobubbles. General issues also emerge such as imprecision in defining what is being studied, bias towards confirmation rather than falsification, and poor standards for reproducibility concerning data originated from investigations on pathological water science.

The chapter by (Gorb and Gorb this volume) discusses microscopical and experimental data that demonstrate the powdery mildew turns the oak leaf surface to a highly hydrophobic state. Intriguingly, the contact angles of polar and non-polar fluids are very high on the leaves covered with the fungal mycelium, which in turn lead to rather high dispersion component of the surface free energy. Since the original wax coverage of intact leaves has strong dispersion (as in favor to polar) component, one may assume that the chemical background of the surface free energy is similar in both circumstances. Put it another way, it means that the fungal surface contains hydrophobic substances that might be similar to the lipophilic coverage of intact leaves. Furthermore, the main differences in contact angles and the surface free energy between intact leaves and those covered by mycelium are presumably based on differences in the surface structure. In fact, the leaves collected in the mid-season bear surfaces covered by three-dimensional wax projections that partially degraded, which is well known for different plant species such as oak leaves.

The chapter by (Tiwari and Persson this volume) argues on the leakage of suction cups both in air and water. The experimental results were analysed using a newly developed theory of fluid leakage valid in diffusive and ballistic limits combined with contact mechanics theory by Persson. In these experiments, the PVC suction cups were pressed against sandblasted PMMA sheets. It is found that the measured failure times of suction cups in air is in good agreement with the theory, except for surfaces with rms-roughness below ca. one micrometer, where diffusion of plasticizer occurred, from the PVC to the PMMA counter-face, ensuing a blockage of critical constrictions. As for experiments in water, it is uncovered that the failure times of suction cup were ca. a hundred times longer than in air, and this could be attributed mainly to the different viscosity of air and water.

In (Strzelewicz et al. this volume), one debates on a description of aquaporins, which are natural channels. Next, the authors' attention was focused on the description of water transport through biological membranes, which are the inspiration for developing new solutions in water transport processes through synthetic membranes. The methods of describing the structure and properties of these membranes were presented. The next part of the work focused on practical examples of using this knowledge to design hybrid membranes applied for ethanol dehydration.

The discovery of the aquaporins explained the selective transport of water through the plasma membranes of cells, while preventing ions from passing through the membrane. The structural models of aquaporins offer remarkable insight into the biophysical functions and give scientists a chance to understand the role of aquaporins in numerous clinical disorders. Aquaporins are involved in some forms of renal vascular diseases, including nephrogenic diabetes insipidus. Aquaporins are also associated with problems of brain edema, loss of vision and with defense against thermal stress. Aquaglyceroporins are involved in the defense against starvation. These proteins are present in the whole natural world.

Natural membranes, like those containing aquaporins are a biomimetic inspiration to develop new artificial membrane materials for different application from desalination to dehydration. The present research evolves in two directions. The first group of studies includes artificial membranes with protein channels, which can be used in filtration processes. Furthermore, research on biomimetic membranes is aimed at replacing aquaporins by imitation synthetic channels. The second approach is based on designing a synthetic hybrid membrane. In these membranes, the role of the channels are the fillings in the form of nanotubes or free areas formed in the polymer matrix as a result of the addition of particles filler.

The chapter by (Dimitrova and Vitanov this volume) is focused on several mathematical results concerning travelling waves connected to arterial wall and blood flow in large arteries, containing virtually structural irregularities in their walls. In order to study these waves the authors used a method for obtaining exact solutions of nonlinear partial differential equations called Simple Equations Method. They present a brief summary of the method and apply it to obtain exact travelling wave solutions of nonlinear partial differential equations which model blood flow pulsations and nonlinearly affected motion of walls of large arteries.

For an artistic depiction of the state-of-the-art presented by this section, cf. Fig. 1.3.

3.3   M-scale

This subsection collects information accumulated about the main findings of four M-scale addressing chapters. First, we wish to start with a chapter by (Gorczyca this volume), when going then over the main arguments presented in chapters by (Futo and Bodnar this volume) and (Cerqueti this volume), and ultimately arriving at the content of the chapter by (Siwek et al. this volume).

The chapter by (Gorczyca this volume) disputes that properties of particle aggregates, filtration cakes and biofilms formed in water/wastewater determine success of most treatment processes. Description of properties of these materials and linking it to the performance of the treatment unit is very challenging. Concepts of fractal geometry with "broken" dimensionalities have been demonstrated to be suitable for characterization of such materials, however, determination of particular fractal dimensions has been proven to be challenging, depending on particular method of determination as well as the type of material. It has been reported that a spectrum of fractal dimensions (grouped into a multifractal analysis), rather than a single fractional number is required to characterize flocs. There is definitely a need for more substantiation in this area, which can have numerous applications in studies of general material structures.

In the chapter by (Futo and Bodnar this volume), it is found that tape drip irrigation of maize is of very low water-use, energy-efficient and generally efficient irrigation technology, which can be a major domestic technical innovation for maize irrigation in future for intensive farming. The yield of maize can be significantly increased by improving the water supply to the plant. In many areas, only little water is available for irrigation. The effect of drip irrigation in their experiment was investigated for maize yields in 2016 and 2017. In 2016, yields increased by 22.3–24.5% compared to the yields of control plots, while the yield gains in the drier year of 2017 reached 46.73–53.46%. In their experiment, the growth of the average yield was economically substantiated (Futo and Bodnar this volume) .

In (Cerqueti this volume), two contexts on the stock of water have been presented. They allow to comprehend the importance of the human activities and of biological and mechanical factors on the vital cycle of this crucial natural resource. On the one hand, this piece of monograph dealt with pollution, and on the other with depletion or a certain lack of water. The presented arguments suggest that the quantitative analysis is capable of fostering good practices when pursuing sustainability. To be specific, decision makers should implement long-term policies for avoiding the irreversible pollution of shallow lakes or the depletion of water in a specific region. The arguments conveyed invoke consciences towards greater responsibility about environmental protection, since the damage caused by external events gets sometimes on irreversible character.

The chapter by (Siwek et al. this volume) unravels issues related to the use of water in agriculture, especially livestock production, which is the major water consumer of the world. Water is found in every cell of the body, as well as in the intercellular spaces, cf. Figs. 1.2 and 1.3. There is a difference between biological (bulk) water and cellular (nonbulk) water, see a discussion in the preceding chapters. At cellular level, water mediates and modulates intermolecular forces, controls the rate of substrate diffusion and conformational changes. In animal physiology, water influences all bodily functions, such as thermoregulation, fluid balance, and salt concentration. Regarding this, water productivity in livestock farming depends to a substantial extent on selection of diets, and the corresponding food production. Since agriculture is the most water-consuming sector of the human activity, the global supply chains are also taken into account when dealing with food security (Cerqueti this volume). It is widely known that the agriculture might be supplied with water in two ways (Fig. 1.4): rainfed or irrigated (Futo and Bodnar this volume). Over the last 50 years the global irrigated area has doubled. Irrigated water competes with water intended for human consumption, as well as producing crops and pulses for human feed. Moreover, the quality of water is seriously affected by the unprecedented climate changes. These and other factors influence agriculture and livestock farming as important sectors of natural production.

For an artistic graphical presentation of the state-of-the-art presented by this subsection, regarding the M-scale argumentation, see Fig. 1.4.

4 Summary

To summarize, it is to be noted first that the underlying monograph aimed at standing for a material that not only is addressed to the community of life scientists and natural scientists, such as physicists, chemists, materials scientists and biologists, but it is equally well intended to reach representatives of (inter)national groups of decision- and policy-makers, to mention but two of them. This explains in part why the argumentation line presented herein is also quite decisively based on the computer graphics unveiled by Figs. 1.1, 1.2, 1.3 and 1.4.

The present author of this chapter wishes to believe that a less rigorous presentation of the material collected by the underlying monograph chapter may result in the so-called longer (than typically) reach of its main thought to a broader milieu of people interested in rational, thus sustainable development of our planet, and the universe, as considered in terms of water treatment and waste, and when regarding the pollution of water too, a substantial ecological concern.

The main thought exploited by the monograph is equivalent to conveying the role and omnipresence of water through a multitude of scales covered by Nature. As long as the subtleties of the scales are switched off the water dynamic behavior is quit foreseeable. If the scale is allowed to manifest with its all subtleties defined as surfaces, interfaces, membranes, cavities, deposits, flocs, and other structural, often fractal-like structural peculiarities, then the water behavior demands a great attention.

As much as the scale-exhibiting dynamic behavior of water molecules, whether in (huge) statistical or in finite quantities, has had an appreciable impact on the overall m-mo-M scalable biosystems, it would merely depend on the mechanical (force and/or pressure) factor. The factor manifests in either explicit or implicit ways; it is an ever present factor, somehow intriguingly hidden behind the force fields tacitly involved in the creation of fantasy engaging subtle chains of sub-effects, yielding the overall depiction of the systems examined by this monograph, as presumably disclosed by those open-minded and popularized artistic views presented by Figs. 1.1, 1.2, 1.3 and 1.4 in the present chapter.

In a final word, let us note that the biomechanical factor manifests as possessing its scalable force-field contribution(s) over the overall M-mo-m scale employed within the monograph. As a simple, albeit instructive example, let us accept that within the M-scale the gravitational force would dominate the system behavior. This force, according to Newton law, scales with a distance R as $1/R^2$. At the mo-scale, in turn, in quite frequent physical circumstances an electric contribution, as represented by the electrostatic force, thus the Coulomb law, would influence the system dynamics; it is also scalable with a distance r as $1/r^2$. Of course, typically $r < R$ applies. In the smallest m-scale Van der Waals force field enters, with its attraction part proportional to $1/l^6$. Here l stands for a mean distance of binary molecular interactions. Typically $l < r < R$ holds. But all distance dependent force fields follow the power laws such $1/L^k$. The exponent k gets either on $k = 2$ (twice) or $k = 6$ when emphasizing the three main force-fields addressed, whereas L indicates the characteristic interaction distance in the overall M-mo-m scale. Put it in short: The (central) force-field characteristics are qualitatively of the same power-law type, obeying $1/L^k$. A simple similarity transformation, making from L another distance, namely aL with a – a real-valued scaling factor, yields the effective force-field expression as being of the same type, namely proportional to $1/(aL)^k$. But it yields an equivalent expression as b/L^k, with $b = 1/a^k$. It then reintroduces the basic property that again the proportionality to $1/L^k$ emerges. Such a simple rationale gives us a privilege to ascertain to a reasonable extent that water dynamic contribution to the entire M-mo-m scale would solely be related with gravitational (M-scale), Coulomb type (mo-scale), and finally, Van der Walls type (m-scale) leading and scalable contributions to the whole biomechanical scenario. Note that the presented basic argumentation line does not discard other contributions, for example, either viscous or hydrophobic, to get in any detailed description of peculiar processes of interest.

Acknowledgements A support by BN-10/2019 is acknowledged. A part of the material has been presented at an international webinar on June 25, 2020, at the UTP University of Science and Technology, Bydgoszcz, Poland, http://zmpf.imif.utp.edu.pl/jsbmw2020/. The author is thankful to the series editor at Springer for helpful advices and useful discussions. The computer graphics has been designed in cooperation with a design studio *https://www.aleksandrabirch.com/* Aleksandra Wesołowska, owner (accessible on January 31, 2021) after translating the scientific contents to a popular graphical style, supposed ultimately to enhance the comprehension of the presented material.

References

Bełdowski, P., Domino, A., Bełdowski, D., & Dobosz, R. (this volume). Analysis of protein intramolecular and solvent bonding on example of major sonovital fluid component. In A. Gadomski (Ed.), *Water in biomechanical and related systems*. Cham: Springer.

Ben-Naim, A. (this volume). Solvent induced effects on protein folding. In A. Gadomski (Ed.), *Water in biomechanical and related systems*. Cham: Springer.

Cerqueti, R. (this volume). External solicitations, pollution and patterns of water stock: Remarks and some modeling proposals. In A. Gadomski (Ed.), *Water in biomechanical and related systems*. Cham: Springer.

Dimitrova, Z. I., & Vitanov, N. K. (this volume). Travelling waves connected to blood flow and motion of arterial walls. In A. Gadomski (Ed.), *Water in biomechanical and related systems*. Cham: Springer.

Elton, D. C., & Spencer, P. D. (this volume). Pathological water science – Four examples and what they have in common. In A. Gadomski (Ed.), *Water in biomechanical and related systems*. Cham: Springer.

Futo, Z., & Bodnar, K. (this volume). Soil hydrology. In A. Gadomski (Ed.), *Water in biomechanical and related systems*. Cham: Springer.

Gorb, E. V., & Gorb, S. N. (this volume). Powdery mildew fungus erysiphe alphitoides turns oak leaf surface to the higly hydrophobic state. In A. Gadomski (Ed.), *Water in biomechanical and related systems*. Cham: Springer.

Gorczyca, B. (this volume). Fractal properties of flocs, filtration cakes and biofilms in water and wastewater treatment processes. In A. Gadomski (Ed.), *Water in biomechanical and related systems*. Cham: Springer.

Johnson, K. (this volume). Water nanoclusters in cosmology, astrobiology, the RNA world and biomedicine: The universe as a biosystem. In A. Gadomski (Ed.), *Water in biomechanical and related systems*. Cham: Springer.

Kruszewska, N., Domino, K., & Weber, P. (this volume). Water behavior near the lipid bilayer. In A. Gadomski (Ed.), *Water in biomechanical and related systems*. Cham: Springer.

Siódmiak, J. (this volume). Water molecules organization surrounding ions, amphiphilic protein residues, and hyaluronan. In A. Gadomski (Ed.), *Water in biomechanical and related systems*. Cham: Springer.

Siwek, M., Sławińska, A., & Dunisławska, A. (this volume). Water on livestock: Biological role and global perspective on water demand and supply chains. In A. Gadomski (Ed.), *Water in biomechanical and related systems*. Cham: Springer.

Strzelewicz, A., Dudek, G., & Krasowska, M. (this volume). Water transport through synthetic membranes as inspired by transport through biological membranes. In A. Gadomski (Ed.), *Water in biomechanical and related systems*. Cham: Springer.

Tiwari, A., & Persson, B. N. J. (this volume). Physics of suction cups in air and in water. In A. Gadomski (Ed.), *Water in biomechanical and related systems*. Cham: Springer.

Yuvan, S., & Bier, M. (this volume). Sense and nonsense about water. In A. Gadomski (Ed.), *Water in biomechanical and related systems*. Cham: Springer.

Chapter 2
Sense and Nonsense About Water

Steven Yuvan and Martin Bier

Abstract Water is present in all of its three phases in our natural world. It carves the landscapes on our planet, it is the solvent for biological activity, and it is central in humankind's physical and intellectual existence. We summarize how water's properties as a liquid and as a solvent are a consequence of the molecule being a strong dipole subject to Brownian motion. Short-lived hydrogen bridges between neighboring water molecules set up a flexible tetrahedral network. Convoluted pseudoscientific theories have been formulated about water. Many of these theories involve elaborate forms of higher organization and quantum physics. Some such theories have been used as a basis for scams and quackeries. We discuss of few of these excesses.

Keywords Liquid water · Water as a solvent · Water phase diagram · Brownian motion · Pseudoscience on water

An over-indulgence of anything, even something as pure as water, can intoxicate.

(Jami 2012).

S. Yuvan (✉)
Department of Physics, East Carolina University, Greenville, NC, USA
e-mail: yuvans16@students.ecu.edu

M. Bier (✉)
Department of Physics, East Carolina University, Greenville, NC, USA

Faculty of Mechanical Engineering and Institute of Mathematics and Physics, University of Technology and Life Sciences, Bydgoszcz, Poland
e-mail: bierm@ecu.edu

© The Author(s), under exclusive license to Springer Nature Switzerland AG 2021
A. Gadomski (ed.), *Water in Biomechanical and Related Systems*,
Biologically-Inspired Systems 17, https://doi.org/10.1007/978-3-030-67227-0_2

1 Introduction

Life left the oceans and went terrestrial on a large scale about half a billion years ago. However, even on land water maintained its role as the great enabler. In any terrestrial life-form the vast majority of molecules is water and the intracellular and extracellular solution are aqueous. Water is a volatile enabler as its availability generally fluctuates and overabundance or shortage of water commonly proves calamitous for entire ecosystems. Water thus came to play a prominent role as humans started to develop and formulate world views.

In all of the major, modern-day religions water plays a role as a cleanser and a purifier. As such it is considered to have a symbolic, if not mystical, ability to close the gap between man and the ideal purity of the divine. The Catholic Church prescribes elaborate rituals for the manufacture and disposal of "holy water." Baptism is central in Christianity. Judaism has a similar approach and has its followers achieve purity through immersion in a "mikveh." The Prophet Muhammed said "cleanliness is half the faith" and, consequently, Muslims are to wash with clean water before prayer and before handling the Quran. Both Judaism and Islam have strict rules about the water and the allowable sources of water to be used in the rituals. In Hinduism and in Buddhism it is believed that the sacred waters of the Ganges will wash away the sins of those that bathe in the river. Deities from Poseidon (Homer's *Iliad* 13:26–31) and Jesus (Mark 6:45–52, Matthew 14:22–33, John 6:15–21, *NIV*) up till Kim Il-Sung purportedly walked on water (McNeill 2011) – an act of dominance that would seem to set them metaphysically, as well as literally, above mere water.

With aqueous magic so firmly entrenched in our collective consciousness, it is hard for an individual scientist to keep a rational and empirical outlook when researching water. Below we will describe how seeing liquid water as mere dipoles performing Brownian motion accounts for properties of water as the solvent that facilitates biological activity. No quantum physics is required and no new phases or higher organization need be postulated. We will also tell the tale of a few pseudoscientific mishaps and scams.

2 The Facts on Water

A major pursuit in chemistry is explaining how the bulk properties of a substance result from the characteristics of the constituent molecules. Many of water's bulk properties are consequences of the strong dipole of the molecule and of the intermolecular hydrogen bonds that ensue from the dipoles.

The H-O-H angle in the water molecule is 104.5 degrees, thereby slightly differing from the 109.5 degrees of a perfect tetrahedron. The strong dipole comes about because the oxygen side of the molecule is a stronger attractor of electrons than the side with the two protons. When water is exposed to an electric field, the dipoles will rotate and orient themselves parallel to the electric field. The result is that the field will be reduced. The factor by which the electric field is reduced is the relative dielectric permittivity. For water the value of the relative dielectric permittivity is 80. Dielectric forces arise when the imposed electric field is not uniform and when, for instance, microparticles are dissolved in the water. The system will then lose energy if the material with the highest dielectric constant is forced towards where the electric field is highest. Dielectric forces are commonly used to move and manipulate microparticles (Rousselet et al. 1994).

The consensus picture is that around a dissolved ion or dipole, the water molecules orient their dipoles and form a nanometer-scale hydration shell (Moore 1974; Israelachvili 2011). Around an ion, there is a rigid inner shell of one layer of water molecules. Immediately beyond this, an outer shell consists of one or two more layers that are less rigid. Further away, Brownian collisions (that carry an average energy of about $k_B T$) overwhelm hydration bonds. Hydration shells around proteins and other biomolecules are the subject of much research. Experimental observation appears to be in good agreement with the results of theory and molecular dynamics simulation (Pettitt et al. 1998; Virtanen et al. 2010). In the references by Ebbinghaus et al. (2007) and Dastidar and Mukhopadhyay (2003) it is shown that the hydration shell around a big protein also extends to no more than about a nanometer away from the protein. The hydration shell effectively screens away the electric field of the underlying ion or dipole. The electric field of an ion in an aqueous solution is already unnoticeable within a nanometer.

The molecules in liquid water are colliding repeatedly as they perform Brownian motion. At room temperature, the speed of a water molecule is about 600 m/s (for each translational degree of freedom $\frac{1}{2}mv^2 = \frac{1}{2}k_B T$). For liquid water, the density is such that the mean free path roughly equals the size of the molecule – about 0.25 nm. From these numbers it is readily derived that there are on the order of a trillion (10^{12}) collisions per water molecule per second. In other words, there is a collision per picosecond for every individual water molecule.

Much of water's bulk behavior is related to the hydrogen bridges that form between neighboring water molecules (see Fig. 2.1). After fluoride, oxygen is the most electronegative element. The ensuing strong attraction of electrons to oxygen leads to the water molecule's strong dipole. Hydrogen bonds are basically dipole-dipole interactions. The strength of the hydrogen bonds in water as well as the large number of hydrogen bonds in a mole of liquid water results in a lot of energy being associated with hydrogen bonds. This energy must be supplied when melting or boiling water and leads to relatively high melting and boiling temperatures. In this context it is instructive to compare water to hydrogen sulfide (H_2S). As a molecule, hydrogen sulfide has an architecture that is almost identical to water. But the electronegativity of sulfur is 2.58, whereas that of oxygen is 3.44. This leads to the H_2S molecule having a smaller dipole than the H_2O molecule. The result is that hydrogen sulfide, in spite of having almost twice the molecular mass of water, is a gas at room temperature. The energy in the hydrogen bridges also leads to liquid water's high specific heat and the high energies associated with melting and evaporation.

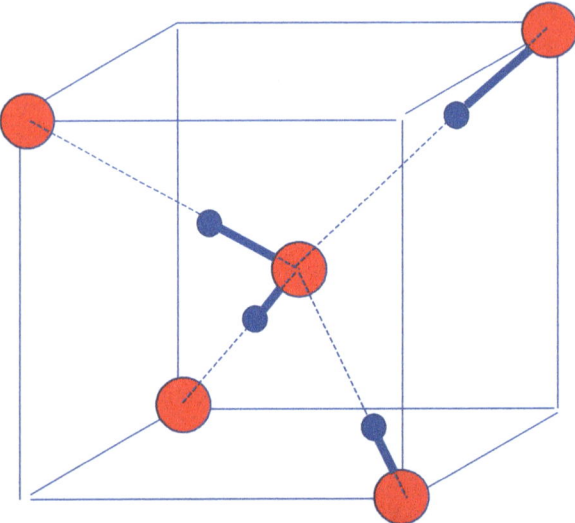

Fig. 2.1 The tetrahedral structure of water. The large red dots represent the oxygen atoms and the smaller blue dots represent the hydrogen atoms. The bold segments represent covalent bonds and the dotted segments represent hydrogen bridges. Hydrogen bridges ensue along the dotted segments as the strongly negative electron cloud around an oxygen attracts protons bound to other water molecules. With the tetrahedral structure, a hydrogen-bound electron can transit to the other side of the hydrogen to next be shared with another oxygen. Note that the same structure is obtained upon an inversion where all the dotted segments become bold and all the bold segments become dotted. Such symmetry adds stability to the structure. For liquid water Brownian collisions have sufficient energy to disrupt a hydrogen bond within picoseconds. But upon freezing the tetrahedral structure becomes rigid

Because of the tetrahedral positioning of the oxygen's four electron pairs, the structure depicted in Fig. 2.1 is the favored lattice arrangement. For liquid water, the energy of Brownian collisions (about $k_B T$) is still such that the hydrogen bond survives for only picoseconds. The tetrahedral structure loses its plasticity upon freezing.

Under atmospheric conditions, water expands by 9% when it freezes. Water has its highest density in its liquid state at 4 °C. This is the reason that ice floats on water. It is also the reason that potholes appear in roads during winter months: water seeps into the cracks in the asphalt when the temperature is above freezing and next fractures the asphalt when it freezes and expands. Expansion upon cooling is counterintuitive as one would think that increased thermal agitation pushes molecules further away from each other. But this behavior is not really anomalous. Silica (SiO_2), the major constituent of sand, also does this and it is the reason that the solid crust of the Earth, with its density of 2750 kg/m^3, effectively floats on the liquid mantle that has a density of 3300 kg/m^3 (Lowrie 1997). In the case of water, it is again hydrogen bridges that are largely responsible for the shrinking upon melting. When water turns liquid, the organized structure that is depicted in Fig. 2.1 literally collapses (Brini et al. 2017). The Van der Waals forces can next pull water molecules closer together than that they are in the lattice of Fig. 2.1. Only above 4 °C is the situation again such that increasing the energy $k_B T$ of the Brownian collisions increasingly overwhelms the energy of the Van der Waals bonding and leads to driving molecules further apart. Alcohol does not expand upon freezing and the addition of a little alcohol to water leads to a mixture that again expands upon heating. Alcohol (C_2H_5OH) also has intermolecular hydrogen bonds, but less of them per unit of mass.

Widespread is the misconception that a microwave oven is generating heat by emitting at a resonance frequency of water. The actual fact is that it is possible to quantitatively account for the operation of a microwave oven with great accuracy by modeling the water molecule as a sphere with an electric dipole that is rotating in a viscous, overdamped medium. Because of the dipole, the microwave oven's GHz radiation applies a torque to the water molecule and makes it rotate. The generated heat is due to the friction with the medium. Imagine a water molecule with its dipole in an arbitrary direction and no electric field present. If an electric field is suddenly applied, the dipole will align. However, there is a characteristic time associated with the relaxation to the new alignment. Peter Debye derived $\tau = 4\pi\eta R^3/k_B T$ for this relaxation time (Debye 1945; Elton 2017). Here η represents the shear viscosity (about 10^{-3} N × s/m^2 for liquid water). With $R = 0.14$ nm for the radius of the water molecule, this equation leads to a remarkably accurate estimate: $\tau \approx 10$ ps. It is obvious that the rotating dipole will keep up with an applied AC field if that field is not changing too fast, i.e. if $\omega \lesssim \tau^{-1}$, where ω is the angular frequency of the applied AC field. For $\omega \gg \tau^{-1}$ the field cycles so fast that the dipole will effectively "feel" the zero-average of the field. No rotation and generation of heat will occur in that case. With the 2.45 GHz of an ordinary microwave oven, we are well below the τ^{-1} cutoff.

There are small discrepancies between Debye's predictions and measured spectra. However, these discrepancies have been resolved by taking the tetrahedral geometry (cf. Fig. 2.1) into account and realizing that much of the molecular motion consists of "hopping" from one such structure to another (Elton 2017). It is not through an invocation of quantum resonances or higher organization that discrepancies have been resolved, but through a more precise assessment of hydrogen bonds and intermolecular geometry.

3 Mishaps

3.1 Polywater

In 1962 the Russian physicists Nikolay Fedyakin and Boris Deryagin reported the discovery of a new type of water (Deryagin and Fedyakin 1962). In narrow capillaries they claimed to have found water that was more dense and more viscous than normal. It was thought to be a polymerized form of water and it therefore came to be called "polywater." Deryagin, who was the head of a prestigious research center in Moscow, showed his results in the West in 1966 and the paranoia of the Cold War soon led to large scale research efforts in the United States and Western Europe. It was suggested that with the proper catalyst, rivers, lakes, and oceans could be turned into molasses. Newspapers frequently reported about polywater in a hysteric and apocalyptic tone. Nevertheless, the international scientific community kept experimenting and theorizing in a free and open manner.

Within a few years it became clear that it had all been a delusion. Polywater was nothing but ordinary water that had been slightly contaminated. The first skeptical voices were heard in 1970 and by 1974 scientists generally agreed that there was no such thing as polywater. Research into the subject died out. Figure 2.2 shows the accumulated number of scientific articles on the subject of Polywater.

Where scientists have left polywater behind and moved on to new subjects, historians and philosophers of science have subsequently written a lot about the polywater affair. Was it a hypothesis that did not survive the testing, i.e., was it science the way it ought to be conducted? Or was it science led astray? Much cited is the book *Polywater* by Felix Franks (1981). Franks takes the latter viewpoint and describes the episode as "pathological science."

Ultimately, 299 papers about polywater were published. After 1974 no more scientific articles on the subject appeared (Ackermann 2006). Figure 2.2 shows the accumulated number of papers as a function of time. What is remarkable is that the logistic curve is an almost perfect fit. The curve is mindful of an epidemic and in the reference by Ackermann (2006) the author indeed characterizes the polywater episode as a "false information epidemic."

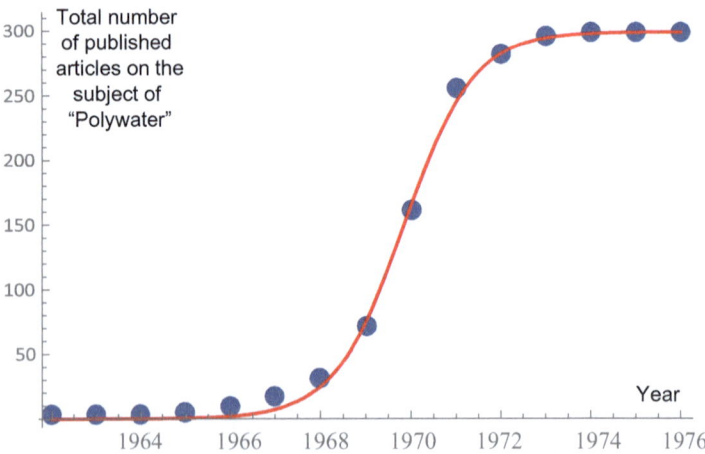

Fig. 2.2 The accumulated number of scientific articles on the subject of polywater as time evolves. Data were taken from the reference by Ackermann (2006). The number of papers that were ultimately published on the subject is $N_0 = 299$. The red curve represents a best fit of the logistic function $n(t) = N_0/(1 + \exp[-\alpha(t - t_0)])$

The logistic function, $n(t) = N_0/(1 + \exp[-\alpha(t - t_0)])$, is the solution of the Verhulst Equation: $\dot{n}(t) = \alpha n(t)(1 - n(t)/N_0)$. The equation describes how for small $n(t)$ there is an exponential increase that is characterized by a rate α. For larger $n(t)$ the solution approaches a horizontal asymptote, i.e. $n(t \to \infty) = N_0$. The free parameter in the solution of the differential equation can be identified with setting the initial time constant t_0. The Verhulst Equation was first formulated for a population of animals that initially increases exponentially but flattens when the carrying capacity of the environment is approached. Half a century ago Bass described how the Verhulst Model also applies to durable goods like lawn mowers, televisions, or clothes dryers (Bass 1969). Sales of such products increase exponentially as the products first enter the market and word about usefulness spreads. But later, as the market approaches its saturation level, sales come to a halt. Ideas and fashions appear to follow a similar dynamic. In the reference by Coulmont et al. (2016) it is shown how Verhulst dynamics also applies to baby names: logistic curves are fitted to data that show how "Diane" and "Seymour" first rose in popularity, but ultimately became outdated.

It is remarkable that the simple Verhulst Equation leads to such a perfect fit to the data shown in Fig. 2.2. It appears as if it were a completely deterministic evolution of half a decade and 299 papers that turned an illusion into a generally-recognized delusion. In the remainder of this section we will see that polywater is the exception rather than the rule. Many of the misapprehensions about water have turned out more long-lasting. Such misapprehensions have also become gratefully adopted and exploited by quacks and scammers.

3.2 Homeopathy and "Water Memory"

Homeopathy was developed about 200 years ago and it was most likely inspired by the proliferating practice of smallpox inoculation. Smallpox used to be fatal in about 25% of all cases. In an inoculation a very small amount of pus or scabs from someone infected with smallpox would be administered. It would generally lead to mild symptoms, but upon recovery the inoculated individual would have immunity against the disease. Homeopathy extrapolated the idea: a strong cup of coffee may lead to sleeplessness but give someone a very diluted caffeine solution and it will work as a *remedy* against sleeplessness. However, once Avogadro's number was established, it became obvious that the dilutions that are used in homeopathic practice are so extreme that no molecule of the original substance is ultimately left.

In an attempt to take their practice out of the realm of quackery and obtain scientific credibility, advocates of homeopathy have concocted a number of theories as to why a substance can have an effect even after it has been completely diluted away. The issue came to a head in June of 1988. In that year Jacques Benveniste and his group published a paper in *Nature* (Davenas et al. 1988). The paper reported the results of a simple bench experiment. Basophils are the white blood cells that are responsible for much of the immune response. The *Nature* article reported how homeopathic dilutions of antibody triggered a response in such basophils. At the end of the article the authors briefly address the fundamental ontological problem of a molecule having an effect without being physically present. In the abstract it writes: "Since dilutions need to be accompanied by vigorous shaking for the effects to be observed, transmission of the biological information could be related to the molecular organization of water." The earliest use of the word "memory" that we found in this context was in a news column in *Nature* that was written in August of 1988 in the midst of the uproar that followed the publication of the article of Benveniste's group (Coles 1988).

After the publication of the article by Davenas et al., *Nature* set up a team to oversee replications. These replications failed. However, Benveniste passionately stood by his results. In the polemics that ensued, he often backed up the "water memory" claim with an article that, also in 1988, had appeared in *Physical Review Letters*. The article was by Del Giudice, Preparata, and Vitiello and it was titled "Water as a Free Electric Dipole Laser" (Del Giudice et al. 1988).

One of the basic outcomes of Quantum Mechanics is that action, the product of energy and time, comes in packages of $\hbar = 1.05 \times 10^{-34}$ J $\times$ s. The $\hbar$ is also known as the Reduced Planck's Constant. Angular momentum has the same Joule×second-dimension as action and this means that rotation can only occur with an angular momentum, $|\mathbf{L}|$, that is an integer multiple of $\hbar$ (Merzbacher 1970). Del Giudice et al. consider water molecules as dipoles that are rotating around their own center of mass and going back and forth between the ground state ($|\mathbf{L}| = 0$) and the first excited state ($|\mathbf{L}| = \hbar$). A photon is emitted when a water molecule falls back to the ground state. But that same photon can next make another water molecule jump to the first excited state. It can also, through so-called stimulated emission, induce another molecule in the excited state to drop to the ground state and emit. Quantum Field Theory is the appropriate framework to describe the interaction between photons and dipoles (Merzbacher 1970). This theory has a much higher level of complexity than the ordinary quantum mechanics that is basic to much of chemistry. After a short introduction the authors set up a quantum-field Lagrangian. Next a dynamical system for the evolution of the three relevant populations (ground state water, excited state water, and photons) is derived from the Lagrangian. The authors do not realize that the derived dynamical system is actually the well-known Euler System to describe rigid body rotation (Goldstein et al. 2002; Shnir 2005). Next, their analysis of the dynamical system is littered with errors (Bier and Pravica 2018).

However, already the setup of the paper is mistaken. The Lagrangian in the reference by Del Giudice et al. (1988) would be legitimate if the water molecules were standing still and if the exchanged photons were the only "contact." But as was mentioned in the previous section, in liquid water an average molecule collides every picosecond with another molecule. Energy is exchanged in these collisions and a realistic Lagrangian has to include terms to this effect.

Brownian collisions underlie friction and let a system relax towards a Boltzmann equilibrium. In the previous section we saw that treating water as spheres that interact through electrostatic forces and perform overdamped motion leads to quantitatively accurate descriptions. The reference by Elton (2017) shows how such descriptions apply until well into far-infrared frequencies.

Almost as an afterthought it is stated at the end of the reference by Del Giudice et al. (1988) that a permanent electrical polarization can form around an "impurity that carries a sizable electric dipole" from, for instance, a macromolecule. Again there is mention of "ordered structures in macroscopic domains." The possibility of such a polarization rolls out of long and complicated formulae and it is presented on the last page of the paper. The convoluted mathematics, however, disguise how the authors are essentially already assuming what they are proving. They postulate that the impurity or macromolecule causes a constant and homogeneous electric field that extends over hundreds of micrometers. That an electric field causes a polarization in a polarizable medium is almost a tautology. Debye's formulae are already successful in establishing a quantitatively accurate relationship between an electric field and its ensuing polarization (Moore 1974; Israelachvili 2011). The problem is, of course, that no ion or polymer in an aqueous solution will ever give rise to a constant, homogeneous electric field that extends over macroscopic distances. As was mentioned before, fields are screened away within nanometers. Moreover, macromolecules are subject to Brownian fluctuations and the electric fields that they exude will not be constant.

The reference by Del Giudice et al. (1988) ends with an appeal to further develop the ideas of water as a free electric dipole laser. But not much subsequent fundamental work has been undertaken in that direction. Most of the citations of this article simply invoke the alleged long-range order as a possible explanation for otherwise inexplicable phenomena (see e.g. Liu et al. 2011). The Journal of Alternative and Complementary Medicine is responsible for a very significant part of the citations. As was mentioned before, especially homeopathic medicine has gratefully picked up on the claimed long-range order. Google Scholar indicates that the article has been cited 480 times (accessed May 2020). Over the last 15 years the paper has received uncritical citations at a steady rate of about 20 per year. More than a "false information epidemic" it looks like we are dealing here with a genetic disorder that is transferred from one generation to the next without serious threat of proliferation.

In the immediate aftermath of the "cold fusion" uproar of 1989, "water as a free electric dipole laser" was advanced as an explanation for the alleged phenomenon (Bressani et al. 1989). Even after cold fusion was discredited (Miskelly et al. 1989; Price et al. 1989), work to this effect continued (Fleischmann et al. 1994).

A brief Google search furthermore reveals how there is a large number of more popularly oriented books and websites where the reference by Del Giudice et al. (1988) is invoked as providing the scientific legitimacy behind a variety of unconventional approaches to life, science, and health (e.g. McTaggart 2002). The reference by Del Giudice et al. (1988) has also featured in advertisements to purportedly explain the efficacy of products with names like "Reverse Aging" and "Divine Alignment" (Trinfinity8 2018).

It is especially the "permanent electric polarization around an impurity"-suggestion in the reference by Del Giudice et al. (1988) that has been cited in the last few decades in order to support claims that liquid water could keep and carry a "memory." Structures in liquid water, it is purported, would be able to retain a "memory" of an originally dissolved compound even if the dilutions are such that no molecule of the original compound is left in the water. But in actual fact, none of the conclusions of the reference by Del Giudice et al. (1988) justifies the idea of "water memory." The reference by Del Giudice et al. (1988) makes no statements about what happens when the macromolecules are diluted away. There is nothing in this reference that suggests an "imprint" that is left by a substance that is no longer there. Even if coherent domains were real, it is impossible to see how they would create a "memory" and constitute a validation for homeopathic practices.

Liquid water is generally not in the realm of interest of physicists working in Quantum Field Theory. It is therefore that those most qualified have not pursued the Quantum Coherent Domains in liquid water. Water, of course, is all-important in Condensed Matter Physics and in the Biosciences. But researchers in these field are generally unfamiliar with Quantum Fields. It is the enigmatic nature of the reference by Del Giudice et al. (1988) that has been most instrumental in building its popularity. Quantum Coherent Domains carry the respectability of basic physics research on matter at its most fundamental level. They thus constitute the ultimate *argumentum ab auctoritate*. Those that are most zealously invoking the reference by Del Giudice et al. (1988) and "water memory" in the peddling of pseudoscience and quackery are generally the least qualified to assess the underlying validity. Writing about homeopathy, water memory, and the reference by Del Giudice et al. (1988), Philip Ball formulated it as follows in a 2007 *Nature News* column: "This 'field' has acquired its own *deus ex machina*, an unsubstantiated theory of 'quantum coherent domains' in water proposed in 1988 that is vague enough to fit anything demanded of it" (Ball 2007).

There is currently an almost universal consensus that a liquid water environment is too hot and too wet for quantum entanglement to play any role of significance (Tegmark 2000). A collision between two water molecules localizes these molecules and implies that the involved quantum mechanical wave functions collapse onto position eigenfunctions (Unruh and Zurek 1989). As was already discussed, molecules in liquid water collide with terahertz frequency. A cubic micrometer of liquid water contains about 10^{10} molecules. As a coherent domain, such a cubic micrometer will therefore not last beyond an insignificant 10^{-22} s. Most importantly, until today science has needed no recourse to quantum mechanics or quantum field theory in order to explain the bulk behavior of water (Brini et al. 2017; Elton 2017).

3.3 Polywater Reiterated

It was already mentioned in the Introduction that the intracellular and extracellular solution are aqueous. The first living cells evolved in the oceans and it is therefore that, even in terrestrial multicellular organisms, the extracellular fluid has an ionic composition that is similar to that of seawater. Blood, sweat, and sperm taste salty.

For sodium and chloride ions, the concentration in the extracellular liquid is generally larger than 10^2 mM. The extracellular potassium concentration is about 10 mM. Inside the cell it is reversed: concentrations of sodium and chloride are about 10 mM and the potassium concentration is about 10^2 mM. The inside of a living cell is, furthermore, negatively charged relative to the outside. The potential difference is about 100 mV (Alberts et al. 1994; Hille 1992; Läuger 1991).

In the years after World War 2, the intricate system of pumps, transporters, and channels that control the electrochemical potentials was slowly getting unraveled (Alberts et al. 1994, Hille 1992, Läuger 1991). Na$^+$,K$^+$-ATPase was discovered in 1957 (Skou 1957). This is the ion pump that maintains the electrochemical gradients for sodium and potassium. It derives its energy from the hydrolysis of ATP (adenosine triphosphate) (Alberts et al. 1994). ATP is the currency of energy in a living cell and it is where the energy that is released in the breakdown of glucose is initially stored.

The transmembrane electrochemical potential of sodium is high, and it is used in a cell as an energy source for transmembrane transport. The sodium-glucose cotransporter (Alberts et al. 1994) is a membrane-spanning protein that was first identified in 1960 (Crane et al. 1961). It binds a glucose molecule on the extracellular side, goes through a number conformational states, and in the end lets that glucose molecule detach on the intracellular side. The scheme works even when the intracellular glucose concentration is larger than the extracellular one. This is because the cotransporter, concurrently with the glucose transport, also transports a sodium ion from the outside to the inside of the cell. The electrochemical gradient of sodium thus provides the driving force for the energetically-uphill transport of glucose.

The genes that are associated with the different proteins that carry out the transmembrane transport have been identified. Also identified have been many mutations in these genes and the malfunctions that these mutations can lead to. Cystic fibrosis, for instance, is caused by a mutation that leads to a defective chloride channel. The 3D atomic structure of most pumps, transporters, and channels is currently known. Nowadays, the knowledge of such 3D structures is often the starting point for drug design. The challenge in such design is to construct a molecule that will bind to the intended membrane protein and elicit the desired effect on that protein's activity. Much of the search for drugs against obesity and diabetes, for instance, is focused on the aforementioned glucose-sodium cotransporter.

In 1947 Gilbert Ling (1919–2019) was prominently involved in the development of a type of microelectrode that facilitated the measurement of transmembrane potentials. The microelectrode greatly helped the advance of our understanding of the cell's electrochemistry. But Ling took issue with the developing consensus in the scientific community. He proposed that a mixture of crosslinked proteins and water ultimately formed a gel inside the living cell. Polarized layers of this gel would give the cell its electrical properties. That there is more potassium than sodium or chloride inside the cell would be a result of the affinities of involved proteins. Consumption of energy to maintain the transmembrane electrochemical potentials would not be necessary. Starting in the 1950s, Ling published books and articles to promote his model. However, as molecular biology developed rapidly, his ideas became ever more marginal. Electron microscopy made cell membranes and embedded proteins visible. With radioactive and fluorescent markers, with atomic force microscopy, with patch clamp, and with optical tweezers the movement and activity of biomolecules could be probed and followed (Barkai et al. 2012). Ling's model lacked evidence and became a sideshow.

Nevertheless, Ling's ideas did find a small following. Prominent is the work of Gerald Pollack. Pollack has focused on a more basic surface effect. When water with dissolved molecules or microspheres has an interface with a hydrophilic surface, a so-called exclusion zone (EZ) forms (Zheng and Pollack 2003). The solutes or microspheres move away from the interface. The EZ can measure a few hundred micrometers. Pollack has proposed that the water in the exclusion zone has polymerized and entered another phase: planar sheets of hexagonal lattices with $(H_3O_2)^-$ as the basic unit (see Fig. 2.3) (Pollack 2013). This would explain the observed negative charge near the interface. The membrane of a living cell consists of phospholipids. These molecules consist of strongly hydrophilic phosphate headgroups and apolar hydrocarbon tails. The cell membrane is a bilayer in which the apolar hydrophobic tails point toward one another and the hydrophilic phosphate headgroups stick out into the aqueous solutions. Living cells would thus consist in large part of water in the EZ phase. This would in Pollack's view also explain why living cells are negatively charged.

The major problem with the theory is that the configuration depicted in Fig. 2.3 has never been observed. Because it concentrates so much negative charge, the structure would be very unstable if it were to exist. Postulating a new phase to explain the observed EZ breaches the spirit of Occam's Razor. Surface effects can be very complex (Israelachvili 2011), but it is the appropriate direction to look into

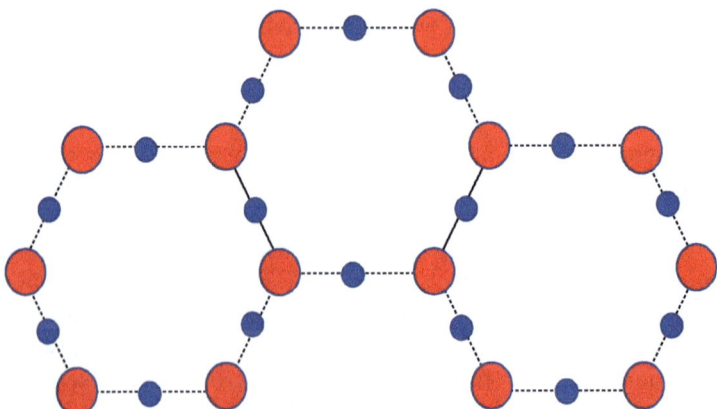

Fig. 2.3 The planar sheet of hexagons that was proposed as a polymeric form of water by Pollack. Water would presumably go to this phase in the close vicinity of a hydrophilic surface. The basic unit is $(H_3O_2)^-$. However, because of the net negative charge the phase would be highly unstable and it is actually unlikely to ever materialize

for an explanation of the exclusion zone. It is likely because of a "polywater déjà vu" that not many researchers have paid serious attention to Pollack's theory. However, there is no reason to *a priori* mistrust the published EZ measurements. These measurements moreover, have been replicated (see, e.g., Elton et al. 2020) and it is a matter of scientific integrity to get to understand the phenomenon. Furthermore, the EZ can be important in understanding biological phenomena and it can possibly find technological applications in microfluidics and filtration.

The EZ effect has also been observed when polar liquids other than water have been used (Chai and Pollack 2010). This is already an indication that the explanation may not lie in specific molecular structures and rearrangements thereof, but instead in more generally applying dynamics of solutes and solvents.

The observed negative charge of the EZ could be explained as follows. Because of the excluded solutes or microparticles, water has an effectively higher concentration in the EZ layer. Water molecules are perpetually dissociating into hydrogen and hydroxide and then again re-associating, i.e., $H_2O \leftrightarrow H^+ + OH^-$. The diffusivities of water, hydrogen, and hydroxide have been well-studied. Hydrogen has a higher diffusion coefficient than hydroxide (Lee and Rasaiah 2011). If the water concentration is higher in the EZ layer, then more H^+ and OH^- will be formed there. As the H^+ will more rapidly diffuse away into the bulk outside the EZ layer, a net negative charge will remain behind.

3.4 Spin-off

"Water" with adjectives like "EZ," "hexagonal," "structured," etc., is the snake oil of the twenty-first century. A small Google search suffices to recognize that. Special mention in this context should be made of Shui-Yin Lo.

Soon after publishing two papers in *Modern Physics Letters B* on a newly discovered form of ice (I_E crystals) in 1996 (Lo 1996; Lo et al. 1996), Shui-Yin Lo was involved in dubious marketing operations involving the I_E crystals. A blue laundry ball of the size of a tennis ball was supposed to make detergent unnecessary. The 75-dollar laundry ball would create I_E crystals outside the ball and the "nanotricity" of these crystals would remove dirt from cloth. State attorneys general and the Federal Trade Commission were very quick to recognize this for the fraud that it was and put a stop to it. The same company also sold a device to be put into a car's air filter. There the I_E-technology would improve gas mileage and engine performance. Also this product was taken of the market through interference of the criminal justice system.

Lo, however, persevered in spite of these setbacks. In 2009 he published another article in a scientific journal on stable water clusters at room temperature (Lo et al. 2009). Doubts about this article were later expressed in a comment in the same journal (Kožíšek et al. 2013). Next Lo and his co-workers discovered that a double helix shape was common among these clusters. They then went on to focus on the "immunological and enzymatic effects" and moved into the health care business. A paper was published in *Forum on Immunopathological Diseases and Therapeutics* (Lo et al. 2012) and a website "www.doublehelixwater.com" was set up. Many forms of questionable science coalesce in the 2012 article of Lo et al. as both the EZ water of Pollack and the Quantum Coherent Domains of Del Giudice are proposed as possible explanations of "Double Helix Water." On the website the "Double Helix Water" is offered for sale and many health benefits are described. "Promising and startling" preliminary results are reported in curing autism, but it appears to also work as an antibiotic and as a remedy for ailments like shoulder pain and arthritis. At the bottom of every page on the website it writes: "These statements have not been evaluated by the Food and Drug Administration. This product is not intended to diagnose, treat, cure, or prevent any disease." The website has been active for almost a decade and it is sad to have to conclude that the disclaimer makes it possible to indefinitely peddle the same rip-off as a health care product after it was almost immediately terminated as a laundry-ball or fuel-additive.

3.5   *"Magnetic Water"*

Water is not a magnetic material. Nevertheless, there are many claims surrounding "magnetic water" – water that has merely been passed through a strong magnetic field. A variety of devices for household and industrial use is for sale. According to the advertisements, routing the water along a series of magnets leads to benefits such as: reduced corrosion and mineral deposition, increased or decreased pH, increased bioavailability of dissolved compounds, improved kidney and digestive health, increased crop yields, and enhanced setting of concrete.

The inspiration for these devices appears to derive from a nebulous industrial technique for reducing mineral buildup in pipes and tanks. There is some evidence of magnetism being investigated to this end even as early as the 1880s for use in steam boilers (Ambashta and Sillanpää 2010). From an engineering perspective, the process is straightforward. Passing supersaturated solution through a strong magnetic field supposedly facilitates precipitation or growth of suspended particles. Particles can then settle at the bottom of reservoirs and be flushed out, rather than adhering to interior surfaces and requiring more costly chemical removal.

Several reviews have been published, yet there is no agreed upon mechanism for how and why a magnetic field would possibly enhance precipitation. The rate of chemical precipitation is sensitive to a variety of factors (temperature, pH, contaminants, etc.). Most studies have not been rigorous in this regard and this is probably why results have been inconsistent (Chibowski and Szcześ 2018). In industrial practice, bulk precipitation is commonly induced by simply cooling a saturated solution (Keister 2008). Even if magnetic fields were to trigger bulk precipitation, any such application to household tap-water pipes is undermined by the fact that any precipitant simply remains suspended all the way to the drinking glass.

In the marketing of the devices, claims of altered structures and memory are once again common in order to produce an appealing mystique. Magnets supposedly possess the ability to induce a distinct "North" or "South" character into molecules or reduce the size of water "clusters" surrounding ions (see, for example, *Magnetized Water* 2020). We have already discussed Brownian motion, the short duration of hydrogen bonds, and the limited reach of hydration shells. In this context it is obvious that even if there were magnetically induced alterations in order, they would disappear again once the magnetic field is gone. Water itself is actually weakly diamagnetic, a property owed precisely to the molecule's lack of an intrinsic magnetic moment. But the effect is so weak as to make water practically immune to magnetic effects. Long-lasting, extremely large field gradients (to the order of $100\ T^2/m$) are needed to produce any significant effects (Ueno and Iwasaka 1994), requiring expensive superconductor-based magnets. Here we are reminded of a more direct application of strong magnetic fields to ingested water: Magnetic Resonance Imaging (MRI). Yet despite the hundreds of millions of exams performed yearly (OECD 2020), we are still left wanting for tales of their miraculous effects on human health.

The "polywater outbreak" lasted less than a decade (cf. Figure 2.2). But for coherent-domain-water, hexagonal water, and magnetic water the dynamics have been a lot slower. Figure 2.4 shows the number of accumulated publications on the subject of "magnetic water treatment" up to the year 2019. If a Verhulstian course is again going to be followed, then there is still a long way to go. So far, even the inflection point is not in sight.

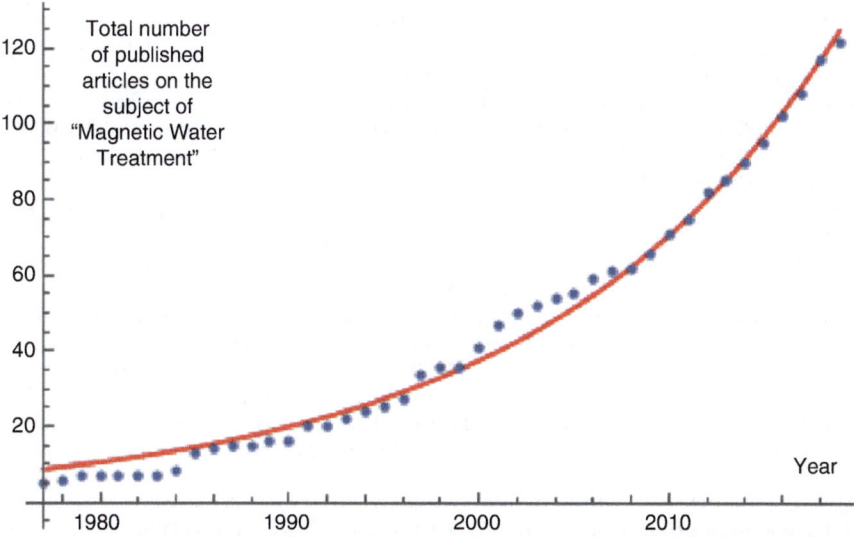

Fig. 2.4 The accumulated number of scientific articles on the subject of "Magnetic Water Treatment" as time evolves. Data was retrieved from Elsevier's Scopus (http://www.scopus.com/). The red curve represents the best fit of an exponential $n(t) = \exp[\alpha(t - t_0)]$

4 Conclusions

The Career of Nikodem Dyzma was first published in 1932. It was the debut novel of the Polish author Tadeusz Dołęga-Mostowicz (2014). The entertaining story describes how a former postal clerk from a small rural town catches a lucky break as he is struggling to make a living in Warsaw. He finds an envelope containing an invitation to an exclusive upper-class party. He goes there in the hope to finally eat a decent meal. In a confrontation with other guests, his cluelessness and coarseness are taken as signs of profound insight and aphoristic wisdom. Next the protagonist does not have to take a lot of initiative to become a rising star in Warsaw high-society. He is seen as a wish-come-true in spite of indications to the contrary. He is made the head of a national bank and, ultimately, he is offered the prime-ministership.

There is a community of researchers for whom water has become an inanimate Nikodem Dyzma. Guided by a desire for Transcendence and Redemption and in search of a Theory-of-Everything, they have turned a simple, common, stable and small molecule into something that it is not.

Sometimes a solvent is just a solvent.

References

Ackermann, E. (2006). Indicators of failed information epidemics in the scientific journal literature: A publication analysis of polywater and cold nuclear fusion. *Scientometrics, 66*(3), 451–466. https://doi.org/10.1007/s11192-006-0033-0.

Alberts, B., Bray, D., Lewis, J., Raff, M., Roberts, K., & Watson, J. D. (1994). *Molecular biology of the cell* (3rd ed.). New York: Garland Science.

Ambashta, R. D., & Sillanpää, M. (2010). Water purification using magnetic assistance: A review. *Journal of Hazardous Materials, 180*(1), 38–49. https://doi.org/10.1016/j.jhazmat.2010.04.105.

Ball, P. (2007, August 8). Here lies one whose name was writ in water. *Nature News.* https://doi.org/10.1038/news070806-6.

Barkai, E., Garini, Y., & Metzler, R. (2012). Strange kinetics of single molecules in living cells. *Physics Today, 65*(8), 29–35. https://doi.org/10.1063/PT.3.1677.

Bass, F. M. (1969). A new product growth for model consumer durables. *Management Science, 15*(5), 215–227. https://doi.org/10.1287/mnsc.15.5.215.

Bier, M., & Pravica, D. (2018). Limits on quantum coherent domains in liquid water. *Acta Physica Polonica B, 49*(9), 1717. https://doi.org/10.5506/APhysPolB.49.1717.

Bressani, T., Del Giudice, E., & Preparata, G. (1989). First steps toward an understanding of «cold» nuclear fusion. *Il Nuovo Cimento A, 101*(5), 845–849. https://doi.org/10.1007/BF02844877.

Brini, E., Fennell, C. J., Fernandez-Serra, M., Hribar-Lee, B., Lukšič, M., & Dill, K. A. (2017). How water's properties are encoded in its molecular structure and energies. *Chemical Reviews, 117*, 12385–12414.

Chai, B., & Pollack, G. H. (2010). Solute-free interfacial zones in polar liquids. *The Journal of Physical Chemistry B, 114*(16), 5371–5375. https://doi.org/10.1021/jp100200y.

Chibowski, E., & Szcześ, A. (2018). Magnetic water treatment–A review of the latest approaches. *Chemosphere, 203*, 54–67. https://doi.org/10.1016/j.chemosphere.2018.03.160.

Coles, P. (1988). Benveniste controversy rages on in the French press. *Nature, 334*(6181), 372–372. https://doi.org/10.1038/334372a0.

Coulmont, B., Supervie, V., & Breban, R. (2016). The diffusion dynamics of choice: From durable goods markets to fashion first names. *Complexity, 21*(S1), 362–369. https://doi.org/10.1002/cplx.21748.

Crane, R. K., Miller, D., & Bihler, I. (1961). The restrictions on possible mechanisms of intestinal transport of sugars. In A. Kleinzeller & A. Kotyk (Eds.), *Membrane transport and metabolism. Proceedings of a symposium held in Prague, August 22–27, 1960* (pp. 439–449). Prague: Czech Academy of Sciences.

Dastidar, S. G., & Mukhopadhyay, C. (2003). Structure, dynamics, and energetics of water at the surface of a small globular protein: A molecular dynamics simulation. *Physical Review E, 68*(2), 021921. https://doi.org/10.1103/PhysRevE.68.021921.

Davenas, E., Beauvais, F., Amara, J., Oberbaum, M., Robinzon, B., Miadonna, A., Tedeschi, A., Pomeranz, B., Fortner, P., Belon, P., Sainte-Laudy, J., Poitevin, B., & Benveniste J. (1988). Human basophil degranulation triggered by very dilute antiserum against IgE. *Nature, 333*, 816–818.

Debye, P. (1945). *Polar molecules* (pp. 89–77). New York: Dover Publications.

Del Giudice, E., Preparata, G., & Vitiello, G. (1988). Water as a free electric dipole laser. *Physical Review Letters, 61*(9), 1085–1088. https://doi.org/10.1103/PhysRevLett.61.1085.

Deryagin, B. V., & Fedyakin, N. N. (1962). Special properties and viscosity of liquids condensed in capliaries. *Proceedings of the Academy of Sciences of the USSR Physics Chemistry, 147*(2), 808–811.

Dołęga-Mostowicz, T. (2014). *Kariera Nikodema Dyzmy*. Warszawa: Bellona SA. English translation to appear in September 2020. https://nupress.northwestern.edu/content/career-nicodemus-dyzma

Ebbinghaus, S., Kim, S. J., Heyden, M., Yu, X., Heugen, U., Gruebele, M., Leitner, D. M., & Havenith, M. (2007). An extended dynamical hydration shell around proteins. *Proceedings of the National Academy of Sciences, 104*(52), 20749–20752. https://doi.org/10.1073/pnas.0709207104.

Elton, D. C. (2017). The origin of the Debye relaxation in liquid water and fitting the high frequency excess response. *Physical Chemistry Chemical Physics, 21*(14), 5041.

Elton, D. C., Spencer, P. D., Riches, J. D., & Williams, E. D. (2020). Exclusion zone phenomena in water—A critical review of experimental findings and theories. *International Journal of Molecular Sciences, 21*(14), 5041. https://doi.org/10.3390/ijms21145041.

Fleischmann, M., Pons, S., & Preparata, G. (1994). Possible theories of cold fusion. *Il Nuovo Cimento A (1965–1970), 107*(1), 143. https://doi.org/10.1007/BF02813078.

Franks, F. (1981). *Polywater*. Cambridge, MA: MIT Press.

Goldstein, H., Poole, C. P., & Safko, J. L. (2002). *Classical mechanics*. San Francisco: Addison-Wesley.

Hille, B. (1992). *Ionic channels of excitable membranes*. Sunderland: Sinauer.

Israelachvili, J. N. (2011). *Intermolecular and surface forces*. San Diego: Academic.

Jami, C. (2012). *Venus in arms* (1st ed.). CreateSpace Independent Publishing Platform. Scotts Valley, California: US.

Keister, T. (2008). Non chemical devices: Thirty years of myth busting. *Water Conditioning & Purification, 490*, 11–12.

Kožíšek, F., Auerbach, D., Gast, M. K. H., & Lindner, K. (2013). Comment on: "Evidence for the existence of stable-water-clusters at room temperature and normal pressure" [Phys. Lett. A 373 (2009) 3872]. *Physics Letters A, 377*(39), 2826–2827. https://doi.org/10.1016/j.physleta.2013.07.060.

Läuger, P. (1991). *Electrogenic ion pumps*. Sunderland: Sinauer.

Lee, S. H., & Rasaiah, J. C. (2011). Proton transfer and the mobilities of the H+ and OH− ions from studies of a dissociating model for water. *The Journal of Chemical Physics, 135*(12), 124505. https://doi.org/10.1063/1.3632990.

Liu, Z.-Q., Li, Y.-J., Zhang, G.-C., & Jiang, S.-R. (2011). Dynamical mechanism of the liquid film motor. *Physical Review E, 83*(2), 026303. https://doi.org/10.1103/PhysRevE.83.026303.

Lo, S.-Y. (1996). Anomalous state of ice. *Modern Physics Letters B, 10*(19), 909–919. https://doi.org/10.1142/S0217984996001036.

Lo, S.-Y., Lo, A., Chong, L. W., Tianzhang, L., Hua, L. H., & Geng, X. (1996). Physical properties of water with IE structures. *Modern Physics Letters B, 10*(19), 921–930. https://doi.org/10.1142/S0217984996001048.

Lo, S. Y., Geng, X., & Gann, D. (2009). Evidence for the existence of stable-water-clusters at room temperature and normal pressure. *Physics Letters A, 373*(42), 3872–3876. https://doi.org/10.1016/j.physleta.2009.08.061.

Lo, A., Cardarella, J., Turner, J., & Lo, S. Y. (2012). A soft matter state of water and the structures it forms. *Forum on Immunopathological Diseases and Therapeutics, 3*(3–4), 237–252. https://doi.org/10.1615/ForumImmunDisTher.2013007847.

Lowrie, W. (1997). *Fundamentals of geophysics*. Cambridge: Cambridge University Press.

Magnetized Water. (2020). *Omni enviro water systems*. Retrieved June 25, 2020, from www.omnienviro.com/magnetized-water/

McNeill, D. (2011, December 20). *Kim Jong-Il: Leader of North Korea who deepened the cult of*. The Independent. http://www.independent.co.uk/news/obituaries/kim-jong-il-leader-of-north-korea-who-deepened-the-cult-of-personality-in-his-country-following-the-6279399.html

McTaggart, L. (2002). *The field: The quest for the secret force of the universe*. New York: Harper Collins.

Merzbacher, E. (1970). *Quantum mechanics*. New York: Wiley & Sons.

Miskelly, G. M., Heben, M. J., Kumar, A., Penner, R. M., Sailor, M. J., & Lewis, N. S. (1989). Analysis of the published calorimetric evidence for electrochemical fusion of deuterium in palladium. *Science, 246*(4931), 793–796. https://doi.org/10.1126/science.246.4931.793.

Moore, W. J. (1974). *Physical chemistry* (5th ed.). Englewood: Prentice-Hall.

OECD. (2020). *Magnetic resonance imaging (MRI) exams (indicator)*. https://doi.org/10.1787/1d89353f-en. Accessed on 25 June 2020.

Pettitt, B. M., Makarov, V. A., & Andrews, B. K. (1998). Protein hydration density: Theory, simulations and crystallography. *Current Opinion in Structural Biology, 8*(2), 218–221. https://doi.org/10.1016/S0959-440X(98)80042-0.

Pollack, G. H. (2013). *The fourth phase of water: Beyond solid, liquid, and vapor*. Seattle: Ebner & Sons.

Price, P. B., Barwick, S. W., Williams, W. T., & Porter, J. D. (1989). Search for energetic-charged-particle emission from deuterated Ti and Pd foils. *Physical Review Letters, 63*(18), 1926–1929. https://doi.org/10.1103/PhysRevLett.63.1926.

Rousselet, J., Salome, L., Ajdari, A., & Prost, J. (1994). Directional motion of brownian particles induced by a periodic asymmetric potential. *Nature, 370*(6489), 446–447. https://doi.org/10.1038/370446a0.

Shnir, Y. M. (2005). *Magnetic monopoles* (p. Ch. 6). Berlin: Springer. https://doi.org/10.1007/3-540-29082-6.

Skou, J. C. (1957). The influence of some cations on an adenosine triphosphatase from peripheral nerves. *Biochimica et Biophysica Acta, 23*, 394–401. https://doi.org/10.1016/0006-3002(57)90343-8.

Tegmark, M. (2000). Importance of quantum decoherence in brain processes. *Physical Review E, 61*(4), 4194–4206. https://doi.org/10.1103/PhysRevE.61.4194.

Trinfinity8 Research – Dr. Glen Rein. (2018). *Trinfinity8*. https://trinfinity8.com/research-dr-glen-rein/ and https://trinfinity8.com/our-products/

Ueno, S., & Iwasaka, M. (1994). Properties of diamagnetic fluid in high gradient magnetic fields. *Journal of Applied Physics, 75*(10), 7177–7179. https://doi.org/10.1063/1.356686.

Unruh, W. G., & Zurek, W. H. (1989). Reduction of a wave packet in quantum Brownian motion. *Physical Review D, 40*(4), 1071–1094. https://doi.org/10.1103/PhysRevD.40.1071.

Virtanen, J. J., Makowski, L., Sosnick, T. R., & Freed, K. F. (2010). Modeling the hydration layer around proteins: HyPred. *Biophysical Journal, 99*(5), 1611–1619. https://doi.org/10.1016/j.bpj.2010.06.027.

Zheng, J., & Pollack, G. H. (2003). Long-range forces extending from polymer-gel surfaces. *Physical Review E, 68*(3), 031408. https://doi.org/10.1103/PhysRevE.68.031408.

Chapter 3
Water Nanoclusters in Cosmology, Astrobiology, the RNA World, and Biomedicine: The Universe as a Biosystem

Keith Johnson

Abstract Laboratory generation of water nanoclusters from amorphous ice and strong terahertz (THz) radiation from water nanoclusters ejected from water vapor into a vacuum suggest the possibility of water nanoclusters ejected into interstellar space from abundant amorphous ice-coated cosmic dust produced by supernovae explosions. Cosmic water nanoclusters offer a hypothetical unified solution to major mysteries of our universe: dark matter (Sect. 3), dark energy (Sect. 4), cosmology (Sect. 5), and the origin of life on Earth and other habitable planets throughout the universe as a connected biosystem (Sects. 6 and 7). Despite their expected low density in space compared to hydrogen, their quantum-entangled diffuse Rydberg electronic states make cosmic water nanoclusters a candidate for baryonic dark matter that can also absorb, via the microscopic dynamical Casimir effect, the virtual photons of zero-point-energy vacuum fluctuations above the nanocluster cut-off vibrational frequencies, leaving only vacuum fluctuations below these frequencies to be gravitationally active, thus leading to a possible common origin of dark matter and dark energy. This scenario offers novel explanations of the small cosmological constant, the coincidence of energy and matter densities, possible contributions of the red-shifted THz radiation from cosmic water nanoclusters at redshift $z \cong 10$ to the CMB spectrum, the Hubble constant crisis, the role of water as a known coolant for rapid early star formation, and ultimately how life may have originated from RNA protocells on Earth and exoplanets and moons in the habitable zones of developed solar systems. Together they lead to a novel cyclic universe model instead of a multiverse based on cosmic inflation theory. Finally, from the quantum biomechanics of water nanoclusters interacting with prebiotic organic molecules, amino acids, and RNA protocells on early Earth and habitable exoplanets, as well as with contemporary pathogenic RNA viruses such as COVID-19, for which an antiviral compound is proposed from these principles (Sect. 8), this scenario is consistent with the

K. Johnson (✉)
Massachusetts Institute of Technology, Cambridge, MA, USA
e-mail: kjohnson@mit.edu

A. Gadomski (ed.), *Water in Biomechanical and Related Systems*, Biologically-Inspired Systems 17, https://doi.org/10.1007/978-3-030-67227-0_3

anthropic principle that our universe must have those properties which allow life, as we know it – based on water, to develop at the present stage of its history.

Keywords Water nanoclusters · Cosmological scale · Force and pressure scales · High frequency system · Astrobiological role of water · Water in biosystems · RNA vs. water relationship · Coronavirus

1 Introduction

In the standard model of Big Bang cosmology, the matter and energy resources of our universe are controlled respectively by *dark matter* (Weinberg 2008) – a *non-baryonic* substance of undetermined nature – and *dark energy* (Riess et al. 1998) – a negative-pressure field of exotic physical origin. Over 30 years of searching for dark matter, proposed exotic elementary particles such as *weakly interacting massive particles* (WIMPS) and AXIONS thus far have not been observed experimentally, even in the latest Large Underground Xenon (LUX) and MIT ABRACADABRA detectors, respectively (Akerib et al. 2017; Ouellet et al. 2019). Nor have the WIMPS predicted from supersymmetry theory been created in the CERN LHC or ATLAS detector. The dark energy believed to be responsible for the accelerating expansion of our universe is usually considered to be a separate problem from dark matter and to be associated with zero-point-energy fluctuations of the cosmic vacuum. However, quantum field theory predicts a vacuum energy density that is too large by a factor of 10^{120}, which is the well-known cosmological constant problem (Weinberg 1989). I propose that nanoclusters of water molecules ejected by cosmic rays from amorphous ice layers on ubiquitous cosmic dust produced from exploding supernovae (Matsuura et al. 2019), albeit at low density compared to elemental hydrogen and oxygen, excited to their diffuse *Rydberg states* (Herzberg 1987), are a possible candidate for *baryonic* dark matter. The cut-off terahertz (THz) vibrational frequencies of such water nanoclusters are close to the $\nu_c \cong 1.7$ THz cut-off frequency of zero-point-energy vacuum fluctuations proposed to account for the small value of vacuum energy and cosmological constant (Beck and Mackey 2005, 2007). Cosmic water nanoclusters are postulated to capture via the *microscopic dynamical Casimir effect* (Souza et al. 2018) the high-frequency vacuum zero-point-energy virtual photons, leaving only the low-frequency ones to be gravitationally active. Water constitutes approximately 70% of our body weight, much of it as water nanoclusters called "structured" water (Chaplin 2006; Johnson 2012). This fact adds scientific and philosophical support based on the simplest *anthropic principle* (Weinberg 1987) to the proposal that water nanoclusters distributed as a low-density "dark fluid" throughout our universe are a possible common origin of dark matter and dark energy along with or instead of thus-far undiscovered exotic elementary particles.

Following a discussion of water nanocluster electronic and vibrational structures in Sects. 2, 3, 4, 5, 6, 7 and 8 are devoted respectively to their possible relevance to dark matter, dark energy, cosmology, astrobiology, the RNA world, and a proposed biomedical COVID-19 antiviral compound.

2 Water Nanoclusters

2.1 Cosmic Implications

Since hypothetical WIMP and AXION elementary particles, after many years of costly experiments, have still not been found, and *baryonic* dark matter has been largely ruled out by standard cosmological theory, why would one expect water nanoclusters to exist in the cosmos, and why would they have anything to do with dark matter and/or dark energy? Hydrogen and oxygen are the most abundant chemically reactive elements in our universe, oxygen slightly beating carbon. Water vapor plays a key role in the early stages of star formation, where it is an important oxygen reservoir in the warm environments of star-forming regions, and is believed to contribute significantly to the cooling of the circumstellar gas, thereby removing the excess energy built up during proto-stellar collapse (Bergin and van Dishoeck 2012). For example, the star-forming region of the Orion nebula produces enough water in a day to fill up earth's oceans many times over (Glanz 1998). The largest and farthest reservoir of water ever detected in the universe has been reported to exist in a high-redshift ($z = 3.91$) quasar approximately 12 billion light-years away (Bradford et al. 2011). The quasar water vapor mass is at least 140 trillion times that of all the water in the world's oceans and 100,000 times more massive than the sun. It has been proposed recently that water could have been abundant during the first billion years after the Big Bang (Bialy et al. 2015). In Sect. 5.2 of this paper, the possible role of water nanoclusters as a coolant catalyst for rapid early star formation in high-redshift clouds is discussed.

On planet Earth, through hydrogen bonding between water monomers, stable water nanoclusters, both neutral and ionized, are easily formed in molecular beams (Carlon 1981), occur naturally in the water vapor of earth's atmosphere (Aplin and McPheat 2005), and are produced from amorphous ice by energetic ion bombardment (Martinez et al. 2019). In the cosmos, therefore, a natural route to water nanocluster formation would be via the ejection from amorphous water-ice coatings of cosmic dust grains, which are believed to be abundant in interstellar clouds because they are a product of supernovae explosions (Matsuura et al. 2019). As a prime example, cosmic ray ionization of H_2 molecules adsorbed on amorphous ice-coated dust grains can lead to the reaction (Duley 1996)

$$H_2^+ + nH_2O + grain \rightarrow H_3O^+ \left(H_2O\right)_{n-1} - +grain. \tag{3.1}$$

Interstellar population of protonated water-nanocluster ions released by this process has been estimated to approach 10^{-6} of the average atomic hydrogen population but could likely be significantly greater (Martinez et al. 2019). Due to their large electric dipole moments ($\geq$ 10D) oscillating at THz frequencies, such water nanoclusters are believed to be responsible for the observed strong THz emission from water vapor into a vacuum under intense U.V. optical stimulation (Johnson et al. 2008) (Fig. 3.1) and therefore should be relatively stable under similar cosmic radiation.

While $H_3O^+(H_2O)_{n-1}$ nanoclusters ejected by ion bombardment from amorphous ice have been observed over a range of n-values (Martinez et al. 2019), the "magic-number" n = 21 pentagonal dodecahedral $H_3O^+(H_2O)_{20}$ or equivalent protonated $(H_2O)_{21}H^+$ nanocluster (Fig. 3.2) is exceptionally stable in a vacuum and is of potential cosmic importance because it can be viewed as a H_3O^+ (hydronium) ion caged ("clathrated") by an approximately pentagonal dodecahedral cage of twenty water molecules (Miyazaki et al. 2004; Shin et al. 2004). Interstellar H_3O^+ has been observed recently (Lis et al. 2014) and its discovery points to the challenge of trying to identify spectroscopically larger cosmic water nanoclusters such as $H_3O^+(H_2O)_{20}$ because both spectra fall into the same THz region. The occurrence of stable pentagonal dodecahedral water nanocluster clathrate hydrates in the interstellar medium has recently been predicted (Ghosh et al. 2019). References (Miyazaki et al. 2004; Shin et al. 2004; Lis et al. 2014) could be the starting point for attempts to identify the cosmic presence of $H_3O^+(H_2O)_{20}$.

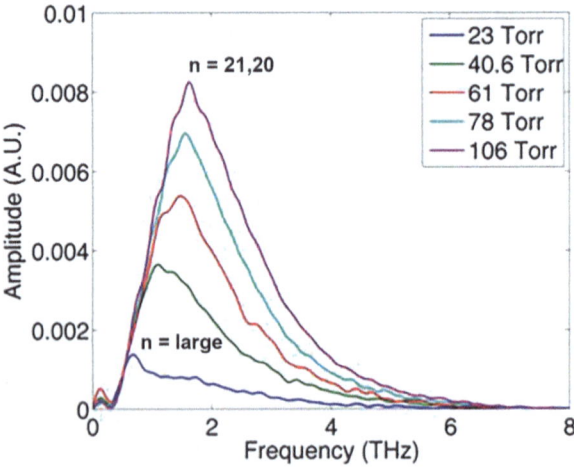

Fig. 3.1 Pressure dependence of the THz wave generation amplitudes (in arbitrary units) from water nanoclusters produced from water vapor in (Johnson et al. 2008)

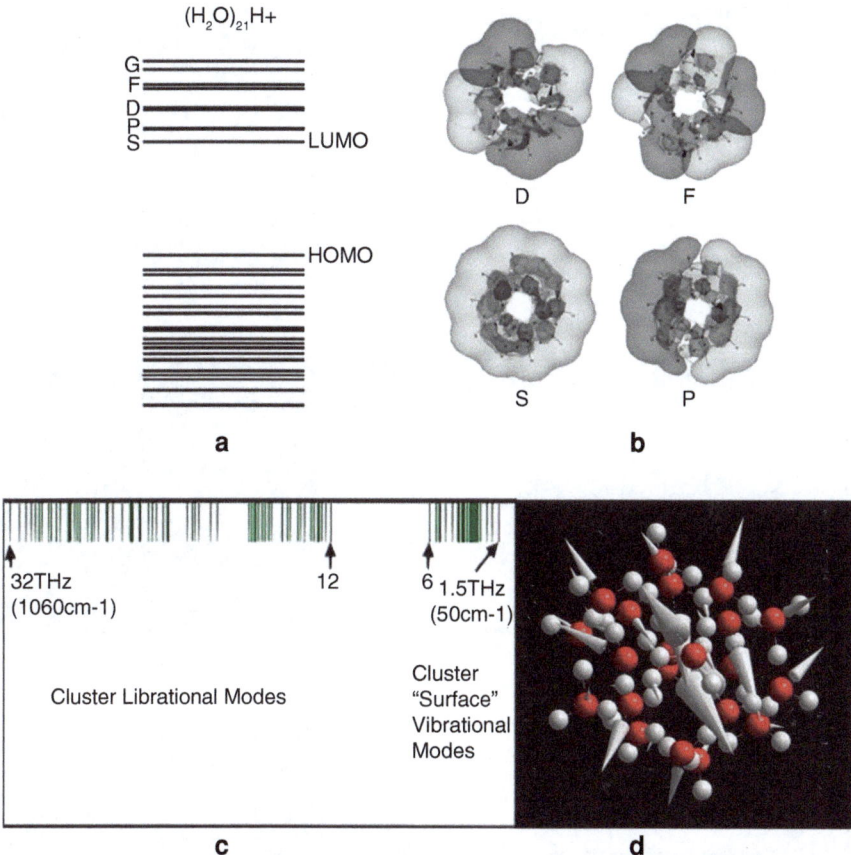

Fig. 3.2 Density-functional molecular-orbital states and vibrational modes of the $(H_2O)_{21}H^+$ protonated water cluster. (**a**). Molecular-orbital energy levels. (**b**). Wavefunctions of the lowest unoccupied molecular orbitals. (**c**). THz vibrational spectrum. (**d**). Lowest-frequency THz vibrational mode and relative vibrational amplitudes

2.2   Electronic and Vibrational Structures

Figure 3.2 shows the ground-state molecular-orbital energies, wavefunctions, and vibrational modes of the pentagonal dodecahedral $(H_2O)_{21}H^+$ or $H_3O^+(H_2O)_{20}$ cluster computed by the SCF-Xα-Scattered-Wave density-functional method co-developed by the author (Slater and Johnson 1972, 1974). Molecular dynamics simulations yield results qualitatively unchanged at temperatures well above 100C, where the cluster remains remarkably intact. Similar calculations for the neutral pentagonal dodecahedral water cluster, $(H_2O)_{20}$ and arrays thereof have also been performed, yielding the THz vibrational modes displayed in Figs. 3.3 and 3.4.

These results are qualitatively similar to those shown in Fig. 3.2, but they indicate a gradual decrease of the cluster cut-off vibrational frequency with increasing

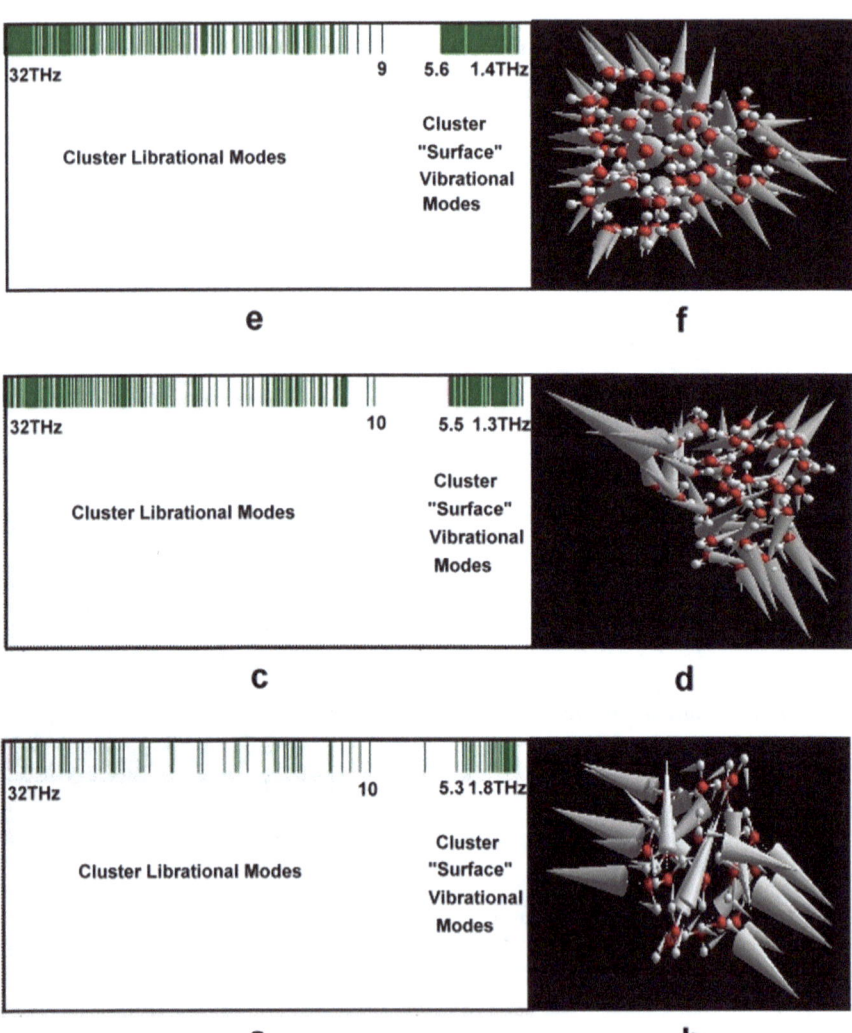

Fig. 3.3 (**a**). THz vibrational spectrum of a pentagonal dodecahedral $(H_2O)_{20}$ cluster. (**b**). Lowest-frequency THz vibrational mode. (**c**). THz vibrational spectrum of an array of three dodecahedral water clusters. (**d**). Lowest-frequency THz vibrational mode. (**e**). THz vibrational spectrum of an array of five dodecahedral water clusters. (**f**). Lowest-frequency THz vibrational mode

cluster size. The latter trend correlates with the experimental studies of THz radiation emission from water vapor nanoclusters (Johnson et al. 2008), showing in Fig. 3.1 the shift in the cluster THz emission peaks toward lower frequencies and intensities – corresponding to a trend toward larger clusters – with decreasing vapor ejection pressure into the vacuum chamber where the radiation was measured. Relating this finding to Eq. 3.1 would suggest decreasing THz emission cut-off

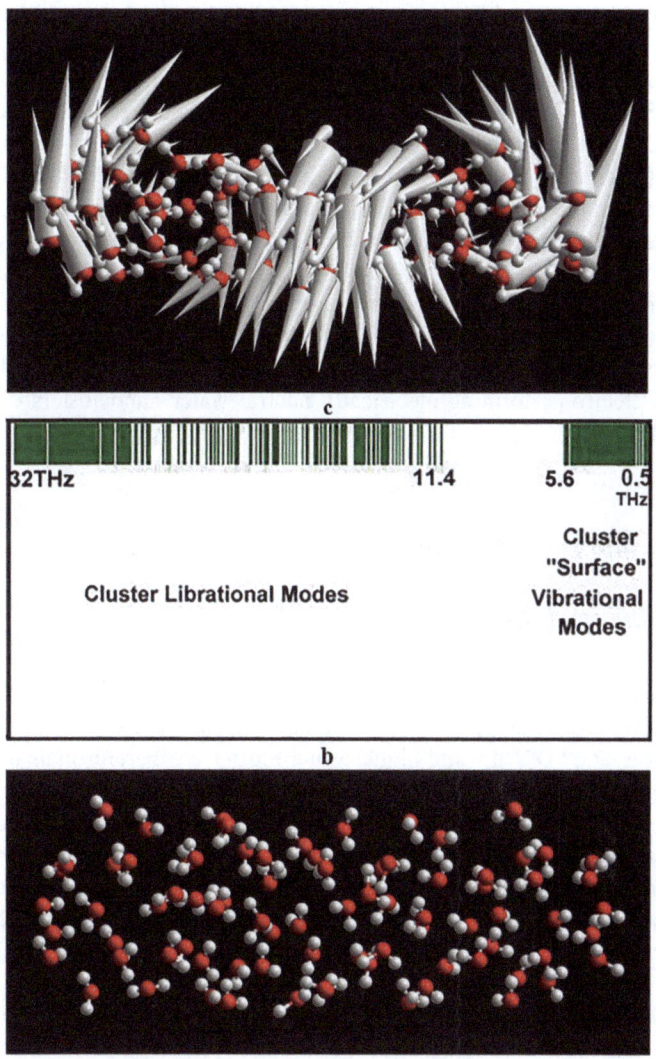

Fig. 3.4 (**a**). Linear array of five pentagonal dodecahedral water nanoclusters. (**b**). THz vibrational spectrum. (**c**). Lowest-frequency THz vibrational mode

frequencies and intensities with the increasing sizes (increasing n-values) of water nanoclusters ejected from ice-coated cosmic dust.

Common to all these water clusters are: (1) lowest unoccupied (LUMO) energy levels like those in Fig. 3.2a, which correspond to the diffuse Rydberg "S"-, "P"-, "D"- and "F"-like cluster "surface" molecular-orbital wavefunctions shown in Fig. 3.2b, and (2) bands of vibrational modes between 0.5 and 6 THz (Figs. 3.2, 3.3 and 3.4), due to O-O-O "squashing" (or "bending") and "twisting" motions between adjacent hydrogen bonds. The vectors in Figs. 3.2, 3.3 and 3.4 represent the directions and relative amplitudes of the lowest THz-frequency modes corresponding to the O-O-O "bending" (or "squashing") motions of the water-cluster "surface" oxygen ions. Surface O-O-O bending vibrations of water clusters in this energy range have indeed been observed under laboratory conditions (Brudermann et al. 1998). Ultraviolet excitation of an electron from the HOMO to LUMO (Fig. 3.2a) can put the electron into the Rydberg "S"-like cluster molecular orbital mapped in Fig. 3.2b. Occupation of this orbital produces a bound state, even when an extra electron is added, the so-called "hydrated electron" (Jordan 2004). In contrast, a water monomer or dimer has virtually no electron affinity. Therefore, in space – especially within dense interstellar clouds – $(H_2O)_{21}H^+$ or $H_3O^+(H_2O)_{20}$ and larger water-nanocluster ions ejected from ice-coated cosmic dust according to Eq. 3.1 are likely to capture electrons, forming electrically neutral water nanoclusters of the types shown in Fig. 3.3.

3   Baryonic Dark Matter

3.1   Rydberg Matter

Starlight energy can stimulate electronic excitations from the HOMO of $(H_2O)_{21}H^+$ (Fig. 3.2a) (or from the LUMO with a captured hydrated electron) to the increasingly diffuse "P", "D", "F" and higher water cluster Rydberg orbitals in Fig. 3.2b. These states have vanishing spatial overlap with the lower-energy occupied ones, have long lifetimes that increase with increasing excitation energy and effective principal quantum number, and thus are candidates for *Rydberg Matter* (RM) – a *low-density* condensed phase of weakly interacting individual Rydberg-excited molecules with long-range effective interactions (Badei and Homlid 2002). RM can interact or become *quantum-entangled* over long effective distances, causing it to be transparent to visible, infrared, and radio frequencies, and thus qualifies as *baryonic dark matter* (Badei and Homlid 2002). Two water nanoclusters of the forms shown in Figs. 3.2 and 3.3a, but each holding an excited or "hydrated" electron in the Rydberg "S" LUMO, are like giant hydrogen atoms, which for short distances between the clusters will form an overlapping "Sσ-bonding" molecular orbital like that shown in Fig. 3.5a holding two spin-paired electrons in analogy to a giant hydrogen molecule.

Promotion of electrons to the Rydberg "P" LUMO produces bonding "Pσ"- and "Pπ"-bonding molecular orbitals like those illustrated in Fig. 3.5b, c. However, the approach of two water nanoclusters to each other in interstellar space should be a rare occurrence because of their relatively low density. For much larger distances between the clusters, the diffuse molecular orbitals of their highest-energy Rydberg states "overlap" sufficiently to permit quantum entanglement of water nanoclusters over long distances in space, thus qualifying cosmic water nanoclusters as possible baryonic dark matter.

Fig. 3.5 (**a**). "Sigma-bond" molecular-orbital overlap of water-nanocluster Rydberg "S" LUMOs shown in Fig. 3.2. (**b**). "Sigma-bond" molecular-orbital overlap of Rydberg "P" LUMOs (**c**). "Pi-bond" molecular-orbital overlap of Rydberg "P" LUMOs

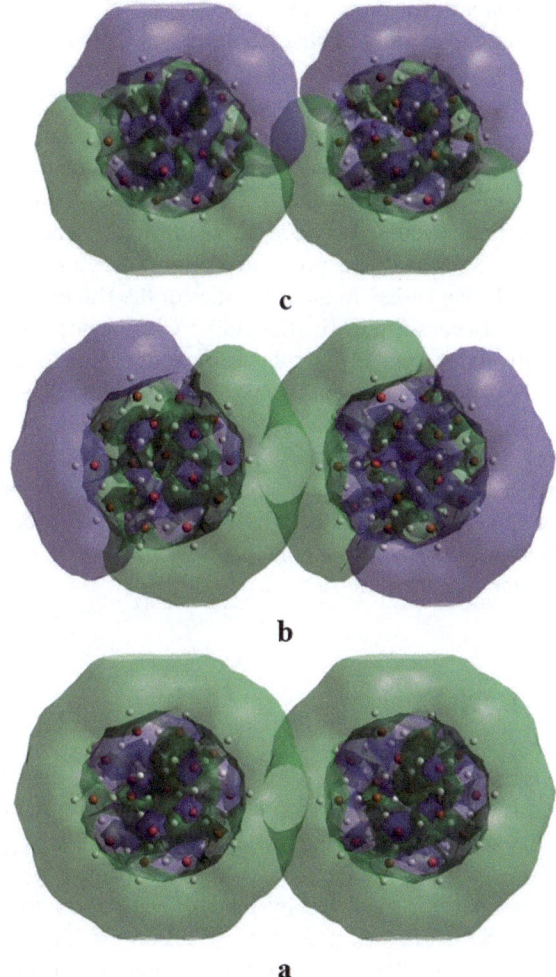

c

b

a

3.2   The Bullet Cluster and Galactic Halos

Can low-density, quantum-entangled water-nanocluster RM account for at least part of the dark matter estimated from inflationary Big Bang theory? The consensus of standard cosmology is that the unknown dark matter cannot be baryonic. Gravitational lensing observations of the *Bullet Cluster* revealing the separation of normal luminous matter and dark matter have been said to be the best evidence to date for the existence of nonbaryonic dark matter (Clowe et al. 2004). Protonated water nanoclusters produced by Eq. 3.1 and shown in Fig. 3.2 for n = 21 are positively charged, although, as pointed out above, such clusters are likely to pick up a "hydrated" electron (Jordan 2004) from space once ejected from ice-coated cosmic dust, forming electrically neutral clusters like those pictured in Fig. 3.3. It is believed that nonbaryonic dark matter is uncharged. Nevertheless, it has recently been argued that a small amount of charged dark matter could cool the baryons in the early universe (Munoz and Loeb 2018). While magnetic fields associated with celestial objects should interact with water nanocluster ions, which are motional sources of magnetism, the origin and relevance of intergalactic magnetic fields is still debated (Jedamzik and Pogosian 2020). Magnetic fields may be important to the possible role of water nanoclusters in water vapor as a coolant catalyst for rapid early star formation, discussed in Sect. 5.2 of this paper. The electric charge of the ice-coated cosmic dust that is the postulated origin of cosmic water nanoclusters according to Eq. 3.1 may be key to the properties of the Bullet Cluster dark matter (Clowe et al. 2004). Because much of the Bullet Cluster normal matter is likely composed of positively charged cosmic dust (Mann 2001), its electrical repulsion of protonated water nanoclusters would enhance the ejection of water nanoclusters described by Eq. 3.1, thus explaining the observed separation of normal luminous matter and dark matter. Despite uncertainties about electric fields on the galactic scale (Bally and Harrison 1978; Chakraborty et al. 2014), it is possible that such fields could cause cosmic water nanocluster RM to aggregate around the peripheries of galaxies, thereby possibly explaining *galactic dark matter halos* similarly to the Bullet Cluster dark matter.

4 Dark Energy

Quantum field theory predicts a vacuum energy density that is too large by a factor of 10^{120}, which is the well-known cosmological constant problem (Weinberg 1989). It has been suggested that if the observed dark energy responsible for the accelerated expansion of the universe is equated to the otherwise infinite cosmic vacuum energy density predicted by theory, then gravitationally active zero-point-energy vacuum fluctuations must have a cut-off frequency of $v_c \cong 1.7$ THz (Beck and Mackey 2005, 2007). In other words, the virtual photons associated with vacuum fluctuations should be of gravitational significance only below this frequency to be consistent with the observationally small magnitude of dark energy. A $v_c \cong 1.7$ THz vacuum fluctuation cut-off frequency is the same order of magnitude as the cut-off vibrational frequencies of prominent water nanoclusters, although these frequencies decrease with increasing cluster size (Figs. 3.2, 3.3 and 3.4) or increasing n-value in Eq. 3.1. Other molecules in space, such as hydrogen, water monomers, carbon buckyballs, and various observed organic molecules do not have vibrational cut-off frequencies in this THz region. Thus as low-density Rydberg Matter spread diffusely through space, water nanoclusters like the one in Fig. 3.6 could absorb the virtual photons of vacuum fluctuations at frequencies greater than $v_c \cong 1.7$ THz via the *microscopic dynamical Casimir effect* (Souza et al. 2018), which converts vacuum virtual photons to real ones, leaving only the low-frequency ones to be gravitationally active. The absorbed photons can then decay via emitted THz radiation (Johnson et al. 2008) (Fig. 3.1).

Fig. 3.6 Microscopic dynamical Casimir absorption of zero-point-energy vacuum fluctuations between THz vibrational states of a cosmic water nanocluster

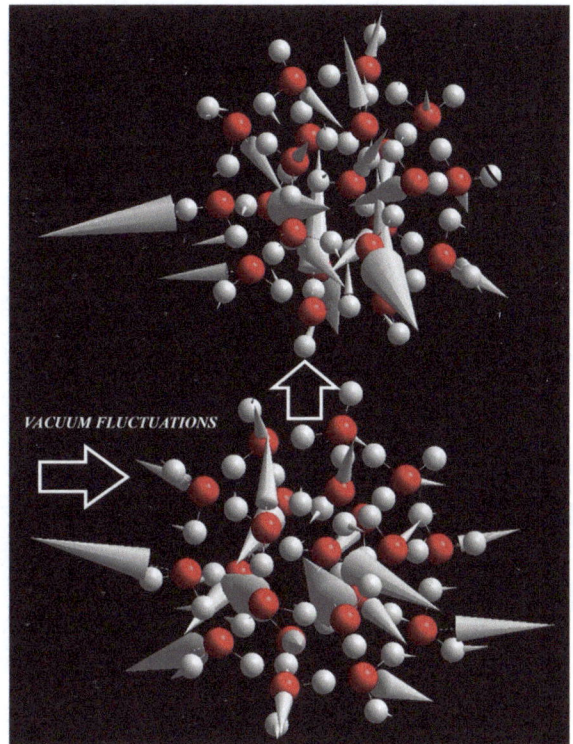

VACUUM FLUCTUATIONS

To quantify this, we conventionally view the vacuum electromagnetic field (excluding other fields) as a collection of harmonic oscillators of normal-mode frequencies ν_k, summing over the zero-point energies of each oscillator mode, leading to the following energy density (energy E per unit volume V)

$$\rho_{vac} = \frac{E}{V} = \frac{1}{V}\sum_k \frac{1}{2}h\nu_k = \frac{4\pi h}{c^3}\int_0^\infty \nu^3 d\nu \qquad (3.2)$$

where the wave vector k signifies the normal modes of the electromagnetic field that are consistent with the boundary conditions on the quantization volume V. As V approaches infinity, one obtains the right-hand side of Eq. 3.2.

The divergent integral in Eq. 3.2 can be avoided by replacing the upper limit by a cut-off frequency set by the Planck scale (Weinberg 1989). However, this results in a huge vacuum energy that exceeds the cosmologically measured value by 120 orders of magnitude.

If instead we subtract from (3.2) the energy density

$$\rho_c = \frac{4\pi h}{c^3}\int_{\nu_c}^\infty \nu^3 d\nu \qquad (3.3)$$

of the virtual photons of zero-point vacuum fluctuations captured by the water clusters through the *microscopic dynamical Casimir effect* (Souza et al. 2018), the divergent integral in Eq. 3.2 is largely cancelled, leaving the finite quantity, Eq. 3.4 to be identified with the present cosmologically observed dark energy density

$$\frac{4\pi h}{c^3}\int_0^{v_c} v^3 dv = \frac{\pi h v_c^4}{c^3} = \rho_{dark}. \tag{3.4}$$

The PLANCK observations have concluded that dark energy presently constitutes 68.3% of the total known energy of the universe (Ade et al. 2016), as compared to the earlier WMAP result of 73% (Bennett et al. 2013), giving

$$\rho_{dark} = 3.6\ 4\frac{\text{GeV}}{\text{m}^3}. \tag{3.5}$$

Equations 3.4 and 3.5 then require a cut-off frequency v_c of approximately 1.66 THz, which is indeed the same order of magnitude as the cut-off frequencies of the smallest pentagonal dodecahedral water clusters shown in Figs. 3.2d and 3.3a. However, since v_c *decreases with increasing water-cluster size* (Figs. 3.3 and 3.4) or with increasing n-value in Eq. 3.1, a trend toward the ejection of larger water clusters from cosmic dust over time would imply a *decrease of dark energy density over time* according to Eq. 3.4.

5   Cosmology

5.1   The CMB Spectrum

The consensus of standard inflationary cosmology (Guth 1981, 2007; Linde 2008) is that the measured cosmic microwave background (CMB) spectrum of the universe has its origin at approximately 380 thousand years after the Big Bang (Ade et al. 2016; Bennett et al. 2013). Is there a possible and credible additional contribution to the CMB that is consistent with its spectrum and the THz vibrational properties of water nanoclusters discussed in the previous sections? It was suggested long ago that the CMB might be attributable to thermalization by "cosmic dust" in the form of hollow, spherical shells of high dielectric constant or conducting "needle-shaped grains" (Layzer and Hively 1973; Wright 1982). Layzer and Hively (1973) argued that a relatively low density of high-dielectric-constant dust could thermalize the radiation produced by objects of galactic mass at redshift z $\cong$ 10. Because of their computed large electric dipole moments and measured strong THz radiation emission (Johnson et al. 2008) (Fig. 3.1), optically pumped water nanoclusters in water vapor consisting of spherical "shells" of water-cluster O-H bonds (Figs. 3.2 and 3.3) or "strings" of water clusters (Fig. 3.4) satisfy the conditions proposed in (Layzer and Hively 1973; Wright 1982). The thin amorphous water ice that coats the cosmic dust from which water clusters are ejected according to Eq. 3.1 itself consists of disordered water nanoclusters of high dielectric constant and could directly contribute. Astronomical observations have pushed back the epoch of pro-togalaxy formation and reionization to redshifts of z = 8.6 and z = 9.6, respectively, *i.e.* to corresponding times of 600 and 500 million years after the Big Bang (Lehnert et al. 2010; Zheng et al. 2012), although more recently, the Hubble Space Telescope has found a galaxy at z $\cong$ 11, corresponding to just 400 million years after the Big Bang (Oesch et al. 2016). At redshift z $\cong$ 10, the distinctive THz vibrational mani-folds of water clusters (Figs. 3.2, 3.3 and 3.4), as well as the laboratory THz emis-sion peaks of Fig. 3.1 are red-shifted to the region of the measured CMB spectrum, suggesting water nanocluster THz emission originating at z $\cong$ 10, where the tem-perature T $\cong$ 30 K is compatible with the existence of such nanoclusters, could indeed contribute to the CMB in addition to photons from the "recombination" period at z $\cong$ 1100, where the temperature is T $\cong$ 4000 K.

Laboratory measurements (Johnson et al. 2008) (Fig. 3.1) of the THz emission from water nanoclusters as a function of ejection pressure into a vacuum chamber indicate emission peaks that decrease in frequency and intensity with decreasing pressure and thus, according to Figs. 3.2, 3.3 and 3.4, with increasing cluster size. The most intense emission peak at approximately 1.7 THz is assigned to water clus-ters of the "magic numbers" n = 21 and 20 shown in Figs. 3.2d and 3.3a, respec-tively, whereas the peaks decreasing in frequency and intensity with lower pressure to approximately 0.5 THz are due to larger clusters like those shown in Figs. 3.3b, c and 3.4a. The effective temperatures of the water nanoclusters scale with this pres-sure trend, with the smallest clusters emitting the "hottest" radiation peaking around 1.7 THz. Applying this power spectrum to the redshifted radiation from water nano-clusters ejected from ice-coated cosmic dust at z $\cong$ 10, the spectrum of the smallest versus larger water nanoclusters in particular regions of space might be similar to the measured CMB power spectrum (Ade et al. 2016; Bennett et al. 2013), but is a subject for future investigation requiring more resources by this author. The labora-tory ejection pressure dependence of the vacuum chamber power spectrum shown in Fig. 3.1 applied to the effective water-nanocluster ejection "pressures" from ice-coated cosmic dust according to Eq. 3.1 further suggests a "pressure wave" created at z $\cong$ 10, or around 500 million years after the Big Bang, in analogy to the CMB acoustic wave assigned to the "recombination" period at z $\cong$ 1100 or 380 thousand years after the Big Bang (Ade et al. 2016; Bennett et al. 2013). In other words, if one were to interpret simplistically the measured CMB power spectrum and its anisot-ropy as due at least partially to the redshifted THz radiation from water nanoclusters ejected from ice-coated cosmic dust at z $\cong$ 10, this would suggest a relatively slow "classical" process extending over millions of years compared to an inflationary hot Big Bang originating from a quantum singularity. Adding to the possible credibility of this scenario is the fact that z $\cong$ 10, or around 500 million years after the Big Bang, is near the first times of suspected early (Population III) star formation, with most of these thus far hypothetical stars having short lives and becoming explosive supernovae that produce the cosmic dust from which cosmic water nanoclusters can be ejected according to Eq. 3.1.

5.2   Early Star Formation

Water vapor is a recognized coolant for star formation (Bergin and van Dishoeck 2012). The report of a huge reservoir of water in a high-redshift ($z \cong 4$) quasar, corresponding to a water vapor mass at least 140 trillion times that of all the water in the world's oceans and 100,000 times more massive than the sun (Bradford et al. 2011), together with the recent proposal that water vapor could have been abundant during the first billion years after the Big Bang (Bialy et al. 2015) suggests the possibility of a significant population of stable water nanoclusters at redshift $z \cong 10$ or around 500 million years after the Big Bang.

The rapidness at which some early stars were created from observed dense gas clouds at $z \cong 6.4$ or 850 million years after the Big Bang (Banandos et al. 2019) can possibly be understood by the presence of water nanoclusters in the cloud's water vapor coolant. Studies of the infrared absorption by water nanoclusters in the laboratory and Earth's atmosphere (Carlon 1981; Aplin and McPheat 2005), including both protonated cluster ions of the type shown in Fig. 3.2 and neutral clusters like those shown in Fig. 3.3, have established their extraordinary heat storage as due mainly to the *nanocluster hydrogen librational modes* – especially those near 32 THz (1060 cm^{-1}) – shown in these figures. The cooling effect of the star-forming gas clouds can consequently be achieved through the release of photons associated with the *cluster surface modes* in the 1–6 THz range (Figs. 3.2 and 3.3). These photons are the same ones that might contribute to the CMB, as described above. In other words, if one accepts a scenario where redshifted THz radiation from cosmic water nanoclusters at $z \cong 10$ contributes to the CMB spectrum, then the CMB also possibly contains information about early star formation.

5.3   The Hubble Constant Crisis

The present model suggests a possible resolution of the Hubble constant "crisis," where the value of this constant deduced from observations of supernovae and cepheids has indicated the universe is expanding significantly faster than the value concluded from measurements of the microwave radiation emitted immediately after the Big Bang. The ice-coated cosmic dust responsible for ejecting the water nanoclusters proposed herein to underlie dark energy and our accelerating universe is a product of stellar evolution. Since the first stars were born only after the reionization phase following the recombination phase of hydrogen formation that began approximately 380,000 years after the Big Bang, there would be no significant cosmic dust and thus no cosmic water nanoclusters until many years after the Big Bang. This conclusion is supported by a recent observation of the oldest cosmic dust 200 million years after the birth of the first stars (Laporte et al. 2017). As the universe expanded over 13.8 billion years, the amount of cosmic dust increased with the formation of more stars and galaxies until one reached the present era where there is enough dust and ejected water-nanocluster Rydberg matter to account for the Hubble constant deduced from recent supernovae and cepheid observations, as well as the current coincidence of dark energy and matter densities. This scenario is consistent with studies of high-redshift quasars showing that dark energy has increased from the early universe to the present (Risaliti and Lusso 2019). It is also somewhat consistent with a claim that baryon inhomogeneities explain away the Hubble crisis but disagrees that they are due to primordial magnetic fields (Jedamzik and Pogosian 2020).

5.4 A Cyclic Cosmology

Inflationary cosmology (Guth 1981, 2007; Linde 2008) leads to the conclusion that dark matter density should decrease faster than dark energy in an accelerating universe, so that eventually dark energy will become dominant, and the ultimate fate of the universe is that all the matter in the universe will be progressively torn apart by its expansion – the so-called "big rip". In contrast, the present model suggests the following hypothetical scenario, albeit qualitative:

As hydrogen is used up in stellar nuclear fusion to produce heavier elements and more water-ice-coated cosmic dust, there will be a tendency for the increasing dust to expel larger water clusters, $i.e.$ larger n-values in Eq. 3.1. A trend to lower vibrational cut-off frequency v_c with increasing water cluster size (increasing n-value) is suggested by the computed results shown in Figs. 3.2, 3.3 and 3.4. With the increasing population of larger water clusters over time, there will be a trend to clusters having a lower vibrational cut-off frequency v_c in Eq. 3.4, resulting in a movement to lower dark energy density ρ_{dark} over time and thus a reduction of the acceleration of the expanding universe. Eventually, the gravity of remaining baryonic mass will take over and the universe will stop expanding, begin to collapse, and slowly return to the $z \cong 10$ period around 500 million years after the Big Bang, where the size of the universe will be approximately 10% of the present one. As the latter happens, pressure of the water vapor expected to exist at $z \cong 10$ (Bialy et al. 2015) will increase and, according to Fig. 3.1, smaller water nanoclusters – especially those with the magic numbers n = 21,20 shown in Figs. 3.2d and 3.3a, respectively – will again be favored over larger ones. Those water nanoclusters will be available again as a coolant catalyst for star formation, as described above. This is a *classical non-singular* starting point for possible *non-inflationary re-expansion*, although further collapse toward the reionization period and decomposition of water vapor to supply hydrogen for star formation and non-inflationary re-expansion is also a possibility. The *cycle period* is an open question to be addressed, but this cyclic universe scenario mimics ones proposed by others (Ijjas et al. 2017; Ijjas and Steinhardt 2019; Penrose 2006) and therefore lessens the credibility of the *multiverse theory* that is an evolutionary part of inflation theory.

6 Astrobiology

The discovery of organic molecules, including prebiotic ones, in interstellar space, dust clouds, comets, and meteorites over the past 50 years has been impressive. While the significance of cosmic carbon compounds to the existence of life throughout the universe has most often been emphasized, it is ultimately water that is the key to the structures and functions of carbon-based biomolecules. The human body, which is approximately 70% water by weight, cannot exist without water, which is essential for the synthesis of RNA, DNA, and proteins. Much of this water is attached to proteins, RNA, and DNA as water nanoclusters, called "structured" water (Chaplin 2006; Johnson 2012). For example, proteins not containing nanostructured water will not fold properly and can lead to degenerative diseases such as Parkinson's, Alzheimer's, and cataracts, while such structured water is also essential to RNA and DNA replication. How water nanoclusters interact with organic molecules in astrobiology can be investigated by first-principles molecular-orbital calculations using the SCF-Xα-Scattered-Wave density-functional method (Slater and Johnson 1972, 1974).

This method was first applied to proteins (Cotton et al. 1973; Yang et al. 1975; Case et al. 1979). The resulting molecular structures and lowest THz-frequency vibrational modes of a pentagonal dodecahedral water nanocluster interacting with *methane, anthracene*, and the *amino acid valine*, respectively, all of which have been found in interstellar space, dust clouds, comets, or meteorites (Lacy et al. 1991; Iglesias-Groth et al. 2010; Kvenvolden et al. 1970), are shown in Fig. 3.7. In all three examples, there is coupling of 1.8 THz water-cluster "surface" oxygen motions to the carbon atomic motions, as represented by the vectors in Fig. 3.7. In the anthracene and valine examples there are carbon-carbon bonds, so that the water-cluster-induced carbon motions at 1.8 THz are "bending" modes of the C-C bonds. In valine, there is also coupling between the water-cluster 1.8 THz vibrational mode to that of the nitrogen atom (Fig. 3.7f). Valine is an α-amino acid that is used in the biosynthesis of proteins by polymerization beginning with the nitrogen atom and ending with a carbon. The point here is that water nanoclusters and prebiotic molecules delivered by cosmic dust and meteorites to Earth could have jump-started life here and on exoplanets in the habitable zones of distant solar systems. This requires the first self-replicating RNA necessary for DNA and the synthesis of proteins from amino acids.

7 The RNA World

7.1 *From Prebiotic Molecules to RNA*

Compelling arguments for the so-called *RNA World* as the origin of life on planet Earth and possibly elsewhere in the universe (Joyce and Orgel 1993) – preceding DNA- and protein-based life – has posed the fundamental problem of explaining

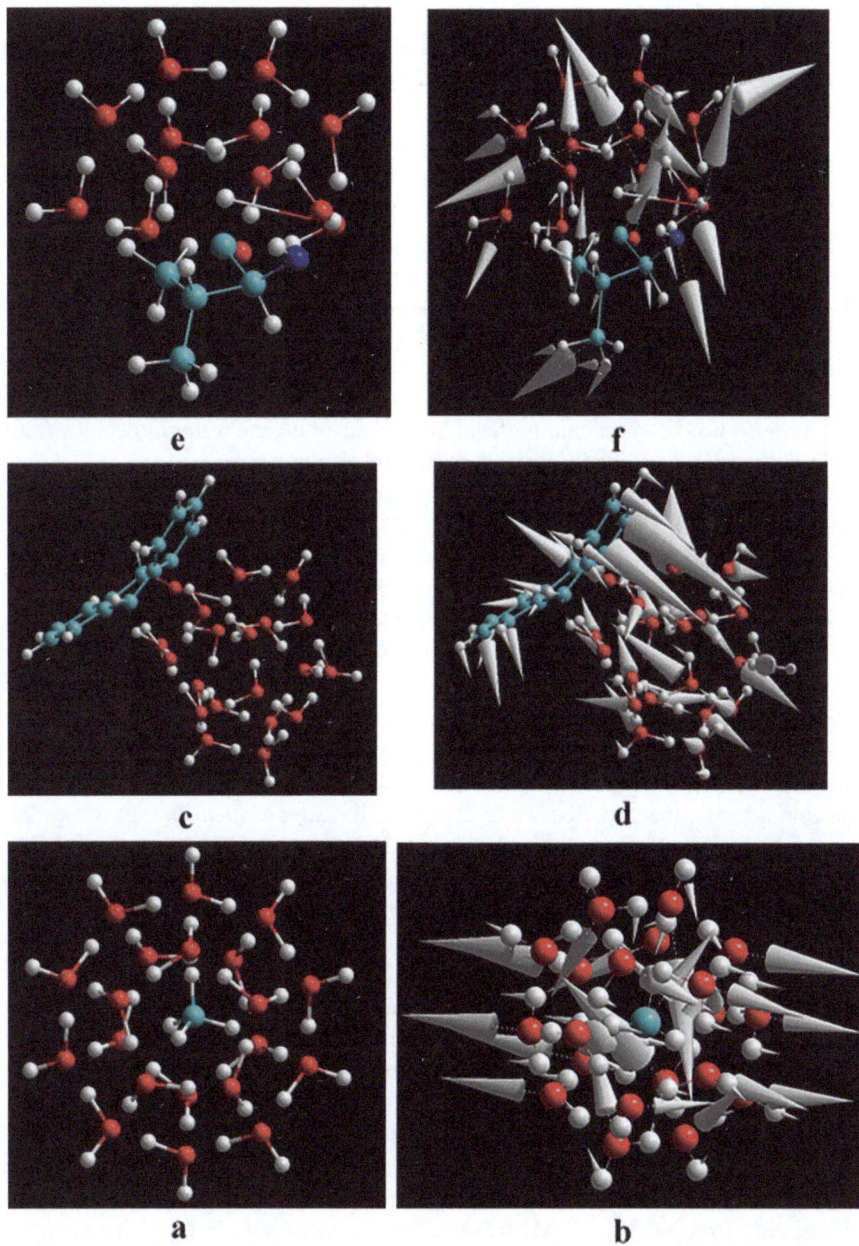

Fig. 3.7 (**a**). Pentagonal dodecahedral $CH_4(H_2O)_{20}$ methane clathrate. (**b**). 1.8 THz vibrational mode. (**c**). $(H_2O)_{20}$ cluster interacting with anthracene. (**d**). The vibrational coupling of the $(H_2O)_{20}$ 1.8 THz vibrational mode to a THZ "bending" mode of anthracene. (**e**). Hemispherical pentagonal dodecahedral water nanocluster clathrating a valine amino acid residue. (**f**). Vibrational coupling of the water clathrate 1.8 THz vibrational mode to the valine amino acid "bending" mode

how the first self-replicating RNA polymers, such as the segment shown in Fig. 3.8, were created chemically from a pool of prebiotic organic molecules, nucleosides, and phosphates. A recent paper (Totani 2020) based on polymer physics and inflationary cosmology proposes that extraterrestrial RNA worlds and thus the emergence of life in an inflationary universe must be statistically rare. Since the polymerization of long RNA chains in water has been demonstrated but not explained definitively (Costanzo et al. 2009), it is proposed here that water nanoclusters comprising liquid water and water vapor can act as catalysts for prebiotic RNA synthesis, increasing the likelihood of RNA worlds and thus the emergence of life wherever water is present in a non-inflationary or cyclic universe of the type described in this paper.

Simply described, the chemical steps of combining the prebiotic molecules of Fig. 3.8 to yield an RNA sequence of four polymerized nucleobases, *guanine*, *adenine*, *uracil*, and *cytosine*, require the effective loss of eleven water molecules from the initial reactants and their effective recombination in RNA.

This is a *dehydration-condensation reaction* (Cafferty and Hud 2014). Cyclic water pentamers (Harker et al. 2005) (Fig. 3.8) have been identified as being key to the hydration and stabilization of biomolecules (Teeter 1984). Such examples indicate the tendency of water pentagons to form closed geometrical structures around amino acids (Fig. 3.7e, f) and nucleosides (Neidle et al. 1980). Studies of supercooled water and amorphous ice have revealed the presence of cyclic and clathrate water pentamers (Nandi et al. 2017; Yokoyama et al. 2008), suggesting the possible delivery of water pentamers by ice-coated cosmic dust and meteorites to the atmospheres of Earth and habitable exoplanets. At the opposite extreme temperatures and pressures of hydrothermal ocean vents (*black smokers*) arising from planetary volcanic activity, the expelled water can be in the *supercritical* phase, where the structure is neither liquid nor vapor but instead isolated water nanoclusters (Sahle et al. 2013). Therefore, at both temperature and pressure extremes, water pentamers interacting with prebiotic molecules could nucleate additional water molecules – eleven in the Fig. 3.8 example – to form the more stable pentagonal dodecahedral cluster $(H_2O)_{21}H^+$, which could then provide via its THz vibrations (Fig. 3.2c, d) the eleven water molecules necessary to yield the RNA sequence of Fig. 3.8. In other words, water nanoclusters delivered by cosmic dust or hydrothermally to planet Earth and habitable exoplanets could have provided a catalytic pathway for the dehydration-condensation-reaction mechanism of RNA polymerization.

How a $(H_2O)_{21}H^+$ cluster could expel water molecules or OH groups when interacting with prebiotic molecules to promote RNA chain growth deserves further analysis. As pointed out in Sect. 2.2, this protonated water cluster readily takes up an extra electron into the LUMO (Fig. 3.2a) – a *hydrated electron* – as noted in Fig. 3.8. The proximity of the resulting electrically neutral $(H_2O)_{21}H$ cluster occupied LUMO "S" orbital to the lowest unoccupied, nearly degenerate cluster "P_x, P_y, P_z" orbitals (Fig. 3.2a, b) suggests the possible coupling between the hydrated electron and the pentagonal dodecahedral cluster THz-frequency "squashing" and "twisting" modes shown in Fig. 3.9a via the *pseudo* or *dynamic Jahn-Teller (JT) effect* (Bersuker and Polinger 1989).

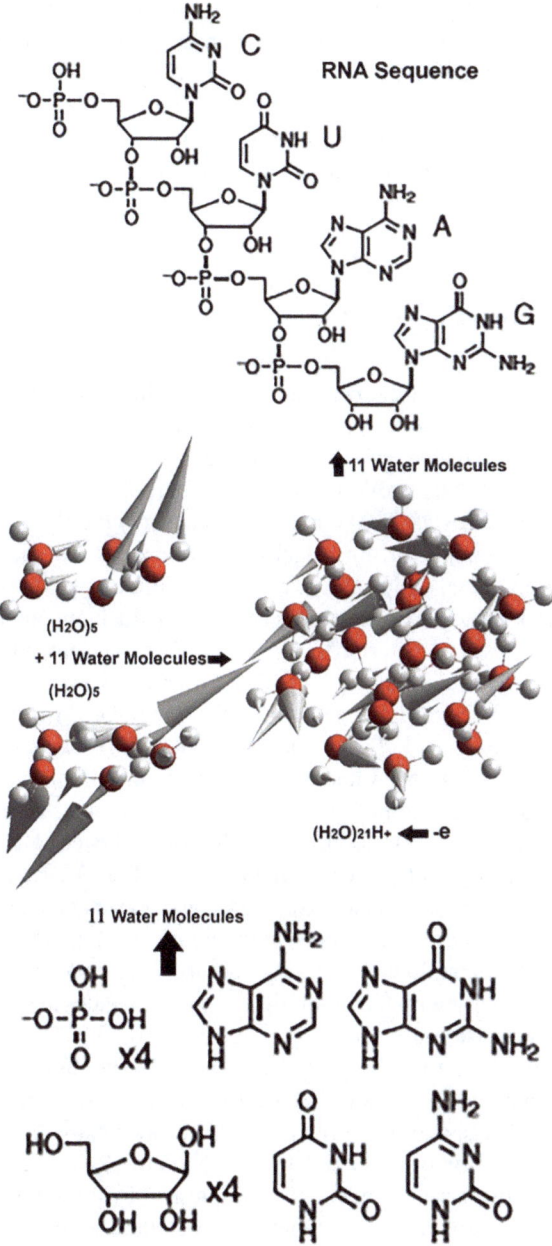

Fig. 3.8 The dehydration-condensation steps from prebiotic molecules to an RNA sequence of four polymerized nucleobases, *guanine*, *adenine*, *uracil*, and *cytosine* via a water nanocluster catalytic pathway for water molecules

JT coupling in $(H_2O)_{21}H$ leads to a prescribed symmetry breaking of the pentagonal dodecahedron along the THz-frequency vibrational mode coordinates Q_s, lowering the cluster potential energy from A to the equivalent minima A' shown in Fig. 3.9b. Because of the large JT-induced vibrational displacements (large Q_s) of water-cluster surface oxygen atoms, the energy barrier for expulsion of water oxygen or OH radicals and their oxidative addition to reactive nucleotides is lowered from $E_{barrier}$ to $E'_{barrier}$ (Fig. 3.9b).

7.2   RNA Protocells

Laboratory investigations of the possible origins of life on Earth have successfully created self-assembling model protocell membranes, the simplest of which are fatty-acid vesicles capable of containing at least short segments of RNA (Meierhenrich et al. 2010; Dworkin et al. 2001; Oberholzer et al. 1995; Chang et al. 2000; Hanczyc and Szostak 2004). Fatty acids are *amphiphilic* molecules, which means that polar and nonpolar functional groups are present in the same molecule. Fatty acids are commonly found in experiments simulating the prebiotic "soup" arising from Earth's early hydrothermal conditions (Milshteyn et al. 2018) and the arrival of extraterrestrial material to the early Earth (Kvenvolden et al. 1970). Quantum-chemical calculations by the SCF-Xα-Scattered-Wave density-functional method (Slater and Johnson 1972, 1974) find that the polar (*hydrophilic*) end of the naturally occurring fatty acid *glycerol monolaurate* ("*GML/monolaurin*") $C_{15}H_{30}O_4$ (Milshteyn et al. 2018) attracts water molecules, forming a stable water-nanocluster-GML molecule, as illustrated in Fig. 3.10a. These calculations reveal: (1) the polar end of the fatty acid donates an electron into the water-cluster LUMO, as shown by the computed molecular-orbital wavefunction Ψ in Fig. 3.10a; (2) the 1.8 THz vibrational mode of the water nanocluster described above resonates with the fatty-acid carbon-chain motions, as represented by the vectors in Fig. 3.10b. This quantum-mechanical coupling of the fatty acid to water nanoclusters can promote their chemical reactivity with the prebiotic organic molecules and phosphates leading to RNA polymerization by the steps shown in Fig. 3.8. Once such couplings occurred naturally during Earth's early hydrothermal conditions and the arrival of extraterrestrial material to the early Earth, the water-nanocluster-fatty-acid molecules tended to aggregate around growing RNA segments, as shown in Fig. 3.10c. The preferred orientation of these molecules around the RNA segment will be recognized as a primitive *reverse micelle*, the simplest self-assembling fatty-acid vesicle that has been demonstrated to be capable of containing at least short segments of RNA (Chang et al. 2000). The laboratory-controlled synthesis of self-assembling water-nanocluster reverse micelles present in *water-in-oil nanoemulsions* has also been demonstrated (Johnson 1998; Daviss 1999).

This scenario, albeit rudimentary, provides at least one possible pathway to cellular life's beginnings on Earth and habitable exoplanets, including the earliest RNA viruses (Moelling and Broecker 2019), while suggesting contemporary applications

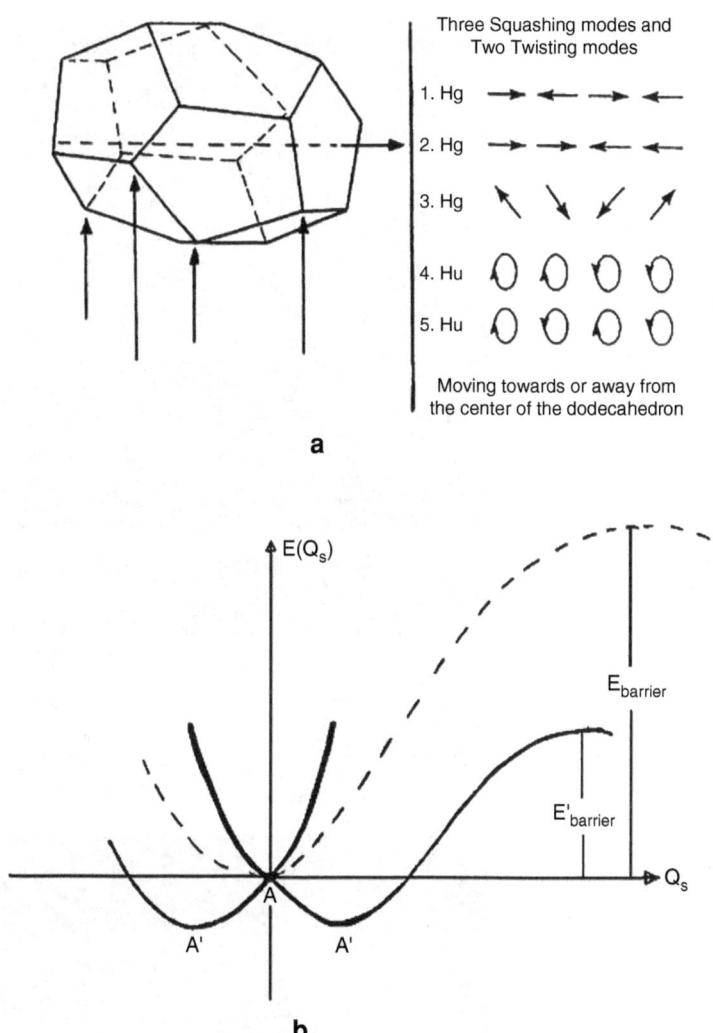

Fig. 3.9 (**a**). "Squashing" and "twisting" vibrational modes of a pentagonal dodecahedron, where Hg and Hu designate the key irreducible representations of the icosahedral point group corresponding to these modes. (**b**). Schematic representation of the double potential energy wells for a Jahn-Teller distorted water pentagonal dodecahedral water nanocluster and the resulting reduction of the energy barrier for the catalytic reaction of the cluster along the reaction path defined by the normal mode coordinates Q_s

to biomedicine, such as pharmaceuticals (Authelin et al. 2014) and *RNA-interference antiviral drugs* (Wu and Chan 2006; Setten et al. 2019). RNA interference induced by small interfering RNA segments like the micellular one shown in Fig. 3.10c can inhibit the expression of viral antigens and so provides a novel approach to the therapy of pathogenic coronaviruses such as COVID-19.

Fig. 3.10 (**a**). Water nanocluster attached to the polar end of the naturally occurring fatty acid *glycerol monolaurate* (*"GML/monolaurin"*) $C_{15}H_{30}O_4$, which donates an electron to the water-cluster LUMO represented by the computed molecular-orbital wavefunction Ψ. (**b**). The 1.8 THz vibrational mode of the combined water-nanocluster-GML molecule. (**c**). The aggregation of water-nanocluster-GML molecules around an RNA segment, forming a primitive reverse micelle

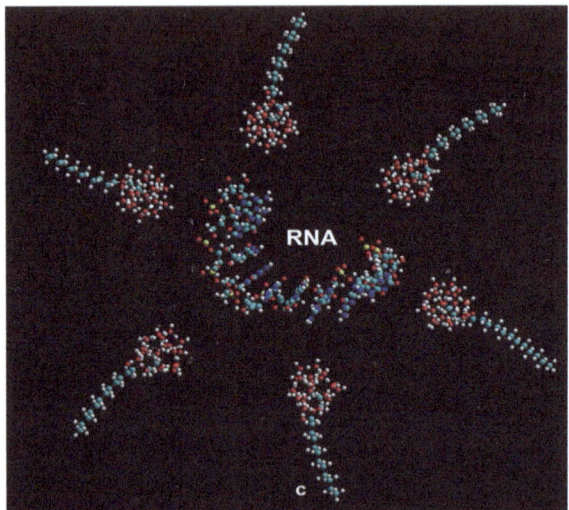

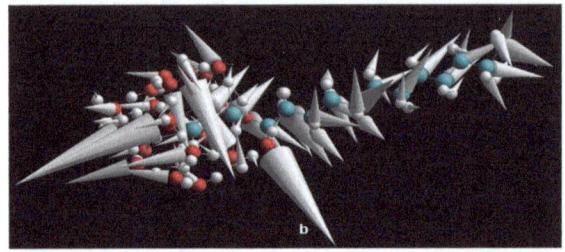

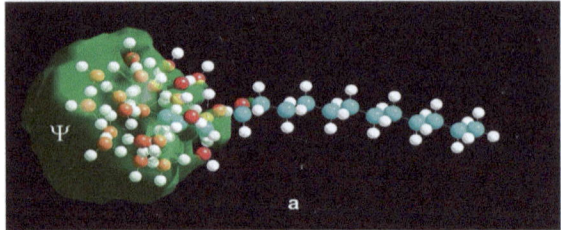

8   A Proposed COVID-19 Antiviral Compound

The relevance of water nanoclusters to protein folding and biomedicine has been discussed previously in (Johnson 2012), Sect. 5. During the wait for vaccines, researchers attempting to defeat the COVID-19 pandemic – caused by an *enveloped single-strand RNA coronavirus* – have been focusing on three areas: (1) disabling the protein "spikes" that attach the virus to cells, (2) destroying the enveloping protein-phospholipid "sheath" that protects the virus, and (3) inhibiting the main protease enzyme that catalyzes the multiplication of the virus in the lung tissues. (Hoffmann et al. 2020). Recent experimental findings (Johnson and Phelan 2020) suggest that the water-nanocluster-GML compound described above can function as a COVID-19 *antiviral* by inhibiting the main protease enzyme, along with enabling GML to attack and disable the enveloping sheath (Welch et al. 2020).

Similar to a lock and key, the COVID-19 protease activity is triggered by the binding of molecules to specific sites in the protease, e.g. *histidine* and *cysteine* amino-acid residues in the reported protease structure (Zhang et al. 2020; Jin et al. 2020) shown in Fig. 3.11a. The binding of a cellular substrate effectively switches the protease on, allowing it to cut the long viral protein strands derived from the viral RNA into smaller segments, leading to virus proliferation in the body. The protease's activity can be blocked by molecules called *inhibitors*. When an inhibitor attaches to an active site, it prevents the binding of substrates – stopping the catalytic action of the protease.

As shown in Fig. 3.11, according to quantum-chemical calculations by the SCF-Xα-Scattered-Wave method (Slater and Johnson 1972, 1974), water-nanocluster-GML does exactly that by bonding with the protease active sites at the water-nanocluster polar end of GML. The computed molecular-orbital wavefunction Ψ localized on the GML-captured water nanocluster in Fig. 3.10a promotes molecular-orbital overlap with the protease active amino acids, as shown by the molecular-orbital bonding wavefunction Ψ in Fig. 3.11b, while the 1.8 THz vibrations of the water-nanocluster-GML molecule in Fig. 3.11c couple with the protease amino-acid bending vibrational modes – these two effects resulting in protease inhibition.

Another proposed protease inhibitor – the highly trumpeted malaria drug *hydroxychloroquine* – should not have been effective in the first place because it lacks these electronic and THz vibrational properties, particularly those of the abundant water-nanocluster oxygens, and indeed hydroxychloroquine has largely failed patient trials (McGinley and Cha 2020). The more oxygens, the better. Moreover, hydroxychloroquine can produce serious side effects, whereas water-nanocluster-GML has none because it is composed of two natural substances already produced by the body – nanostructured water and glycerol monolaurate, a derivative of lauric acid (present in coconut oil). It is also relatively inexpensive to formulate and benefits the immune system. Preliminary tests on patients under treatment by physicians with this compound – taken orally and by nebulizer as a prophylactic for COVID-19, influenza enveloped viruses, and as an antibacterial for chronic Lyme disease, where it inhibits the formation of associated pathogenic bacterial biofilms – have been promising (Johnson and Phelan 2020).

This brief introduction to the biomedical properties of water nanoclusters combined with GML has been presented to stimulate freely basic research on possible COVID-19 therapeutics.

Fig. 3.11 (**a**). Water-nanocluster-GML molecule attacking COVID-19 protease, (**b**). binding to protease active site, and (**c**). the associated 1.8 THz vibrational mode inhibiting the protease

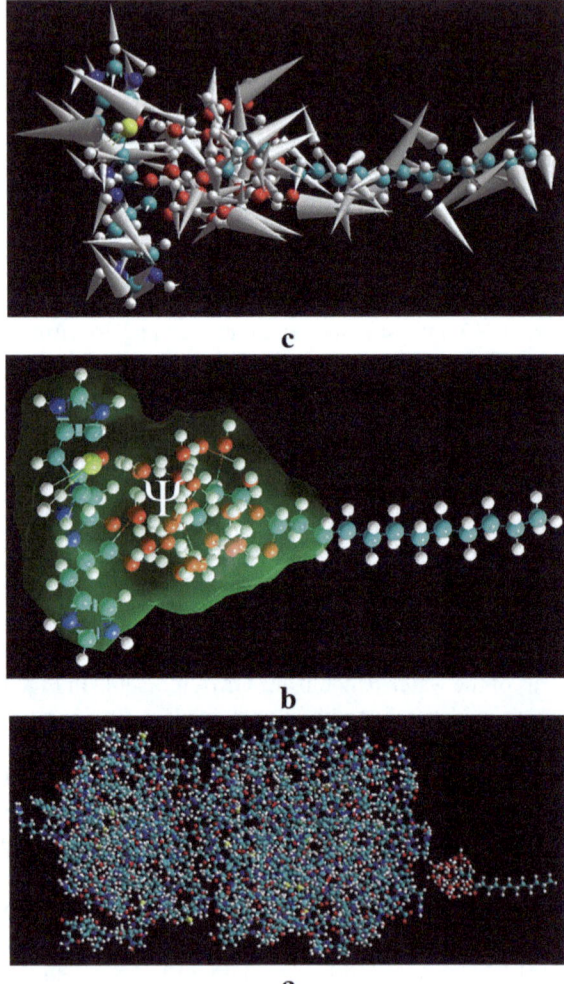

c

b

a

9 Conclusions

9.1 Why Pentagonal Water Clusters?

In the proposed roles of water nanoclusters in cosmology and biology, emphasis has been placed on clusters consisting of pentagonal cyclic rings of water molecules. This is justified because a pentagonal geometry leads to the magic-number water nanoclusters most frequently observed experimentally and discussed in Sect. 2. The higher stability of these clusters is explained simply by the fact that the water molecule bond angle is roughly equal to a regular pentagon angle. Thus, the water molecule hydrogen bonds are only slightly deformed. That said, this scenario does not completely rule out the possible ejection of water nanoclusters of other topologies from cosmic dust according to Eq. 3.1. For example, the calculated cut-off THz vibrational frequency of an icosahedral "water buckyball" is only slightly greater than that of a pentagonal dodecahedral one, and deforming the latter dodecahedron changes that frequency only slightly.

9.2   Can Classical Physics Explain Water Nanoclusters?

In Sect. 5.1 it was pointed out that pentagonal dodecahedral water nanoclusters of the types shown in Figs. 3.2 and 3.3 might be viewed as spherical shells of the type originally proposed by Layzer and Hively (1973) to be a primordial source of the CMB radiation according to classical electromagnetic theory. Likewise, viewing such nanoclusters as tiny spheres could allow one, in principle, to apply the theoretical approach of Gérardy and Ausloos (1983, 1984) to model the infrared absorption spectrum of water-nanocluster arrays from solutions of Maxwell's equations. However, instead of the infrared, here we are focused on the unique THz spectra of such nanoclusters due to the "surface" vibrational modes of the clusters' water-molecule shells.

Since the classical frequency of a thin vibrating spherical shell varies inversely with the radius of shell, it is therefore no surprise that the THz cut-off frequency of an approximately spherical water nanocluster decreases with increasing cluster size or with increasing number of water molecules in Eq. 3.1, as suggested by Figs. 3.2 and 3.3. Likewise, since the classical frequency (lowest harmonic) of a vibrating string is inversely proportional to its length, "strings" of water nanoclusters like the one shown in Fig. 3.4 will tend to have cut-off frequencies decreasing with string length. The numerical values of these frequencies cannot be determined classically, but they can be computed quantum mechanically by the SCF-Xα-Scattered-Wave density-functional method (Slater and Johnson 1972, 1974) described in Sect. 2.2.

9.3   The Key Frequency

My quantum-chemical findings that 1.8 THz water-nanocluster vibrational modes are key to their role in coupling with and activating prebiotic molecules is especially interesting because that frequency is very close to the cut-off frequency $v_c \cong 1.7$ THz which determines, according to Sect. 4 and Eq. 3.4, the dark energy density due to vacuum fluctuations, and that is consistent with the measured cosmological constant at the present time in the expansion of the universe. This result is also consistent with water nanoclusters containing the "magic-numbers" of n = 21 and 20 water molecules (Figs. 3.2d and 3.3a, respectively) to be dominant ones produced by ice-coated cosmic dust according to Eq. 3.1, because with increasing cluster size (Figs. 3.3b, c and 3.4), the cut-off frequency v_c and, according to Eq. 3.4, the dark energy density will be smaller than presently measured. Applying the cyclic cosmological model presented in Sect. 5.4, one concludes that as the universe expands further and larger cosmic water nanoclusters become more dominant from ice-coated cosmic dust, those larger clusters will be less favorable to interact with pre-biotic molecules, suggesting that life will become less probable over astronomical time. In other words, we are likely living at the ideal time in the expansion of the universe for life to exist, and water nanoclusters of the types shown in Figs. 3.2d and 3.3a created on cosmic dust could possibly be "seeds of life" that catalyze the bio-molecules necessary for life.

9.4   Evidence for Cosmic Water Nanoclusters

As discussed in Sect. 2, there is strong experimental evidence for the existence of water nanoclusters and protonated water-nanocluster cluster ions in Earth's atmosphere (Aplin and McPheat 2005), produced in laboratory vacuum chambers (Johnson et al. 2008) (Fig. 3.1), and generated from amorphous ice by energetic ion bombardment (Martinez et al. 2019). The protonated $(H_2O)_{21}H^+$ or $H_3O^+(H_2O)_{20}$ cluster (Fig. 3.2) is exceptionally stable, even at high temperatures and irradiation, and has been identified experimentally from its infrared spectrum (Miyazaki et al. 2004; Shin et al. 2004). Because this species can be viewed as a hydronium ion (H_3O^+) caged by an approximately pentagonal dodecahedron of twenty water molecules, the report of widespread hydronium in the galactic interstellar medium through its inversion spectrum (Lis et al. 2014) is key to the challenge of identifying the presence of $H_3O^+(H_2O)_{20}$ because both spectra fall into the same THz region. Nevertheless, both experiment and theory are relatively strong on the likely occurrence of this water nanocluster in the galactic interstellar medium, which after years of searching, can no longer be said for nonbaryonic subatomic particles such as WIMPS and AXIONS (Haynes 2018).

9.5   Final Conclusions

Although this paper is "out of the box" of generally popular inflationary cosmology and multiverse theory (Guth 1981, 2007; Linde 2008), it is not the only example. Other "cyclic-universe" theories have been widely promoted (Ijjas et al. 2017; Ijjas and Steinhardt 2019; Penrose 2006), while critiquing the multiverse scenario (Steinhardt 2011). This author has offered a unified *interdisciplinary* approach to cyclic cosmology and the origin of life in the universe based on quantum astrochemistry, while still attempting to include complementary relevant astrophysics facts. The proposal that cosmic water nanoclusters may constitute a form of invisible baryonic dark matter does not rule out nonbaryonic dark matter, such as WIMPS and AXIONS, although observational evidence for their existence is lacking

(Haynes 2018). Regarding dark energy, it is difficult to escape the conclusion that some "substance" fills the vacuum of space which largely but not entirely cancels via Eq. 3.3 the otherwise infinite vacuum energy density, Eq. 3.2. This scenario is consistent with the reported "web" of dark matter permeating the cosmos (Heymans et al. 2012). The report of a neutral hydrogen gas bridge connecting the Andromeda (M31) and Triangulum (M33) galaxies (Lockman et al. 2012) suggests the possibility that water nanoclusters, albeit at lower density than pure hydrogen, might similarly be dispersed as intergalactic gas constituting Rydberg dark matter. The striking consistency of the THz cut-off vibrational frequencies of water nanoclusters with the zero-point vacuum energy THz cut-off frequency that produces a dark-energy density in agreement with cosmological data is possibly only coincidental. Nevertheless, a common origin of dark matter and dark energy outside the realm of conventional elementary-particle physics, which has yet to identify conclusively the origins of either dark matter or dark energy, is a tempting idea. Indeed, one might conceptually view water nanoclusters in the vacuum of space as baryonic nanoparticles that break the real-space symmetry of the otherwise isotropic vacuum due to their physical presence. Surprisingly, their masses are within the range of those estimated for WIMPS. Again, there are no other identified baryonic substances in our universe, including hydrogen, water monomers, and organic molecules that exhibit all these characteristics, while possessing the Rydberg-excited electronic states of low-density condensed matter that qualifies also as dark matter. Fullerene buckyballs in planetary nebulae (Cami et al. 2010) are also ruled out as candidates, even though Rydberg states have been observed in the C_{60} molecule (Boyle et al. 2001) because their vibrational frequencies lie beyond the required THz range. Planets, and moons, as well as water vapor in solar atmospheres (Bergin and van Dishoeck 2012), nebulae (Glanz 1998), and distant quasars (Bradford et al. 2011) are widely present throughout the cosmos, and therefore should be included as possible sources of cosmic water nanoclusters. In fact, water clusters have been detected recently in the hydrothermal plume of Enceladus – a moon of Saturn (Coates et al. 2013).

Thus, a low-density gas of cosmic water nanoclusters of the types shown in Figs. 3.2, 3.3 and 3.4, primarily ejected from the abundant cosmic dust population, could be a common astrophysical basis for baryonic dark matter and dark energy, offering simple explanations for the cosmological coincidence of matter and dark energy densities and the Hubble constant crisis. Finally, from the quantum chemistry of cosmic water nanoclusters interacting with prebiotic organic molecules, amino acids and RNA protocells on early Earth and habitable exoplanets, this scenario is consistent with the *anthropic principle* that our universe is a connected biosystem and has those properties which allow life, as we know it – based on water, to develop at the present stage of its history.

Acknowledgement The author is grateful to Franziska Amacher for introducing me to the RNA world and the work of Harvard Professor J. W. Szostak.

References

Ade, P. A. R., et al. (2016). Planck 2015 results-xii. Cosmological parameters. *Astronomy & Astrophysics, 594*, 1–28.

Akerib, D. S., et al. (2017). Results from a search for dark matter in the complete LUX exposure. *Physical Review Letters, 118*, 021303–021311.

Aplin, K. L., & McPheat, R. A. (2005). Absorption of infra-red radiation by atmospheric molecular cluster-ions. *Journal of Atmospheric and Solar - Terrestrial Physics, 67*, 775–783.

Authelin, J.-R., et al. (2014). Water clusters in amorphous pharmaceuticals. *Journal of Pharmaceutical Sciences, 103*, 2663–2672.

Badei, S., & Homlid, L. (2002). Rydberg matter in space: Low-density condensed dark matter. *Monthly Notices of the Royal Astronomical Society, 333*, 360–364.

Bally, J., & Harrison, J. R. (1978). The electrically polarized universe. *The Astrophysical Journal, 220*, 743–744.

Banandos, E., et al. (2019). A metal-poor damped Lyα system at redshift 6.4. *The Astrophysical Journal, 885*, 59–74.

Beck, C., & Mackey, M. C. (2005). Could dark energy be measured in the lab? *Physics Letters B, 605*, 295–300.

Beck, C., & Mackey, M. C. (2007). Measurability of vacuum fluctuations and dark energy. *Physica A, 379*, 101–110.

Bennett, C. L., et al. (2013). Nine-year Wilkinson microwave anisotropy probe (WMAP) observations: Cosmological parameter results. *The Astrophysical Journal Supplement Series, 208*, 1–25.

Bergin, E. A., & van Dishoeck, E. F. (2012). Water in star- and planet-forming regions. *Philosophical Transactions of the Royal Society A, 370*, 2778–2802.

Bersuker, I. B., & Polinger, V. Z. (1989). *Vibronic interactions in molecules and crystals*. Berlin: Springer.

Bialy, S., Sternberg, A., & Loeb, A. (2015). Water formation during the epoch of first metal enrichment. *The Astrophysical Journal, 804*, L29–L34.

Boyle, M., et al. (2001). Excitation of rydberg series in C_{60}. *Physical Review Letters, 83*, 273401–273405.

Bradford, C. M., et al. (2011). The water vapor spectrum of APM 08279+5255: X-ray heating and infrared pumping over hundreds of parsecs. *The Astrophysical Journal, 741*, L37–L43.

Brudermann, J., Lohbrandt, P., & Buck, U. (1998). Surface vibrations of large water clusters by He atom scattering. *Physical Review Letters, 80*, 2821–2824.

Cafferty, B. J., & Hud, N. V. (2014). Abiotic synthesis of RNA in water: A common goal of prebiotic chemistry and bottom-up synthetic biology. *Current Opinion in Chemical Biology, 22*, 146–157.

Cami, J., et al. (2010). Detection of C60 and C70 in a young planetary nebula. *Science, 329*, 1180–1182.

Carlon, H. R. (1981). Infrared absorption by molecular clusters in water vapor. *Journal of Applied Physics, 52*, 3111–3115.

Case, D. A., Huynh, B. H., & Karplus, M. (1979). Binding of oxygen and carbon monoxide to hemoglobin. An analysis of the ground and excited states. *Journal of the American Chemical Society, 101*, 4433–4453.

Chakraborty, K., et al. (2014). Possible features of galactic halo with electric field and observational constraints. *General Relativity and Gravitation, 46*, 1807–1820.

Chang, G.-G., Huang, T.-M., & Hung, H.-C. (2000). Reverse micelles as life-mimicking systems. *Proceedings of the National Science Council ROC(B), 24*, 89–100.

Chaplin, M. (2006). Do we underestimate the importance of water in cell biology? *Nature Reviews Molecular Cell Biology, 7*, 861–866.

Clowe, D., Gonzalez, A., & Markevich, A. (2004). Weak-lensing mass reconstruction of the interacting cluster 1E 0657-558: Direct evidence for the existence of dark matter. *The Astrophysical Journal, 604*, 596–604.

Coates, A. J., et al. (2013). Photoelectrons in the Enceladus plume. *Journal of Geophysical Research, Space Physics, 118*, 5099–5108.

Costanzo, G., et al. (2009). Generation of long RNA chains in water. *The Journal of Biological Chemistry, 284*, 33206–33216.

Cotton, F. A., Norman, J. G., & Johnson, K. H. (1973). Biochemical importance of the binding of phosphate by arginyl groups. Model compounds containing methylguanidinium ion. *Journal of the American Chemical Society, 95*, 2367–2369.

Daviss, B. (1999, March 13). Just add water. *New Scientist*.

Duley, W. W. (1996). Molecular clusters in interstellar clouds. *The Astrophysical Journal, 471*, L57–L60.

Dworkin, J. P., et al. (2001). Self-assembling amphiphilic molecules: Synthesis in simulated interstellar/precometary ices. *Proceedings of the National Academy of Sciences, 98*, 815–819.

Gérardy, J. M., & Ausloos, M. (1983). Absorption spectrum of clusters of spheres from the general solution of Maxwell's equations. IV. Proximity, bulk, surface, and shadow effects (in binary clusters). *Physical Review B, 27*, 6446–6463.

Gérardy, J. M., & Ausloos, M. (1984). Absorption spectrum of clusters of spheres from the general solution of Maxwell's equations. III. Heterogeneous spheres. *Physical Review B, 30*, 2167–2181.

Ghosh, J., Methikkalam, R. J., & Bhuin, R. G. (2019). Clathrate hydrates in the interstellar environment. *Proceedings of the National Academy of Sciences, 116*, 1526–1531.

Glanz, J. (1998). A water generator in the Orion nebula. *Science, 280*, 378–382.

Guth, A. H. (1981). Inflationary universe: A possible solution to the horizon and flatness problems. *Physical Review D, 23*, 347–356.

Guth, A. H. (2007). Eternal inflation and its implications. *Journal of Physics A, 30*, 6811–6826.

Hanczyc, M. M., & Szostak, J. W. (2004). Replicating vesicles as models of primitive cell growth and division. *Current Opinion in Chemical Biology, 8*, 660–664.

Harker, H. A., et al. (2005). Water pentamer: Characterization of the torsional-puckering manifold by terahertz VRT spectroscopy. *The Journal of Physical Chemistry. A, 109*, 6483–6497.

Haynes, K. (2018, September 21). What is dark matter? Even the best theories are crumbling. *Discover Magazine*.

Herzberg, G. (1987). Rydberg molecules. *Annual Review of Physical Chemistry, 38*, 27–56.

Heymans, C., et al. (2012). CFHTLenS: The Canada-France-Hawaii telescope lensing survey. *Monthly Notices of the Royal Astronomical Society, 427*, 146–166.

Hoffmann, M., et al. (2020). SARS-CoV-2 cell entry depends on ACE2 and TMPRSS2 and is blocked by a clinically proven protease inhibitor. *Cell, 181*, 1–10.

Iglesias-Groth, S., et al. (2010). A search for interstellar anthracene towards the Perseus anomalous microwave emission region. *Monthly Notices of the Royal Astronomical Society, 407*, 2157–2165.

Ijjas, A., & Steinhardt, P. J. (2019). A new kind of cyclic universe. *Physics Letters B, 795*, 666–672.

Ijjas, A., Steinhardt, P. J., & Loeb, A. (2017). Pop goes the universe. *Scientific American, 316*, 32–39.

Jedamzik, K., & Pogosian, L. (2020) Relieving the Hubble tension with primordial magnetic fields. arXiv:2004.09487v2 [astro-ph.CO] 28 Apr 2020.

Jin, Z., et al. (2020). Structure of Mpro from SARS-CoV-2 and discovery of its inhibitors. *Nature, 582*, 289–293.

Johnson, K. H. (1998). Water clusters and uses therefor. *U.S. Patent No. 5,800,576*.

Johnson, K. (2012). Terahertz vibrational properties of water nanoclusters relevant to biology. *Journal of Biological Physics, 38*, 85–95.

Johnson, K., & Phelan, E. (2020). *Antibacterial and antiviral uses of water-soluble glycerol monolaurate*. Manuscript in preparation.

Johnson, K. H., et al. (2008). Water vapor: An extraordinary terahertz wave source under optical excitation. *Physics Letters A, 371*, 6037–6040.

Jordan, K. D. (2004). A fresh look at electron hydration. *Science, 306*, 618–619.

Joyce, G. F., & Orgel, L. E. (1993). Prospects for understanding the origin of the RNA world. In R. F. Gesteland & J. F. Atkins (Eds.), *The RNA world* (pp. 1–22). Cold Spring Harbor: Cold Spring Harbor Laboratory Press.

Kvenvolden, K. A., et al. (1970). Amino acids in the Murchison meteorite. *Nature, 228*, 923–926.

Lacy, J. H., et al. (1991). Discovery of interstellar methane – Observations of gaseous and solid CH_4 absorption toward young stars in molecular clouds. *The Astrophysical Journal, 376*, 556–560.

Laporte, N., et al. (2017). Dust in the reionization era: ALMA observations of a $z = 8.38$ gravitationally lensed galaxy. *The Astrophysical Journal, 837*, L21–L27.

Layzer, D., & Hively, R. (1973). Origin of the microwave background. *The Astrophysical Journal, 179*, 361–370.

Lehnert, M. D., et al. (2010). Spectroscopic confirmation of a galaxy at redshift $z = 8.6$. *Nature, 467*, 940–942.

Linde, A. D. (2008). Inflationary cosmology. In *Inflationary cosmology*. Berlin: Springer.

Lis, D. C., et al. (2014). Widespread rotationally hot hydronium ion in the galactic interstellar medium. *The Astrophysical Journal, 785*, 135–144.

Lockman, F. J., Free, N. L., & Shields, J. C. (2012). The neutral hydrogen bridge between M31 and M33. *Astronomy Journal, 144*, 52–59.

Mann, I. (2001). Spacecraft charging technology. In R. A. Harris (Ed.), *Proceedings of the Seventh International Conference, April 23–27, European Space Agency, ESA SP-476* (pp. 629–639).

Martinez, R., et al. (2019). Production of hydronium ion (H3O)+ and protonated water clusters (H2O)nH+ after energetic ion bombardment of water ice in astrophysical environments. *The Journal of Physical Chemistry. A, 123*, 8001–8008.

Matsuura, M., et al. (2019). SOFIA mid-infrared observations of supernova 1987A in 2016 – Forward shocks and possible dust re-formation in the post-shocked region. *Monthly Notices of the Royal Astronomical Society, 482*, 1715–1723.

McGinley, L., & Cha, A. -E. (2020, June 3) Hydroxychloroquine, a drug promoted by Trump, failed to prevent healthy people from getting covid-19 in trial. *Washington Post*.

Meierhenrich, U. J., et al. (2010). On the origin of primitive cells: From nutrient intake to elongation of encapsulated nucleotides. *Angewandte Chemie, International Edition, 49*, 3738–3750.

Milshteyn, D., et al. (2018). Amphiphilic compounds assemble into membranous vesicles in hydrothermal hot spring water but not in seawater. *Life, 8*, 11–26.

Miyazaki, M., et al. (2004). Infrared spectroscopic evidence for protonated water clusters forming nanoscale cages. *Science, 304*, 1134–1137.

Moelling, K., & Broecker, F. (2019). Viruses and evolution – Viruses first? A personal perspective. *Frontiers in Microbiology, 10*, 1–13.

Munoz, J. B., & Loeb, A. (2018). A small amount of mini-charged dark matter could cool the baryons in the early universe. *Nature, 557*, 684–686.

Nandi, P. K., et al. (2017). Ice-amorphization of supercooled water nanodroplets in no man's land. *ACS Earth and Space Chemistry, 1*, 187–186.

Neidle, S., Berman, H., & Shieh, H. S. (1980). Highly structured water network in crystals of deoxydinucleoside-drug complex. *Nature, 288*, 129–133.

Oberholzer, T., et al. (1995). Enzymatic RNA replication in self-reproducing vesicles: An approach to a minimal cell. *Biochemical and Biophysical Research Communications, 207*, 250–257.

Oesch, P. A., et al. (2016). A remarkably luminous galaxy at $z = 11.1$ measured with Hubble space telescope Grism spectroscopy. *The Astrophysical Journal, 819*, 129–140.

Ouellet, et al. (2019). First results from ABRACADABRA-10 cm: A search for sub-μeV axion dark matter. *Physical Review Letters, 122*, 121802–121809.

Penrose, R. (2006). Before the big bang: An outrageous new perspective and its implications for particle physics. In *Proceedings of EPAC 2006, Edinburgh, Scotland* (pp. 2759–2762).

Riess, A. G., et al. (1998). Observational evidence from supernovae for an accelerating universe and a cosmological constant. *Astronomy Journal, 116*, 1009–1038.

Risaliti, G., & Lusso, E. (2019). Cosmological constraints from the Hubble diagram at high redshifts. *Nature Astronomy, 3*, 272–277.

Sahle, C. J., et al. (2013). Microscopic structure of water at elevated pressures and temperatures. *Proceedings of the National Academy of Sciences, 110*, 6301–6306.

Setten, R. L., Rossi, J. J., & Han, S. P. (2019). The current state and future directions of RNAi-based therapeutics. *Nature Reviews Drug Discovery, 18*, 421–446.

Shin, J. W., et al. (2004). Infrared signature of structures associated with $H^+(H2O)n$ (n = 6 to 27). *Science, 304*, 1137–1140.

Slater, J. C., & Johnson, K. H. (1972). Self-consistent-field $X\alpha$ cluster method for polyatomic molecules and solids. *Physical Review B, 5*, 844–853.

Slater, J. C., & Johnson, K. H. (1974). Quantum chemistry and catalysis. *Physics Today, 27*, 34–41.

Souza, R. D., Impens, F., & Neto, P. A. M. (2018). Microscopic dynamical Casimir effect. *Physical Review A, 97*, 032514–032523.

Steinhardt, P. J. (2011). Inflation theory debate: Is the theory at the heart of modern cosmology deeply flawed? *Scientific American, 304N4*, 18–25.

Teeter, M. M. (1984). Water structure of a hydrophobic protein at atomic resolution: Pentagon rings of water molecules in crystals of crambin. *Proceedings of the National Academy of Sciences, 81*, 6014–6018.

Totani, T. (2020). Emergence of life in an inflationary universe. *Scientific Reports, 10*, 1671–1678.

Weinberg, S. (1987). Anthropic bound on the cosmological constant. *Physical Review Letters, 59*, 2607–2610.

Weinberg, S. (1989). The cosmological constant problem. *Reviews of Modern Physics, 61*, 1–23.

Weinberg, S. (2008). *Cosmology* (pp. 185–200). New York: Oxford University Press.

Welch, J. L., et al. (2020). Glycerol monolaurate is virucidal against enveloped viruses. *MBio, 11*, 1–17.

Wright, E. L. (1982). Thermalization of starlight by elongated grains – Could the microwave background have been produced by stars. *The Astrophysical Journal, 255*, 401–407.

Wu, C.-J., & Chan, Y.-L. (2006). Antiviral applications of RNAi for coronavirus. *Expert Opinion on Investigational Drugs, 15*, 89–96.

Yang, C. Y., Johnson, K. H., Holm, R. H., & Norman, J. G. (1975). Theoretical model for the 4-Fe active sites in oxidized ferredoxin and reduced high-potential proteins. Electronic structure of the analog $[Fe_4S^*_4(SCH_3)_4]^{2-}$. *Journal of the American Chemical Society, 97*, 6596–6598.

Yokoyama, H., Kannami, M., & Kanno, H. (2008). Intermediate range O-O correlations in supercooled water. *Chemical Physics Letters, 463*, 99–102.

Zhang, L., et al. (2020). Crystal structure of SARS-CoV-2 main protease provides a basis for design of improved α-ketoamide inhibitors. *Science, 368*, 409–412.

Zheng, W., et al. (2012). A magnified young galaxy from about 500 million years after the big bang. *Nature, 489*, 406–408.

Chapter 4
Solvent Induced Effects on Protein Folding

Arieh Ben-Naim

Abstract This chapter is about the effects of water on the process of protein fold-ing. Specifically, there are essentially two problems within the so-called protein folding problem (PFP); one is the thermodynamic stability of the protein, the sec-ond is the speed and specificity of the folding process. We shall start with a very brief formulation of the PFP. We then survey the fascinating history of the evolution of various solvent induced effects on the PFP. Initially, it was believed that hydrogen bonds (HB) were the dominate factors in the stability of proteins. In 1959 Kauzmann suggested that the hydrophobic ($H\phi O$) effect might be an important factor. This idea reigned supreme in biochemical literatures for over 50 years. However in the past 30 years, the dominance of the $H\phi O$ effects was seriously contested. Instead, new solvent-induced effects were discovered, all involved hydrophilic ($H\phi I$) groups. It was found that the $H\phi I$ effects were more powerful and had a richer repertoire, compare with the $H\phi O$.

It was suggested that while the *$H\phi I$ interactions* offers simple and straightfor-ward answers to the problem of protein stability, the *$H\phi I$ forces* offers an explana-tion of the fast process of protein folding.

Keywords Protein folding · Water induced folding effects · Thermodynamics of aqueous protein systems · Hydrophilicity · Hydrophobic effect

1 Introduction: What Is the Protein Folding Problem?

There are essentially three different problems which together comprise the PFP. These are:

(i) The existence of a folding code: Can we predict the 3D structure, given the sequence of amino acids?

A. Ben-Naim (✉)
Department of Physical Chemistry, The Hebrew University of Jerusalem, Jerusalem, Israel

A. Gadomski (ed.), *Water in Biomechanical and Related Systems*,
Biologically-Inspired Systems 17, https://doi.org/10.1007/978-3-030-67227-0_4

(ii) The stability problem: What are the main factors that make the 3D structure of a protein stable?

(iii) The kinetic problem: What are the main factors that "speed" and "guide" the folding of a protein to a unique 3D structure in a relatively short time?

The first problem (i) was probably inspired by Anfinsen's thermodynamic hypothesis (Anfinsen 1973; Haber and Anfinsen 1961). From the studies of the renaturation of ribonuclease *in-vitro*, Anfinsen concluded: *"The native conformation is determined by the totality of the inter-atomic interactions, and hence by the amino acid sequence, in a given environment."*

An equivalent statement is that the *information* on the native structure of the protein is already contained in the sequence of the amino acids. We shall not discuss any further this question, we shall next turn to the second problem.

The second problem (ii) has been in the biochemical literature ever since the structures of proteins were determined. In essence, the problem is this: on one hand, the unfolded form of the protein occupies a very large part of the configurational space. On the other hand, the stable 3D structure of the protein consists of a very small number of configurations. The energy favors the structured form while the entropy favors the unfolded form. Hence the problem is: What are the factors that compete with the entropy, which favors the unfolded form, so that the folded form becomes favorable? Thus, the problem is to pinpoint the factors that determine the stability of the native structure of a protein. This question, as well as a plausible answer, was thoroughly studied by Schellman (1955a, b) and Kauzmann (Kauzmann 1959; Tanford 1973; Tanford and Reynold 2003; Ben-Naim 1980). We shall discuss this problem in Sect. 3.1.

The third problem (iii) was eloquently stated by Levinthal (1968). Following his rough estimate of the time it would require for the folding of the protein had it been searching at random its configurational space, Levinthal (1968, 1969) reached the conclusion that there must be some preferential pathways along which the protein folds:

> We feel that protein folding is speeded and guided by rapid formation of local interactions which then determine the further folding of the peptide.

Thus, in Levinthal's formulation of the *PFP* the main unknown are the factors that "speed" and "guide" the protein in its folding pathways. This formulation is also contained in the list of 125 big questions of science (Kennedy and Norman 2005):

> Can we predict how proteins will fold? Out of a near infinitude of possible ways to fold, protein picks one in just tens of microseconds. The same task takes 30 years of computer time.

We shall further discuss the Levinthal's question and its answer in Sect. 3.2. Here, we note only that most people have focused on what became to be known as the Levinthal's paradox, while ignoring or overlooking some other more important statements made by Levinthal.

2   The Various Solvent Induced Effects

There is a general agreement that water has an important, if not dominant effect on both the stability of the structure of proteins as well as on the kinetic of the folding process. The main unsettled question is how exactly the water, the main component in the solvent, does that.

2.1   Hydrogen Bonds

A convenient point to begin with is Pauling's book "The Nature of the Chemical Bond" (Pauling 1948, 1960), In the first two editions of the book Pauling discussed the HB, but no mention of proteins or nucleic acids. In the second edition, the chapter on HBs ends with some estimates of the HB energies and HB distances. The third edition contains two new sections on HBs in proteins and HBs in nucleic acids.

Following the works of Anson and Mirsky (1934), and Mirsky and Pauling (1936) on the denaturation of proteins, and later works of Pauling and Corey (1951a, b, c, d), the role of HBs in stabilizing the native structure of proteins was accepted The HBs, with bond energies of the order of 24 kJ/mol, which provided explanation for many anomalous properties of water (Ben-Naim 2009).

Doubts concerning the HB-paradigm started to accumulate in 1955, after the publications of a few articles by Schellman (1955a, b), on the dimerization of urea in aqueous solutions. Urea has two hydrophilic groups that can serve as donor and acceptor for hydrogen bonding, Fig. 4.1. In 1959 Kauzmann (1959) summarized Schellman's work and concluded:

> Hydrogen bonds, taken by themselves, give marginal stability to ordered structures, which may be enhanced or disrupted by interactions of side chains.

Similar conclusions have appeared in many articles and textbooks.

However, it was Fersht (1999) who formulated the HB inventory argument as follows:

> The inventory shows that the net change in the number of hydrogen bonds is zero.

The argument seems simple, straightforward and convincing. Write the stoichiometric reaction between a donor and an acceptor of a HB in the form

$$E - C = O \cdots w + w \cdots HN - S \rightarrow E - C = O \cdots HN - S + w \cdots w \qquad (4.1)$$

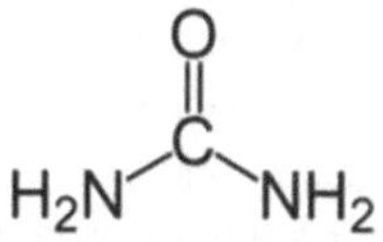

Fig. 4.1 Urea molecule having both a donor and an acceptor for hydrogen bonding

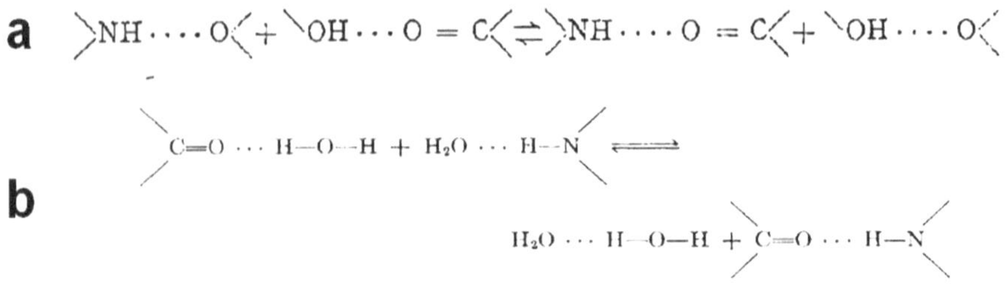

a $\rangle$NH $\cdots$ O$\langle$ + $^\backslash$OH $\cdots$ O = C$\langle$ $\rightleftharpoons$ $\rangle$NH $\cdots$ O $=$ C$\langle$ + $^\backslash$OH $\cdots$ O$\langle$

$^\backslash$C=O $\cdots$ H—O—H + H$_2$O $\cdots$ H—N$\langle$ $\rightleftharpoons$

b

H$_2$O $\cdots$ H—O—H + C=O $\cdots$ H—N$\langle$

Fig. 4.2 The stoichiometric reaction as written by: (**a**) Schellman (1955a, b) and by (**b**) by Kauzmann (1959)

where E and S stand for an enzyme and a substrate, respectively, but can be any two molecules or parts of the same molecule, and w is a water molecule. Equation 4.1 suggests that in the process of formation of a *direct* hydrogen bond between a donor (here amine group) and an acceptor (here a carbonyl group), *two* HBs are *broken* on the left hand side of the equation and *two* HBs are *formed* on the right hand side of the equation. Therefore, ignoring the differences in the various HB energies between the various pairs (carbonyl-water, amine-water, carbonyl-amine and water-water), we can conclude by simply *counting*, that the net effect of the formation of a direct HB is negligibly small.

A stoichiometric equation of the form (4.1) was first written by Schellman for the association between two urea molecules, Fig. 4.2. It was Kauzmann (1959), who in 1959 reached the conclusion quoted above.

2.2 The Rise of the HϕO Effect

Following Kauzmann's conclusion, based on Schellman's experiments, there was a kind of vacuum about the question which factor that is responsible for the stability of proteins; if HBs is not the main factor, what factors are the important ones? This vacuum was filled by Kauzmann in 1959, and became known as the *HϕO* effect. The *HϕO* effect was known long before Kauzmann applied it to the problem of protein folding (Levinthal 1968). It was applied successfully to explain surface tension of certain aqueous solutions of organic molecules, micelle formation and membranes. All these phenomena involve molecules having two moieties; a hydrophobic part, which "fears" water and tries to avoid it, and a hydrophilic part, which "loves" water and mingles with it comfortably. As Tanford and Reynolds quoted from a personal communication with Kauzmann, the idea of the *HϕO* effect had been hovering "in the air" for long a long time (Tanford and Reynold 2003).

In a classical review article "Some Factors in the Interpretation of Protein Denaturation," Kauzmann applied the idea of the *HϕO* effect to protein folding (Pauling 1948). For this purpose he coined the term *HϕO*-bond, and speculated that this "bond" could be the more important factor in the stabilization of the native structure of protein.

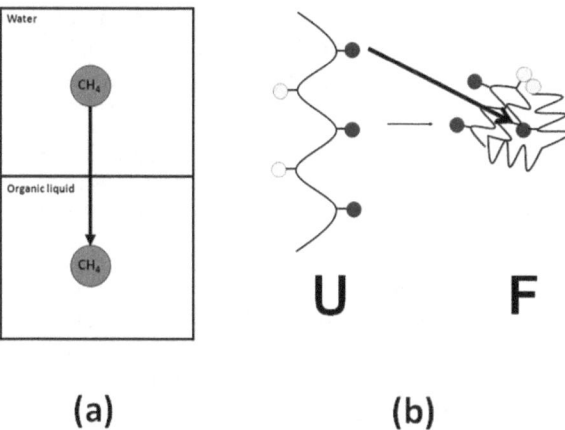

Fig. 4.3 (**a**) transfer of a non polar molecule from water into an organic liquid, and (**b**) The Kauzmann's model for hydrophobic effect

Kauzmann's idea was very simple. It was known that the Gibbs energy of transferring a small non-polar solute such as methane or ethane, from water into an organic liquid involves a large negative change in Gibbs energy. Kauzmann also noticed that there are about one third of amino-acid side-chains which are $H\phi O$, and most of these find themselves in the interior of the folded protein. Kauzmann suggested to use the process of transfer in Fig. 4.3a to represent the process of transferring of a side chain, Fig. 4.3b. If a protein, having about 150 amino acids, has about 50 $H\phi O$ groups, and if each of these contributes between -12 and -16 kJ/mol, then we expect to get a very large "driving force" for the folding process in the process Fig. 4.3b.

Following Kauzmann's brilliant idea, many authors rushed to announce that the $H\phi O$ effect is the *most important* "driving force" in protein folding, without really examining the validity of Kauzmann's model.

Since it was difficult to prove anything on such a complex process as protein folding, carried out in poorly understood solvent, involving a mysterious "entropy driven" concept (Ben-Naim 2009, 2011), it is not surprising that the dominance of the $H\phi O$ effect has prevailed for over half a century. The fact that $H\phi O$ groups are in the *interior* of the *protein*, and the fact that the transfer of $H\phi O$ molecules from water to an organic liquid is large and negative are undeniable. The former lends credibility to Kauzmann's model, while the latter provides the large negative Gibbs energy change.

2.3   Doubt About the Dominance of the HϕO Effects

The first crack in the $H\phi O$ dogma was not on the $H\phi O$ effect itself, but rather on the HB-inventory on which $H\phi O$ arose in the first place.

As mentioned in the previous section, the *HφO* dogma is still alive and thriving. One can find statements about the "dominance" of the *HφO* effect in protein folding even in the most recent reviews and textbooks. In 1980, in the preface of the book, "Hydrophobic Interactions," we find (Pauling and Corey 1951b):

> In spite of my researches in this field over almost 10 years, I cannot confirm that there is at present either theoretical or experimental evidence that unequivocally demonstrates the relative importance of the HφO interactions over other types of interactions in aqueous solutions.

These doubts were based on *lack* of *evidence* in favor of the contention that the *HφO* effect is the *most important* effect in the "driving force" for protein folding. How can one claim that one factor is more important, or most important when one does not have a full *inventory* of *all* the factors involved in protein folding? Remember that Kauzmann's paper was on "some factors in the interpretation of protein denaturation" – not on *all factors* involved. No one knew what were *all the factors* especially those that are solvent-induced. The only factor that could have competed with the *HφO* effect was the HB, but the HB-inventory argument debilitated the effect of the HBs in aqueous media, and rendered them powerless in explaining the driving force for protein folding.

This was the main motivation for the examination of the entire question of the solvent-induced effects on the protein folding and protein-protein association that was undertaken late in the 1980s. The results of this examination were stunning.

2.4   The Fall of the HB Inventory Argument, and Doubts About the Kauzmann Model

In 1990 it was shown that the HB-inventory argument was fundamentally faulty (Pauling and Corey 1951c). Then it was shown that Kauzmann's model was not adequate to the protein folding process (Anson and Mirsky 1934; Mirsky and Pauling 1936; Pauling and Corey 1951d; Fersht 1999; Ben-Naim 1989, 1990a, b, c, 1991, 1992, 2009, 2011).

Recall that the HB-inventory argument was based on the stoichiometric Eq. (4.1). The mere counting of the number of HBs on each side of the equation led to the dismissal of the role of direct hydrogen bonding to the driving force. The first serious challenge to the HB-inventory argument was expressed in 1990 (Fersht 1999). It was shown that the very writing of the stoichiometric equation in the form (4.1) is faulty for two reasons:

1. What one loses on the left hand side are not HB *energies* but *solvation Gibbs energies* of the *Hφl* groups. Actually it was the conditional solvation Gibbs energy that was lost, see Fig. 4.4b.

2. Whatever the water molecules do when they are released from the solvation sphere of the two *HϕI* groups is irrelevant to the driving force. These water molecules "flow" from the solvation sphere into the pool of water at constant chemical potential. Therefore, they cannot contribute anything to the driving force. The stoichiometric reaction must be written instead in the form

$$\left(E - C = O\right)_{solvated} + \left(HN - S\right)_{solvated} \rightarrow \left(E - C = O \cdots HN - S\right)_{solvated} \quad (4.2)$$

Here, each of the solutes (or the groups) involved in the formation of a HB is *solvated* by the water molecule. In this form, the HB-inventory argument does not exist. Therefore, the foundation on which the *HϕO* dominance has risen has now been demolished. This particular *HϕI* effect was estimated to contribute somewhat between −4 and −6 kJ/mol to the driving force of protein folding, for each intramolecular HB formed between two "arms" of the *HϕO* groups.

Second, the analysis of all the solvent-induced factors revealed that Kauzmann's model *does not* feature in the "driving force" for the process of protein folding (Fersht 1999; Ben-Naim 2009). Instead of the Gibbs energy of *solvation* of a *HϕO* molecule in water, the *conditional solvation* Gibbs energy of a *HϕO* group features in the "driving force." These Gibbs energies are very different from the Gibbs energies of solvation in water. The main reason is that a *HϕO* group attached to the backbone (BB) of the protein is surrounded by water molecules which are perturbed

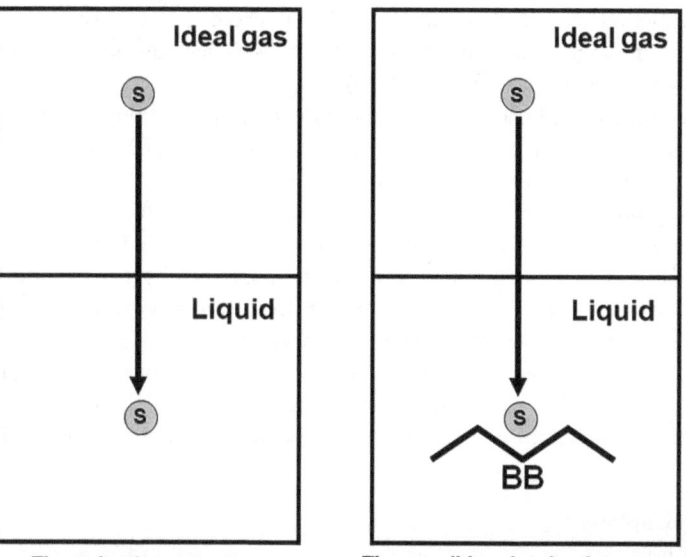

Fig. 4.4 The process of solvation and conditional solvation

by the BB. The difference between *solvation* and conditional solvation is shown in Fig. 4.4.

Thus, not only the basis on which the *HφO* model was built upon was *demolished*, but the Kauzmann *HφO* – model itself was now shown to be inadequate (Ben-Naim 1992, 2009).

Finally, the fact that the *HφO* groups are in the interior of the protein does not necessarily mean that the *HφO* effect is the "driving force" for protein folding. "Can anything be more convincing?" Tanford and Reynolds asked rhetorically (Tanford and Reynold 2003). Such an inference turned out to be only an illusion. This is exactly the same argument invoked when two different ideal gases mixed spontaneously. The mixing is a fact, but the mixing, in itself does not provide the "driving-force" for the mixing process (Ben-Naim 1992, 2007).

2.5   The Discovery of New **HφI** Effects

In the early 1990s we have analyzed all possible solvent-induced effects on protein folding. As a result a host of new solvent-induced effects that were discovered. These effects involved *HφI* rather than *HφO* groups. The most important one, and so far the most studied, was the pairwise *HφI* interactions between pairs of *HφI* groups at a distance between 4 and 5 Å, Fig. 4.5. For this particular *HφI* interaction there is overwhelming evidence that it is far stronger than any of the *HφO* effects. The evidence comes from theoretical estimates (Mirsky and Pauling 1936; Ben-Naim 1992), simulations (Mezei and Ben-Naim 1990; Durell and Ben-Naim 2017), and experimental data (Haberfield et al. 1984). There are also some estimates of *HφI* effects involving three and four *HφI* groups. These are more powerful, but probably less frequent (Ben-Naim 1990b).

The qualitative explanation of the pairwise *HφI* interaction is quite simple. A *HφO* group is characterized by a few "arms" along which HB may be formed. An amine group on the BB of the protein has one arm, a carbonyl group has two arms, a hydroxyl group three arms, and a water molecule itself has four arms. When a *HφI* group is in water its arms are solvated by water molecule. The Gibbs energy of solvation per one arm was estimated to be of the order of about −9.4 kJ/mol (Ben-Naim 1990b). Note that this is quite different from HB *energy*, as one might have erroneously counted in the HB-inventory argument.

When two *HφI* groups approach each other to a distance of about 2.8 Å, they form a direct HB. Thus, the Gibbs energy balance is: loss of the solvation Gibbs energy of two arms costs about 20 kJ/mol, and the formation of a HB provide

Fig. 4.5 Two hydroxyl groups attached to a backbone (BB) forming a hydrogen-bond-bridge at a distance of 4.5 Å

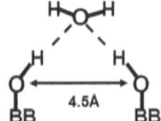

a HB energy of about -24 kJ/mol. Therefore, the net change in Gibbs energy for this particular $H\phi I$ effect at a distance of about 2.8 Å is about -6 kJ/mol (Ben-Naim 1990b).

A more dramatic $H\phi I$ effect was found at a distance of about 4.5 Å, the same distance of the second nearest neighbors in ice (Ben-Naim 1989). When two solvated arms approach each other to this distance, and with the correct orientation, they do not lose their solvation Gibbs energy as in the former $H\phi I$ case. They also do not gain HB *energy*. Instead, the solvation Gibbs energy of the pair of $H\phi I$ groups increase by an amount which was estimated to be between -10 and -12 kJ/mol.

The reason for such a strong $H\phi I$ interaction is that at this particular configuration the two arms of the two $H\phi I$ groups *can* be bridged by a water molecule. It should be stressed however that this effect is not due to a *formation* of permanent HB-bridge, as some have misunderstood. Such a "permanent" bridge could provide two HB *energies*, i.e. about -48 kJ/mol. The real effect is a *mutual solvation* of the two arms of water molecules. This effect involves HB energy, but also involves probability of finding a water molecule that can form a HB-bridge between the two $H\phi I$ groups. The most direct evidence for the existence of such a $H\phi I$ effect is the second peak in the radial distribution function of pure liquid water (Mirsky and Pauling 1936; Ben-Naim 1989). Other experimental evidence comes from the relative solubility of two isomers of the same molecule, having two $H\phi I$ groups at two different distances (Ben-Naim 1990b).

Because of the short range of the HB, there exists a steep gradient of the potential of mean force between two $H\phi I$ groups at a distance about 4.5 Å. This leads to a strong *force* between the two $H\phi I$ groups, a force which plays a crucial role in the process of protein folding, see below.

One can also think of other $H\phi I$ interactions; one involving one water molecule bridging three $H\phi I$ groups, or two water molecules forming one bridge connecting two $H\phi I$ groups. The former is strong, but rare, the second might be more frequent but very weak. Therefore, it is believed that the pairwise $H\phi I$ interaction at a distance of about 4.5 Å is the more important among the $H\phi I$ effects, hence probably the most important in the process of protein folding as well as in the process of protein-protein association or protein binding to DNA. A simple demonstration of the $H\phi I$ effect is shown in Fig. 4.6. Fumaric and maleic acids are two isomers having the same chemical formula; trans and cis butenedioic acid. However, the solubility of the cis-isomer is larger than the trans-isomer by a factor of about 100. This difference can be explained by the pairwise $H\phi I$ effect, see reference (Ben-Naim 1992) for details.

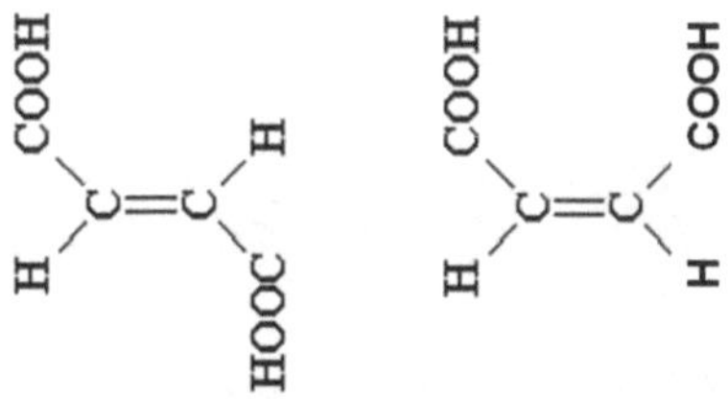

Fig. 4.6 Fumaric (left) and Maleic (right) acids. Which one is more soluble in water?

2.6   *Rough Estimates of the* HφI *Interaction Between Two,*
Three and Four HφI *Groups*

In this section we will present an approximate estimate of the strength of the *HφI* interaction between two "arms" of the *HφI* groups or molecules. Then I shall quote some values of the *HφI* interaction between two, three and four *arms* of *HφI* groups at the right configuration so that they can be bridged by a single water molecule.

Consider first the reaction shown in Fig. 4.7, this can represent either an intra-molecular or an intermolecular HB. We bring two arms of say, two water molecules from infinite separation to the final distance of about 2.8 Å so that they form a direct HB. In this process, we assume that the HB energy is about −27.2 kJ mol⁻¹. The loss of the solvation Gibbs energy of the two arms is about $2 \times 9.4 = 18.8$ kJ mol⁻¹. The net change in Gibbs energy is therefore

$$\Delta G \left(\text{Process in Figure 7} \right) \approx -27.2 + 18.8 = -8.4 \,\text{kJ mol}^{-1} \tag{4.3}$$

The above calculation is based on a correct procedure, except for the numerical values which might not be correct. Nevertheless, I believe that this result is of the correct order of magnitude.

Inspired by the calculation of ΔG in (4.3) we now attempt to estimate the Gibbs energy change for the process of bringing two "arms" of either *HφI* groups or *HφI* molecules from infinite separation to the distance of about 4.5 Å and properly oriented in such a way that the two arms can form a hydrogen-bonded bridge by a water molecule, Fig. 4.8.

Assuming again that the two HBs are formed, contributing about $-2 \times 27.2 \,\text{kJ mol}^{-1}$, and that *four* solvation Gibbs energies of four arms are lost $4 \times 9.4 \,\text{kJ mol}^{-1}$, we estimate

$$\delta G \left(\text{Process in Figure 8} \right) = -2 \times 27.2 + 4 \times 9.4 = -16.8 \,\text{kJ mol}^{-1} \tag{4.4}$$

Note that here we estimate only the solvent induced parts of the Gibbs energy change. We can assume that the direct interaction energy between the two *HφI* at this distance is much smaller than the indirect part of the Gibbs energy change.

The estimate in (4.4) is probably too high. The reason is that we have assumed that the water molecule *actually* forms two HBs with the two arms of the *HφI* group. However, the process that actually occurs when we bring the two *HφI* groups from infinity to the distance 4.5 Å, is not the process depicted in Fig. 4.8, i.e. we do

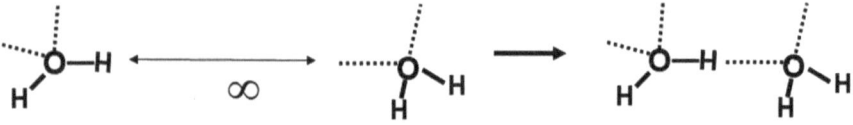

Fig. 4.7 The process of bringing two water molecules. To form a direct HB in the water

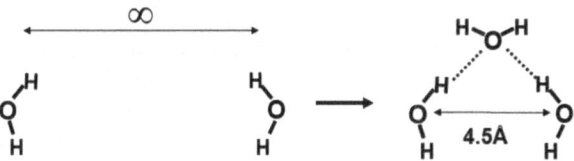

Fig. 4.8 The process of bringing two water molecules. To form an indirect HB-bridge with a water molecule

not *form* a two-hydrogen-bond-bridge, but instead the two arms are *solvated* by water molecules in both the initial and in the final configuration. In other words, we need to take into account the statistical character of the solvation of the two arms in the final configuration. To do this we need a more elaborate argument (see below).

We present here another heuristic argument which is based on the radial distribution function of water (Ben-Naim 2009). First, note that the RDF is actually an average over all possible orientations of the two water molecules

$$\bar{g}(R) = \frac{1}{\left(8\pi^2\right)^2} \int g\left(R,\vec{\Omega}_1,\vec{\Omega}_2\right) d\vec{\Omega}_1, d\vec{\Omega}_2 \tag{4.5}$$

At the distance of $R_1 \approx 2.8$ Å, we find a first peak in the RDF of height about 2.

We now wish to *de-average* the pair correlation function in Eq. (4.5). i.e. want to find out a value of $g\left(R_1,\vec{\Omega}_1,\vec{\Omega}_2\right)$, given $\bar{g}\left(R_1\right)$. To do that, we assume that the integrand $g\left(R_1,\vec{\Omega}_1,\vec{\Omega}_2\right)$ has only two values; a very high value when the orientations of the two molecules are such that a direct HB is formed between the two molecules, and a small value for all other orientations, Fig. 4.7, shows one example of an orientation for which the two water molecules form a HB.

Let x_H be the fraction of all orientations which are favorable for HBing. Hence, we can rewrite the average in (4.5) as:

$$\bar{g}(R_1) = g_H x_H + g_L\left(1 - x_H\right) \tag{4.6}$$

We can estimate the highest value of g_H as follows:

The work of bringing the two water molecules from infinite separation to the final distance R_1, with such orientations that a direct HB is formed is about

$$W\left(R_1\right) \approx -27.2 + 18.83 - 2.51 \approx -10.9 \text{ kJ mol}^{-1} \tag{4.7}$$

The three terms in (4.7) corresponds to the energy of the HB formed by the two molecules, the loss of the solvation of two arms, and some weak van der Waals interaction between the two water molecules at R_1. The value of $W(R_1)$ corresponds to the value of the pair correlation function of water at room temperature 25°C:

$$g_H\left(R_1\right) \approx \exp\left[-\beta W\left(R_1\right)\right] \approx 77 \tag{4.8}$$

The experimental value of the (angle average) pair correlation function of water at $R_1 \approx 2.8$ is about 2.2. This means that the *averaging* over all orientations has reduced the value of the pair correlation function from the highest value of about 77 to the average value of about 2.2, i.e. the reduction by a factor of about 35. We may say that the "deaveraging" of the experimental value of $\bar{g}(R_1)$ should cause an increase in the value of the pair correlation function at the most favorable configuration of forming a HB by a factor of about 35.

Now the heuristic – and very weakly justified – inference regarding the second peak of the pair correlation function of water at $R_2 \approx 4.5 \ \overset{\circ}{A}$ which is about

$$\bar{g}(R_2) \approx 1.2 \tag{4.9}$$

"Deaveraging" this value by multiplying it by 35 we may estimate the highest value of the pair correlation function at a specific configuration, such as shown in Fig. 4.8. Thus, we conclude that

$$g_H(R_2) \approx 1.2 \times 35 = 42 \tag{4.10}$$

which corresponds at room temperature to

$$W_H(R_2) \approx -k_B T \ln g_H(R_2) \approx -9.4 \text{ kJ mol}^{-1} \tag{4.11}$$

This is of course a very crude estimate. It is much smaller than the value we calculated in (4.4). A more elaborate calculation of the $H\phi I$ at the distance $R_2 \approx 4.5 \ \overset{\circ}{A}$ is given in references (Ben-Naim 1989, 1990a, b, 2009, 2011). The value obtained there is about $W(R_2 \approx 4.5) \approx -11.7 \text{kJ mol}^{-1}$, which is in good agreement with some experimental data and simulated calculation.

Thus, assuming that the direct interaction between the two arms at the configuration of Fig. 4.8 is much smaller than the indirect, solvent induced effect, we may conclude that

$$\delta G(1,2 \ / \ LJ) \approx -11.7 \text{ kJ mol}^{-1} \tag{4.12}$$

We quote here more estimates of the $H\phi I$ interactions from Ben-Naim (2009). The $H\phi I$ interaction between three $H\phi I$ arms at the correct distances and orientations, Fig. 4.9b, such that a water molecule can form three HB with three $H\phi I$ groups is estimated as:

$$\delta G(1,2,3 \ / \ LJ) \approx -26.4 \text{ kJ mol}^{-1} \tag{4.13}$$

The $H\phi I$ interaction between four $H\phi I$ arms at the correct distances and orientations, as in Fig. 4.9c, such that a water molecule can form four HBs with these groups is estimated as:

$$\delta G(1,2,3,4 \ / \ LJ) \approx -47.3 \text{ kJ mol}^{-1} \tag{4.14}$$

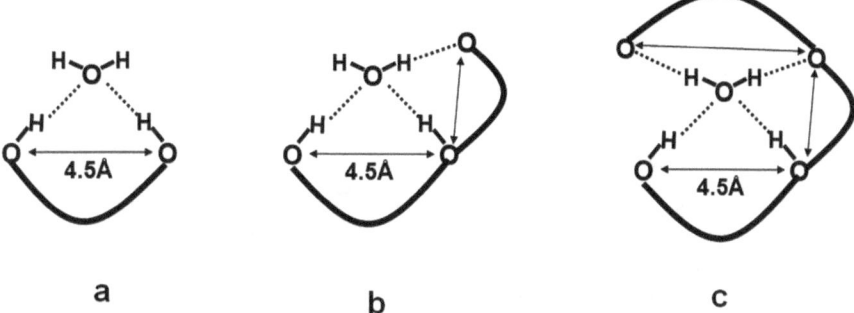

Fig. 4.9 Hydrophilic interaction between (**a**) two, (**b**) three and (**c**) four groups

3   The Protein Folding Problem (PFP)

As we have noted in the Introduction, there are several problems associated with protein folding. The most general question asked in this connection is that of the existence of a "code" that translates from the sequence of amino acids into a three dimensional structure. Following the works of Anfinsen (Anfinsen 1973; Haber and Anfinsen 1961), and others on the spontaneous folding of a denatured protein into the original native structure, it was speculated that the "information" about the folding pathway is already contained in the sequence of amino acids. We shall not discuss here the existence of a folding code. Instead, we shall focus on two aspects of the protein folding problem, in which $H\phi I$ *interactions,* and $H\phi I$ *forces* play a crucial role. One is the hydrophilic *interactions,* which is the most important factor in the stabilization of the protein structure, and the second is the hydrophilic *force.* The solvent-induced *forces* are probably the most important factor that governs the speed and the specificity of the process of protein folding.

It is believed that water not only "reads" the information contained in the sequence of amino acids but also translates the instruction into executable orders. These "orders" are the *forces* that are exerted on each of the atoms of the protein that causes the motion of the entire protein towards the end product. Because of the statistical character of these forces the motion of the protein is not along a unique deterministic route, but along a narrow range of routes or pathways.

Figure 4.10 shows a schematic process of folding of a segment of a protein. In this process we consider only two kinds of $H\phi O$ and $H\phi I$ effects. In Fig. 4.10 we show two de-solvation effects of a $H\phi O$ and a $H\phi I$ groups, and two solvent-induced interaction between two $H\phi O$ groups and between two $H\phi I$ groups. We note here that each $H\phi I$ effect is about an order of magnitude larger than the corresponding $H\phi O$ group. In addition, there are far more $H\phi I$ groups in a protein than $H\phi O$ groups. Therefore, it is clear that the $H\phi I$ effects (for both solvation and interactions) are dominant in determining the stability of the 3D structure of the protein.

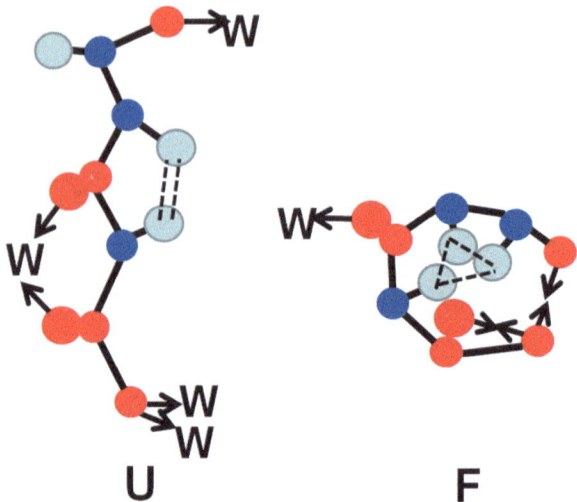

Fig. 4.10 Schematic process of folding a segment of a protein. Hydrophobic and hydrophilic groups are represented by blue and red circles respectively

It should be noted that, contrary to what was initially believed, the *solvation* and *pair-interaction* of either *HϕO* or *HϕI molecules* are totally irrelevant to the PFP. Instead, the real factors are the *conditional solvation* and the *conditional pair-interaction* between either *HϕO* or *HϕI groups* hung on the protein.

To summarize, when a random coil of a protein folds, the main two contributions to the driving force (i.e. to the negative change in Gibbs energy) are: (1) formation of direct intramolecular HB (see Fig. 4.10), and (2) solvent-induced interaction between two *HϕI* groups which are exposed to the solvent.

We note here that recently we found that pair- *HϕI* interactions are probably the most important factor that determines the solubility of globular proteins in water. The main factor in the solubility is the solvation Gibbs energy of the protein in water (Wang and Ben-Naim 1996, 1997).

Finally, we also note that the phenomenon of cold denaturation can also be explained by the different temperature dependence of the two *HϕI* effect (direct hydrogen bonding and solvent-induced interactions between *HϕI* effects).

3.1   The HϕI *Effect on the Stability of the Proteins*

In this section, we briefly discuss how the *HϕI* effects contribute to the stability of the protein. In an article entitled "The Problem of How and Why Proteins Adopt Folded Conformations," Creighton (1985) discusses the questions of "How?" and "Why?" as if they were one. Clearly, if we knew all the forces acting on all the atoms at each intermediate state of a specific protein, we could answer the question of "Why" and thereby the answer to the question of "How." This answer is pertinent to that *specific* protein.

An analysis of all the contributions to the solvent-induced effects on the driving forces for the process of protein folding reveals that $H\phi I$ interactions at a distance of about 4.5 Å are probably the strongest (Ben-Naim 2011).

The first indication that such correlation might be important came from some experimental data published by Haberfield et al (1984). These data were used to extract the quantity we initially called the correlation between two $H\phi I$ groups.

Soon, more data became available, as well as some simulation of these $H\phi I$ effects, (Mezei and Ben-Naim 1990; Durell et al. 1994; Durell and Ben-Naim 2017; Haberfield et al. 1984), and a theoretical estimate of the strength of these effects (Ben-Naim 1989, 1990a, b, c, 1991). All these data, led to the conclusion that correlation between two $H\phi I$ groups (at the correct distance and orientation) are quite significant, and their role in the process of protein folding should be taken more seriously.

The conclusion reached here has some overlaps with the conclusion reached by Rose et al. (2006). Rose *et al* proposed an inversion of the "side-chain/backbone paradigm." We advocate the inversion of the $H\phi O/H\phi I$ paradigm. Clearly, since most of the $H\phi I$ groups are in the backbone, it follows that the role of the backbone should be more important than the role of the side chains. In this sense there is an overlap between our proposal and Rose *et al* proposal. However, the two proposals are quite different.

In reference (Ben-Naim 1990a) we present all the evidence in favor of the contention that the various $H\phi I$ interactions play a dominant role in the stability of the protein. It should also be noted that these $H\phi I$ interactions also explain the high solubility of proteins, the association between proteins, as well as binding of drugs to proteins. This conclusion is contrary to the common accepted view that the hydrophobic effect is the most important one in protein folding (Dill 1990).

3.2   The HϕI *Effect on the Speed and Specificity of the Protein Folding*

Most reviews on protein folding focus on the thermodynamic "driving forces" rather than the *forces* themselves (Ben-Naim 1990a; Dill 1990). These "driving forces" are not really forces. In this section we discuss the *solvent-induced forces* in protein folding.

In early 1973, Levinthal asked a simple question: What are the factors that *speed* and *guide* the proteins in the process of protein folding? This question became one of the "big questions" of science (Kennedy and Norman 2005). Levinthal tried to answer this question, but did not give an answer. Assuming that a protein walks randomly in its configurational space, it would take eons to reach the native structure. Therefore, Levinthal concluded that there must be some "local interactions" that speeded and guided the protein toward the native structure. In his writings, there is no hint of any paradox. Nevertheless, for many years, Levinthal's *question* and his tentative answer were considered to be a paradox. Many scientists tried to resolve the paradox, which actually never existed. This consumed a great deal of effort. The apparent paradox is that proteins fold in a very short time – not in eons, as estimated in a random walk. However, there is no paradox and there was never a paradox. The immediate answer to Levinthal's question is simple: the rapid and guided folding of the protein must result from some *strong forces* that are exerted on the groups of the protein. The main problem is, therefore, to find out what the origin of these strong forces is. Unfortunately, people stuck to the paradigm of the "dominant *HφO* forces," which could not provide any strong *forces* resulting from the *HφO* interactions. Here we discuss the origin of the *strong forces* that speed and guide the folding of the protein. The missing factor in Levinthal's question is now available – the strong solvent-induced force operating on *Hφl* groups. The discovery of the *Hφl* forces essentially resolves the dynamic aspect in the protein folding problem.

We shall briefly discuss here the nature of the hydrophilic forces. These are solvent induced forces exerted on hydrophilic groups, such as OH, C=O, NH on the protein. Once we recognize the existence of the strong hydrophilic forces, then the answer to the dynamics of the protein folding problem becomes straightforward.

There are several kinds of hydrophilic (*Hφl*) forces. The most obvious one is the intramolecular hydrogen-bonds (HB) between *Hφl* groups of the protein. In this section we shall discuss several solvent induced forces between *Hφl* groups. As we shall soon see the strongest solvent-induced forces are exerted *on* the various *Hφl* groups rather than on *HφO* groups.

Before we describe the strong *Hφl* force it should be emphasized that although the *force* is derived from the potential of mean force, it is in general not true that strong force imply strong interactions, or strong interactions imply strong force. Figure 4.11 shows an example of strong force derived from relatively weaker interactions.

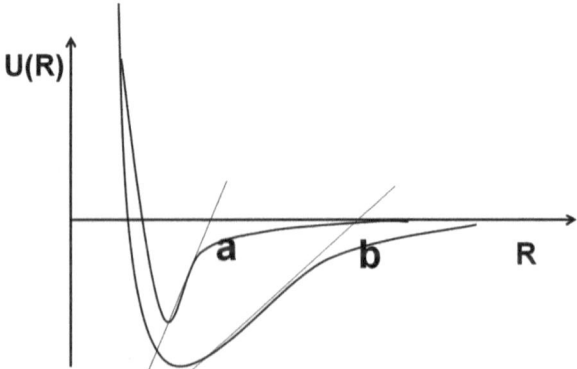

Fig. 4.11 The interaction energy at (**a**) is weaker than at (**b**), but the force (the slope) is stronger in (**a**) than in (**b**)

We start with a brief *definition* of the solvent-induced effects, and then present the evidence for the relative importance of the hydrophilic forces in protein folding.

Let $G(\boldsymbol{R}^M)$ be the Gibbs energy of a system of N water molecules at a given temperature T and pressure P, and a single protein molecule being at a specific conformation denoted by $\boldsymbol{R}^M = (\boldsymbol{R}_1, \boldsymbol{R}_2, \boldsymbol{R}_3, \cdots, \boldsymbol{R}_M)$ where $\boldsymbol{R}_i$ is the locational vector of atom i, or group i of the protein. It is convenient to choose the groups, such as methyl, ethyl, hydroxyl, carbonyl, etc. rather than each atom separately to describe the conformation of the protein. The function $G(T, P, N; \boldsymbol{R}^N)$ may also be referred to as the Gibbs energy landscape.

The statistical mechanical expression for the Gibbs energy of such system is:

$$\exp\left[-\beta G\left(\boldsymbol{R}^M\right)\right] = C\int dV \exp\left[-\beta V\right]\int dX^N \exp\left[-\beta U\left(\boldsymbol{R}^M, X^N\right)\right] \quad (4.15)$$

Here, $\beta = (k_B T)^{-1}$, k_B is the Bolzmann constant, and T the absolute temperature. C is a constant having the dimensions of length to the power $-3(N + 1)$. This is necessary to render the right-hand side of (4.15) dimensionless.

The quantity $U(\boldsymbol{R}^M, X^N)$ in (4.15) is the total interaction energy between all the molecules in the system being at a fixed configuration. We write this as

$$U\left(\boldsymbol{R}^M, X^N\right) = U\left(\boldsymbol{R}^M\right) + U\left(X^N\right) + B\left(\boldsymbol{R}^M, X^N\right) \quad (4.16)$$

where $U(\boldsymbol{R}^M)$ is the potential energy of the protein being at a specific conformation $\boldsymbol{R}^M$. $U(X^N)$ is the total interaction energy among all water molecules being at a fixed configuration $X^N = (X_1, X_2, \cdots, X_N)$, and $B(\boldsymbol{R}^M, X^N)$ is the binding energy of the protein, i.e. the interaction between the protein at $\boldsymbol{R}^M$, and the solvent molecules at X^N.

To examine the probable direction, the protein will move in its configurational space. We need to know the *forces* acting on each of the M groups of the protein being at the conformation $\boldsymbol{R}^M$. This force is obtained by taking the gradient of the Gibbs energy with respect to each of the $\boldsymbol{R}_i$.

We take an infinitesimal change in $\boldsymbol{R}_1$, keeping all other $\boldsymbol{R}_i (i \neq 1)$ unchanged and averaging over all the configurations of the solvent molecules. Thus,

$$-F\left(\boldsymbol{R}_1\right)d\boldsymbol{R}_1 = G\left(\boldsymbol{R}_1 + d\boldsymbol{R}_1\right) - G\left(\boldsymbol{R}_1\right) = \left[\nabla_1 U\left(\boldsymbol{R}^M\right) + \nabla_1 \delta G\left(\boldsymbol{R}^M\right)\right]d\boldsymbol{R}_1 \quad (4.17)$$

Note that in (4.17), G is a function of $\boldsymbol{R}^M$, but we have omitted from the notation all the $\boldsymbol{R}_i$ which do not change in this process. The first gradient on the right-hand side of (4.17) is simply the force acting on group 1 being at $\boldsymbol{R}_1$, by all other groups of the protein. We refer to this as the *direct force*.

The second gradient on the right-hand side of (4.17) is the *solvent-induced force* on group 1, given a fixed conformation of the protein and averaged over all possible configurations of the solvent molecules. For any fixed conformation $\boldsymbol{R}^M$ of the protein, we can write the equality:

$$\delta G\left(\boldsymbol{R}^M\right) = \Delta G^*\left(\boldsymbol{R}^M\right) - \Delta G^*\left(\infty\right) \quad (4.18)$$

where $\Delta G^*(\mathbf{R}^M)$ is the Gibbs energy of solvation of the protein at the conformation $\mathbf{R}^M$, and $\Delta G^*(\infty)$ denotes the Gibbs energy of solvation of all the M groups when they are at infinite separation from each other. Since the latter is not a function of $\mathbf{R}_1$, we can write the solvent induced force as

$$F_1^{SI}\left(\mathbf{R}_1\right) = -\nabla_1 \delta G\left(\mathbf{R}^M\right) = -\nabla_1 \Delta G^*\left(\mathbf{R}^M\right) \qquad (4.19)$$

Thus, the solvent-induced force is determined by the gradient of the *solvation Gibbs energy* of the protein at a given conformation $\mathbf{R}^M$.

From Eq. (4.19) we can make the following general statement: For any given conformation $\mathbf{R}^M$ of the protein, a solvent-induced force will be exerted on group i, if and only if, a small change in $\mathbf{R}_i$ causes a change in the solvation Gibbs energy of the protein (keeping T, P, N as well as $\mathbf{R}_j(j \neq i)$ constants). Thus, if we believe that the solvent is important in the folding process we should focus on the indirect force in (4.19). This is exactly the component of the force which is the most relevant to the folding process.

From Eq. (4.19), it is difficult to see the various factors that contribute to the solvent-induced force. A more useful expression which is also easier to analyze and interpret is the following: For details, see references (Ben-Naim 1990a, 2011).

$$F_1^{SI}\left(\mathbf{R}_1\right) = \int\left[-\nabla_1 U\left(\mathbf{R}_1, \mathbf{X}_W\right)\right]\rho\left(\mathbf{X}_W\middle|\mathbf{R}^M\right)d\mathbf{X}_W \qquad (4.20)$$

In Eq. (4.20) we see that the integrand contains two factors [note that the gradient operates only on the potential function $U(\mathbf{R}_1, \mathbf{X}_W)$, and not on $\rho(\mathbf{X}_W|\mathbf{R}^M)$]. The first is the force exerted on group 1 by a water molecule at $\mathbf{X}_W = (\mathbf{R}_W, \mathbf{\Omega}_W)$. The second is the conditional density of water molecules at $\mathbf{X}_W$ given the protein at conformation $\mathbf{R}^M$.

Figure 4.12 shows schematically the force $-\nabla_1 U(\mathbf{R}_1, \mathbf{X}_W)$ by a full line, and the effect of all the groups of the protein on the density of water molecules at $\mathbf{X}_W$, by a dashed line. The quantity $\rho(\mathbf{X}_W|\mathbf{R}^M)d\mathbf{X}_W$ may also be interpreted as the conditional probability of finding a water molecule within the element of "volume" $d\mathbf{X}_W$ given the conformation $\mathbf{R}^M$.

Looking at the expression for the solvent-induced part (4.20), we can say that whenever there exists a strong gradient there is a strong *genuine* force. However, a strong force does not imply an instantaneous motion in the direction of the force. First, because there might be some constraints that prevent the motion in the direction of the force, and second, the solvent-induced force is statistical in nature. Therefore, all we can say is that there is a certain probability that within the time scale of our measurements, the group will move in the direction of the gradient: the stronger the force, the higher the probability of moving in that direction.

Note also that the integrand in (4.20) is a product of two factors; a *force* and a *density*. In order to have a significant contribution to the integral, both factors in the integrand must not be too small. Therefore, the integration in (4.20) is effectively over that region of space where both factors in the integrand are not negligible.

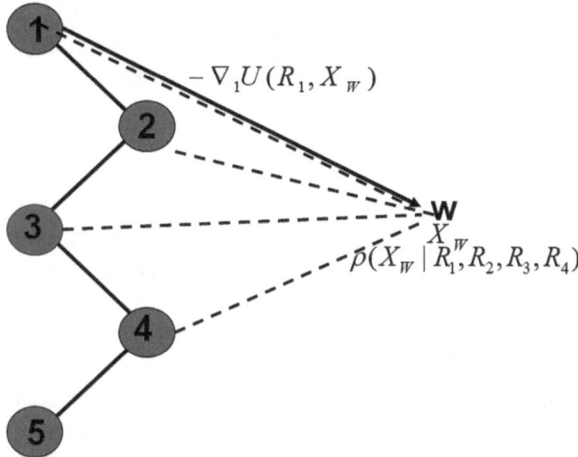

Fig. 4.12 The force exerted on group 1 by a water molecule, W is show by a full arrow. The contributions to the local density of water molecules are indicated by the dashed lines

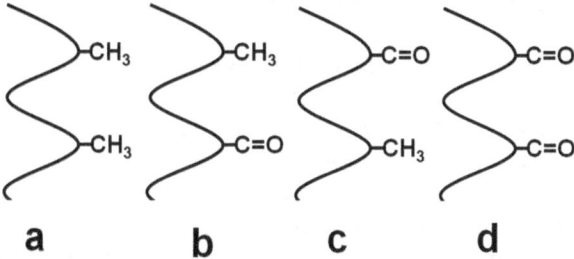

Fig. 4.13 Four cases of two groups on a protein

We now examine a simple case of a segment of a protein having groups of two kinds: methyl group representing $H\phi O$ groups and carbonyl to represent $H\phi I$ groups. In this case we have four possible cases, Fig. 4.13.

(a) **Group 1, $H\phi O$ and Group 2, $H\phi O$**

In this case the force $-\nabla_1 U(R_1, X_W)$ (between a water molecule and the $H\phi O$ group) is expected to be weak. Furthermore, since the interaction between a water molecule and the two $H\phi O$ groups is weak even at the most favorable configuration, for the triplet at R_1, R_2 and R_W, we expect that the conditional density at X_W will be only slightly higher than the bulk density ρ_W.

(b) **Group 1, $H\phi O$ and Group 2, $H\phi I$**

In this case the force $-\nabla_1 U(R_1, X_W)$ is expected to be weak as in case (a). However, because of the presence of the $H\phi I$, group 2, there exists a distance from this $H\phi I$ group where the conditional density might be significantly enhanced.

Therefore, we expect in this case, to obtain a solvent induced force larger than in case (a). Note that the larger the conditional density of water molecules is expected only at certain distance and direction from the $H\phi O$ group.

(c) <u>Group 1, $H\phi I$ and Group 2, $H\phi O$</u>

In this case the force $-\nabla_1 U(\mathbf{R}_1, \mathbf{X}_W)$ is expected to be larger than in cases (a) and (b). Because of the presence of one $H\phi I$ group, the conditional density will also be enhanced. This has the same effect as the corresponding term in case (b), presuming the correct orientation of the $H\phi I$ group. Thus, in this case we expect to get a solvent induced force larger than case (b).

(d) <u>Group 1, $H\phi I$ and Group 2, $H\phi I$</u>

In this case the force $-\nabla_1 U(\mathbf{R}_1, \mathbf{X}_W)$ will be as large as in case (c). However, because of the presence of two $H\phi I$ groups, we might, under the right conditions, get a higher conditional density. The right conditions mean that the two hydrophilic groups are at a distance of about 4.5 Å, and the correct orientation of the two $H\phi I$ groups, Fig. 4.14. At these right conditions we expect large conditional density of water molecules. Thus, in this case both the force and the conditional density will be large and the resulting solvent-induced force is expected to be larger than in case (c).

Figure 4.15 shows schematically the relative order of magnitudes of the solvent induced force from case (a) to (d). We can conclude that the solvent induced force on the $H\phi I$ group in case (d) will be the largest of the four cases described above. It should be noted that we have discussed the *force* exerted on group 1, in the presence of different environments. We also stress that we have also examined the condition under which we expect the *maximum* forces exerted on group 1. In each case the maximum solvent-induced force might be in different directions. We could have also taken the gradients with respect to the rotational angles ϕ_1 in the Gibbs energy function $G(T, P, N; \phi_1, \cdots, \phi_m)$. In this case we would have obtained the torque rather than the force.

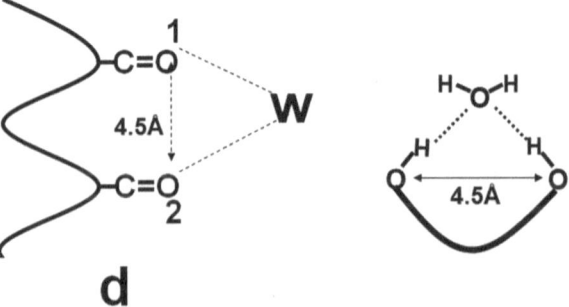

Fig. 4.14 The possibility of water molecule forming a Hydrogen-bond-bridge between two hydrophilic groups

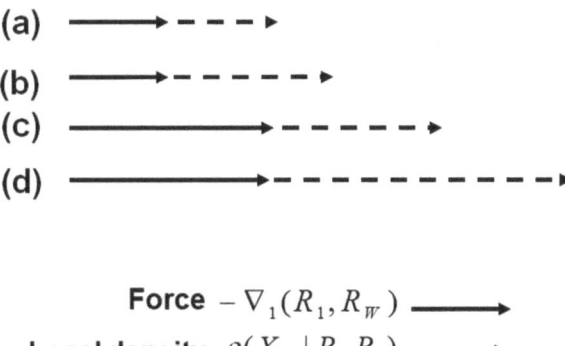

$$\text{Force} \; -\nabla_1(R_1, R_W) \longrightarrow$$
$$\text{Local density} \; \rho(X_W \mid R_1, R_2) \; --\;-\blacktriangleright$$

Fig. 4.15 The relative strength of the maximum forces in cases (a) to (d) in Fig. 4.14

Thus, the general conclusion is that the strongest solvent induced force is expected to be exerted on a *HφI* group, when this group is also surrounded by other *HφI* groups in its immediate neighborhood. We have discussed in details only the case of *two HφI* groups, but clearly, the argument may be extended to include the effect of the presence of more *HφI* groups. The force produced by such a one-water bridge is operative in the range of distances between the two *HφI* groups of about 4.5 Å. Long range *HφI* forces are also possible (Ben-Naim 1992).

3.3 Answering the PFP

Based on the data on the hydrophobic forces, we envisage the process of protein folding to be as follows: Starting from an arbitrary conformation of the protein chain, there will be forces acting on each of the groups of the protein. As we have seen, the forces exerted on the *HφI* groups are significantly stronger than the forces on the *HφO* groups. Add to this that on average each protein has many more *HφI* than *HφO* groups, we can safely conclude that the forces on all the *HφI* groups will dominate the speed of the folding. If the protein is initially in an extended conformation, we expect each *HφI* group to have on average another, single *HφI* group in its vicinity. In this case the forces on these groups will be similar to the results for two hydrophilic solutes, discussed in Sect. 3.2. As the protein becomes more compact, each *HφI* group is expected to be surrounded by a larger number of *HφI* groups, which will result in even greater forces and stabilization. Hence, we expect that the folding process will be accelerated as the protein conformation becomes more and more compact. This scenario is very different from the so-called "hydrophobic collapse," where the tendency of the *HφO* groups to aggregate is presumed to be the main "driving force" for the folding process. As for the "guidance" of the folding process, once we recognize the importance of the *HφI* forces, we can conclude that the specific pattern of amino acids in the chain provides a pattern of strong forces, hence also a preferable pathway for the folding. In this sense, the $H\phi I$ force answers the two questions raised by Levinthal regarding the speed and guidance of the folding process.

4   Conclusion

The paradigm change from $H\phi O$ to $H\phi I$ effects has brought us as close as one can hope for the solution of the problem of protein folding. The $H\phi I$ *interactions* provide explanation of the stability of proteins. The $H\phi I$ *forces* provide the answer to the speed and specificity of the folding process.

References

Anfinsen, C. B. (1973). Principles that govern the folding of protein chains. *Science, New Series, 181*, 223–230.

Anson, M. L., & Mirsky, A. E. (1934). The equilibria between native and denatured hemoglobin in salicylate solutions and the theoretical consequences of the equilibrium between native and denatured protein. *The Journal of General Physiology, 17*, 393–408.

Ben-Naim, A. (1980). *Hydrophobic interactions*. New York: Plenum Press.

Ben-Naim, A. (1989). Solvent-induced interactions: Hydrophobic and hydrophilic phenomena. *Journal of Chemical Physics, 90*, 7412–7525.

Ben-Naim, A. (1990a). On the role of hydrogen-bonds in protein folding and protein association. *Journal of Physical Chemistry, 95*, 1444–1473.

Ben-Naim, A. (1990b). Solvent effects on protein association and protein folding. *Biopolymers, 29*, 567–596.

Ben-Naim, A. (1990c). Solvent induced forces in protein folding. *Journal of Chemical Physics, 94*, 6893–6895.

Ben-Naim, A. (1991). Strong forces between hydrophilic macromolecules; implications in biological systems. *Journal of Chemical Physics, 93*, 8196–8210.

Ben-Naim, A. (1992). *Statistical thermodynamics for chemists and biochemists*. New York: Plenum Press.

Ben-Naim, A. (2007). *Entropy demystified. The second law reduced to plain common sense*. Singapore: World Scientific.

Ben-Naim, A. (2009). *Molecular theory of water and aqueous solutions part I: Understanding water*. Singapore: World Scientific.

Ben-Naim, A. (2011). Molecular theory of water and aqueous solutions. In *Part II: Hydrophilic effects in protein folding, self-assembly and molecular recognition*. Singapore: World Scientific.

Creighton, T. E. (1985). The problem of how and why proteins adopt folded conformations. *Journal of Physical Chemistry, 89*, 2452.

Dill, K. A. (1990). Dominant forces in protein folding. *Biochemistry, 29*, 7133–7155.

Durell, S. R., & Ben-Naim, A. (2017). Hydrophobic-hydrophilic forces in protein folding. *Biopolymers, 2017*(10), 23020.

Durell, S. R., Brooks, B. R., & Ben-Naim, A. (1994). Solvent-induced forces between two hydrophilic groups. *Journal of Physical Chemistry, 98*, 2198–2202.

Fersht, A. (1999). *Structure and mechanism in protein science*. New York: W. H. Freeman and Comp.

Haber, E., & Anfinsen, C. B. (1961). Studies on the reduction on reformation of protein disulfide bonds. *Journal of Biological Chemistry, 236,* 1361–1363.

Haberfield, P., Kivuls, J., Haddad, M., & Rizzo, T. (1984). Enthalpies, free energies, and entropies of transfer of phenols from nonpolar solvents to water. *Journal of Physical Chemistry, 88,* 1913–1916.

Kauzmann, W. (1959). Some factors in the interpretation of protein denaturation. *Advances in Protein Chemistry, 14,* 1–63.

Kennedy, A., & Norman, C. (2005). What don't we know? *Science, 309,* 75.

Levinthal, C. (1968). Are there pathways for protein folding? *Journal de Chimie Physique, 65,* 44.

Levinthal, C. (1969). Mossbauer spectroscopy. In T. Debrunner, J. C. M. Tsibris, & E. Munck (Eds.), *Biological systems* (p. 22). Urbana: University of Illinois Press.

Mezei, M., & Ben-Naim, A. (1990). Calculation of the solvent contribution to the potential of mean force between water molecules in fixed relative orientation in liquid water. *Journal of Chemical Physics, 92,* 1359–1361.

Mirsky, A. E., & Pauling, L. (1936). On the structure of native, denatured and coagulated proteins. *Proceedings of the National Academy of Sciences of the United States of America, 22,* 439–449.

Pauling, L. (1948). *The nature of chemical bond* (2nd ed.). Ithaca/New York: Cornell University Press.

Pauling, L. (1960). *The nature of chemical bond* (3rd ed.). Ithaca/New York: Cornell University Press.

Pauling, L., & Corey, R. B. (1951a). Atomic coordinates and structure factors for two helical configurations of polypeptide chains. *Proceedings of the National Academy of Sciences of the United States of America, 37,* 235–248.

Pauling, L., & Corey, R. B. (1951b). The pleated sheet, a new layer configuration of polypeptide chains. *Proceedings of the National Academy of Sciences of the United States of America, 37,* 251–256.

Pauling, L., & Corey, R. B. (1951c). The structure of fibrous proteins of the collagen-gelatin group. *Proceedings of the National Academy of Sciences of the United States of America, 37,* 272–281.

Pauling, L., & Corey, R. B. (1951d). Configurations of peptide chains with favored orientations around single bonds: Two new pleated sheets. *Proceedings of the National Academy of Sciences of the United States of America, 37,* 729–740.

Rose, G. D., Fleming, P. J., Banavar, J. R., & Maritan, A. (2006). A backbone based theory of protein folding. *Proceedings of the National Academy of Science, 103,* 16623–16633.

Schellmann, J. A. (1955a). The thermodynamics of urea solutions and the heat of formation of the peptide hydrogen bond. *Comptes Rendus des Travaux du Laboratoire Carlsberg. Série Chimique, 29,* 223–230.

Schellmann, J. A. (1955b). The stability of hydrogen-bonded peptide structures in aqueous solution. *Comptes Rendus des Travaux du Laboratoire Carlsberg. Série Chimique, 29,* 230–259.

Tanford, C. (1973). *The hydrophobic effect: Formation of micelles and biological membranes.* New York: Wiley-Interscience.

Tanford, & Reynold, J. (2003). *Nature's robots, a history of proteins.* Oxford: Oxford University Press.

Wang, H., & Ben-Naim, A. (1996). A possible involvement of solvent induced interactions in drug design. *The Journal of Medicinal Chemistry, 39,* 1531–1539.

Wang, H., & Ben-Naim, A. (1997). Solvation and solubility of globular proteins. *Journal of Physical Chemistry B, 101,* 1077–1086.

Chapter 5
Analysis of Protein Intramolecular and Solvent Bonding on Example of Major Sonovital Fluid Component

Piotr Bełdowski, Krzysztof Domino, Damian Bełdowski, and Robert Dobosz

Abstract Molecular interactions within proteins are fundamental to maintain their conformation and role in biological systems. Understanding the nature and dynamics of the interactions is crucial as it can help understand phenomena occurring during physiological processes, drug design, and delivery, etc. This chapter presents the analysis of the dynamics of molecular interactions inside proteins on an example of albumin. We have performed computer simulations of albumin protein at its native/equilibrium state to understand the dynamics of bonding/interactions inside the protein and with water: hydrogen bonds and hydrophobic interactions. Furthermore, we extracted the data to look into interactions between particular amino acids (AA). As expected, charged AA, such as the glutamic acid (GLU) and lysine (LYS) one, form most intermolecular hydrogen bonds and bind most water.

Further, in hydrophobic contacts, the significant role plays phenylalanine (PHE) and leucine (LEU) amino-acids. We have also found that although its hydrophobic nature, the LEU forms the most stable contacts. This can be attributed to its position in the core of the protein. The presented results may help better understand the dynamics of a complex structure such as protein.

Keywords Protein intramolecular interactions · Protein intermolecular interactions · Water assisted effects · Protein water nanointerface · Hydrogen bonding · Hydrophobicity · Computer simulation of albumin-water interface

P. Bełdowski (✉) · R. Dobosz
Institute of Mathematics and Physics, UTP University of Science and Technology, Bydgoszcz, Poland
e-mail: piobel000@utp.edu.pl; robertd@utp.edu.pl

K. Domino
Institute of Theoretical and Applied Informatics, Polish Academy of Sciences, Gliwice, Poland
e-mail: kdomino@iitis.pl

D. Bełdowski
Institute of Mathematics, Jagiellonian University, Kraków, Poland
e-mail: damian.beldowski@student.uj.edu.pl

© The Author(s), under exclusive license to Springer Nature
Switzerland AG 2021
A. Gadomski (ed.), *Water in Biomechanical and Related Systems*,
Biologically-Inspired Systems 17, https://doi.org/10.1007/978-3-030-67227-0_5

1   Introduction

On page 1123, the *Glossary of terms used in physical organic chemistry (IUPAC Recommendations 1994)* (IUPAC 1997) explains two terms that are the main problem of this chapter. We will start by quoting these definitions. Namely:

"hydrogen bond

The hydrogen bond is a form of association between an electronegative atom and a
   hydrogen atom attached to a second, relatively electronegative atom. It is best
   considered as an electrostatic interaction, heightened by the small size of hydrogen, which permits proximity of the interacting dipoles or charges. Both electronegative atoms are usually (but not necessarily) from the first row of the Periodic
   Table, i.e., N, 0, or F. Hydrogen bonds may be intermolecular or intramolecular.
   With a few exceptions, usually involving fluorine, the associated energies are
   less than 20–25 kJ mol^{-1} (5–6 kcal/mol)."

"hydrophobic interaction

The tendency of hydrocarbons (or of lipophilic hydrocarbon-like groups in solutes)
   to form intermolecular aggregates in an aqueous medium and analogous intramolecular interactions. The name arises from the attribution of the phenomenon
   to the apparent repulsion between water and hydrocarbons. However, the phenomenon ought to be attributed to the effect of the hydrocarbon-like groups on
   the water-water interaction. The misleading alternative term "hydrophobic bond"
   is discouraged."

These are terms related to water and the impacts that arise from it. These interactions are inseparable from many fields of chemistry, biochemistry and molecular biology. In our study, we would like to show that they play an essential role in maintaining protein conformation, where hydrogen bonds stabilize protein structure (Rose et al. 2006). Whereas hydrophobic interactions are the main strength of the protein folding process (Compiani and Capriotti 2013). Understanding the nature and dynamics of these interactions is crucial in the drug design and transport and explaining many biochemical problems, such as intercellular transport or the joint lubrication system. The protein Human Serum Albumin (HSA) in a cartoon-like representation is shown in Fig. 5.1. HSA is the main component of synovial fluid and constitutes about 50% of the proteins found in blood plasma (Evans 2002). HSA is an abundant multifunctional non-glycosylated, negatively charged plasma protein (Peters 1996). It is relatively small to other plasma proteins (66 kDa). It is composed of 585 amino acids with a few tryptophan and methionine residues but an abundance of charged residues such as lysine and aspartic acids and no prosthetic groups or carbohydrate (Quinlan et al. 2005). It can bind an incredibly diverse range of drugs, metabolites, and fatty acids, which underlines its great importance in the functioning of organisms (Bhattacharya et al. 2000).

The HSA, like any other polymer, consists of monomers, the amino-acids. They are combined in a specific sequence and interact with each other. To understand the structure of the albumin (and other proteins), we need to know what influence

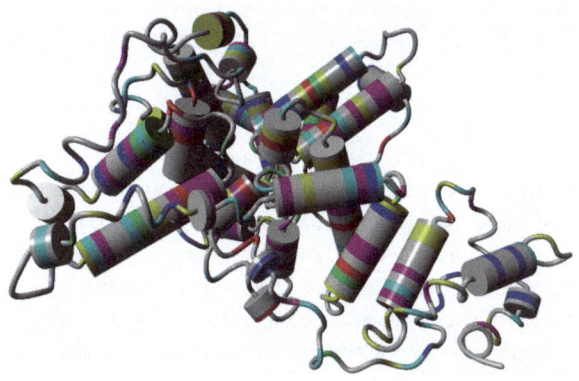

Fig. 5.1 Visualization of albumin molecule in a cartoon form. Several amino-acids have been colored to show their position in the molecule. Colors read as follows: pink-LEU, yellow-LYS, green-TYR, cyan-GLU, blue-PHE, red-ARG. Tubes reflects alpha helices

non-covalence interactions such as hydrogen bond (HB), hydrophobic contacts (HP) have on the formation of the protein chain. An important aspect is also which amino acids most often interact with each other and what types of interactions are they (Subha Mahadevi and Narahari Sastry 2016; Cerny and Hobza 2007; Dedinaite et al. 2019; Bełdowski et al. 2018, 2019). Thus, these interactions are studied in the chapter.

This study aims to present a statistical analysis of basic interaction occurring in albumin containing aqueous solution. Namely, hydrogen bonding between the solute and solvent was examined as well as intramolecular solute bonding. Moreover, two other major interactions were analyzed: hydrophobic effect, hydrogen bond. We discuss preliminary studies aimed at understanding protein dynamics under equilibrium conditions.

2   Materials and Methods

Molecular dynamics simulations were performed. Simulation box containing albumin (downloaded from the protein data bank https://www.rcsb.org/structure/1E78) was filled with water (TIP3P model) containing 0.9% NaCl (physiological solution) in the amount corresponding to established water density. System minimalization was performed with a time step 1 fs for 1000 steps. For the presented configuration, the volume box is about 864 nm³ (11.2 nm × 8.3 nm × 9.3 nm), with a periodic boundary condition applied. All-atom isothermal–isobaric (NPT) ensemble simulations were performed under the following conditions: temperature 300 K, with a time step of 2 fs, pressure = 1 bar. Berendsen barostat and thermostat with a relaxation time of 1 fs were used to maintain constant temperature and pressure. The simulation results were sampled every 50 ps, whereas the simulation lasted 100 ns. A total of 2000 data points were collected to analyze. An Assisted Model Building with Energy Refinement (AMBER) force field has been employed to mimic the interactions between molecules. The functional form of the AMBER03 force field is:

$$E_{total} = E_{bonds} + E_{angles} + E_{dihedrals} + E_{VdW} + E_{electrostatic} \qquad (5.1)$$

where E_{bonds} represents the energy between covalently bonded atoms, E_{angles} represents the energy due to the geometry of electron orbitals involved in covalent bonding, $E_{dihedrals}$ represents the energy for twisting a bond due to bond order (e.g., double bonds) and neighboring bonds or lone pairs of electrons, E_{VdW} represents the non-bonded energy between all atom pairs (van der Waals energy), and $E_{electrostatic}$ represents the electrostatic interactions between all atom pairs.

Hydrogen bond (HB) is defined by YASARA as follows. Hydrogen bond energy is strictly not less than 6.25 kJ/mol (or 1.5 kcal/mol) (weak and very weak bonds are omitted). This number is only 25% of the optimal 25 kJ/mol energy value. The energy in kJ/mol is given by the following formula:

$$E_{HB} = 25 \cdot \frac{\chi^{(D-A-H)} \cdot \chi^{(H-A-X)} \cdot \left[2.6 - max\left(dis_{(H-A)}, 2.1\right)\right]}{0.5} \qquad (5.2)$$

where in the first from above scaling factors depends from the angle formed by Donor-Hydrogen-Acceptor, and the second scaling factor is obtained from the angle formed by Hydrogen-Acceptor-X. The latter X stands for the atom covalently bond to the acceptor. The first and the second scaling factor can take values from the range [0,1] (angles in both formulas are expressed in radians angles). The first scaling factor is:

$$\chi^{(D-A-H)} = \begin{cases} 0 & \text{if angle in betewen } \left[0, \dfrac{5\pi}{9}\right) \text{ radians,} \\ x \in [0,1] & \text{if angle in betewen } \left[\dfrac{5}{9}\pi, \dfrac{11\pi}{12}\right] \text{ radians,} \\ 1 & \text{if angle in betewen } \left[\dfrac{11\pi}{12}, \pi\right) \text{ radians,} \end{cases} \qquad (5.3)$$

In the case when X is a heavy atom, the second scaling factor attains following values:

$$\chi^{(H-A-X)} = \begin{cases} 0 & \text{if angle in betewen } \left[0, \dfrac{17\pi}{36}\right) \text{ radians,} \\ x \in [0,1] & \text{if angle in betewen } \left[\dfrac{17}{36}\pi, \dfrac{19\pi}{36}\right] \text{ radians,} \\ 1 & \text{if angle in betewen } \left[\dfrac{9}{36}\pi, \pi\right) \text{ radians,} \end{cases} \qquad (5.4)$$

In the case when X is a hydrogen atom, the smaller angles are allowed. In such scenario the scaling factor has the following form:

$$\chi^{(H-A-X)} = \begin{cases} 0 & \text{if angle in beteween } \left[0, \dfrac{5\pi}{12}\right) \text{radians,} \\ x \in [0,1] & \text{if angle in beteween } \left[\dfrac{5}{12}\pi, \dfrac{17\pi}{36}\right] \text{radians,} \\ 1 & \text{if angle in beteween } \left[\dfrac{17\pi}{36}, \pi\right) \text{radians,} \end{cases}$$ (5.5)

In the case when the Acceptor forms has more than one covalent bond, the one which has the lowest scaling factor is taken into calculation.

The hydrophobic interaction strength between hydrophobic atoms (carbon based) is calculated as follows. Hydrophobic atoms are identified and the atom type is assigned to them. Type 1 is for carbon atoms with three or more bond hydrogen atoms ($-CH_3$), type 2 is for carbon atoms with two hydrogen atoms or one hydrogen atom plus three carbon atoms bond ($-CH_2-$, HCC_3), and type 3 is for aromatic ring carbon atoms with only carbon and hydrogen atoms bond (Bełdowski et al. 2019).

For each experimental setting simulation were performed for 5 distinct realizations in order to account for an impact of random factors on simulations and collect statistics.

The radial distribution function (or RDF), g(r), is an example of a pair correlation function, which describes how, on average, the atoms in a system are radially packed around each other.

$$g(r) = n(r) / \left(\rho 4\pi r^2 \Delta r\right)$$ (5.6)

In which n(r) is the mean number of atoms in a shell of width Δr at distance r, ρ is the mean atom density. The method is not restricted to one atom but all atoms in the system can be considered, leading to an improved determination of the RDF as an average over many atoms.

3 Results and Discussion

The presence of 585 amino acids which build albumin in the environment of water molecules gives the possibility of extracting thousands of interactions (Fig. 5.2) that affect the structure. Those are as follows—500 intramolecular hydrogen bonds (HB) between amino acids. Additionally, more than 1000 hydrogen bonds that form amino acids with water (HB H_2O) and more than 2500 hydrophobic contacts (HP), we get a complete picture of more than 4000 weaker or stronger interactions that determine the structure of HSA. Hydrogen bonds (HB and HB H_2O) are stable.

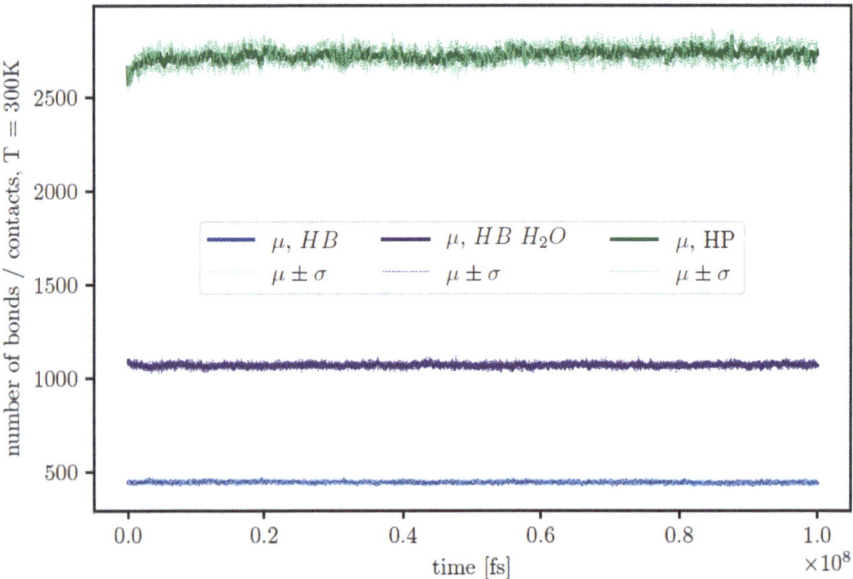

Fig. 5.2 Number of hydrogen bonds and HP contacts, for an exemplary temperature of 300 K. Mean and standard deviation were calculated over 5 realizations of experiments

During the whole time of the simulation, their number has not changed. Interestingly at the first part of simulations number of HP contacts rises, next it stays constant.

In Fig. 5.3 we present aggregate histograms of interactions and contacts lengths. Aggregations were preformed over simulation time moments and realizations. Observe the negative skewness and some sort of the cut-off at $5.0*10^{-10}$ m for HP contacts. Histograms of hydrogen bonds lengths appear to be positively skewed, with significantly different skewness between two particular types of hydrogen bonds.

On the basis of Fig. 5.3 we can see that hydrogen bonds between amino acids (HB) and amino acid-water (HB H_2O) have different length and consequently different strength: from the weakest with the length of approximately 2.4 Å, to the strongest with the length 1.4–1.6 Å. Obviously, the most are HB and HB H_2O with a length of 1.8–2.0 Å.

3.1 Analysis of Particular Amino-Acids Interactions

There are 20 amino-acids building the albumin. Following Siódmiak (Siódmiak and Bełdowski 2019), in this sub-section, we analyze between which amino-acids bonds or contacts are formed.

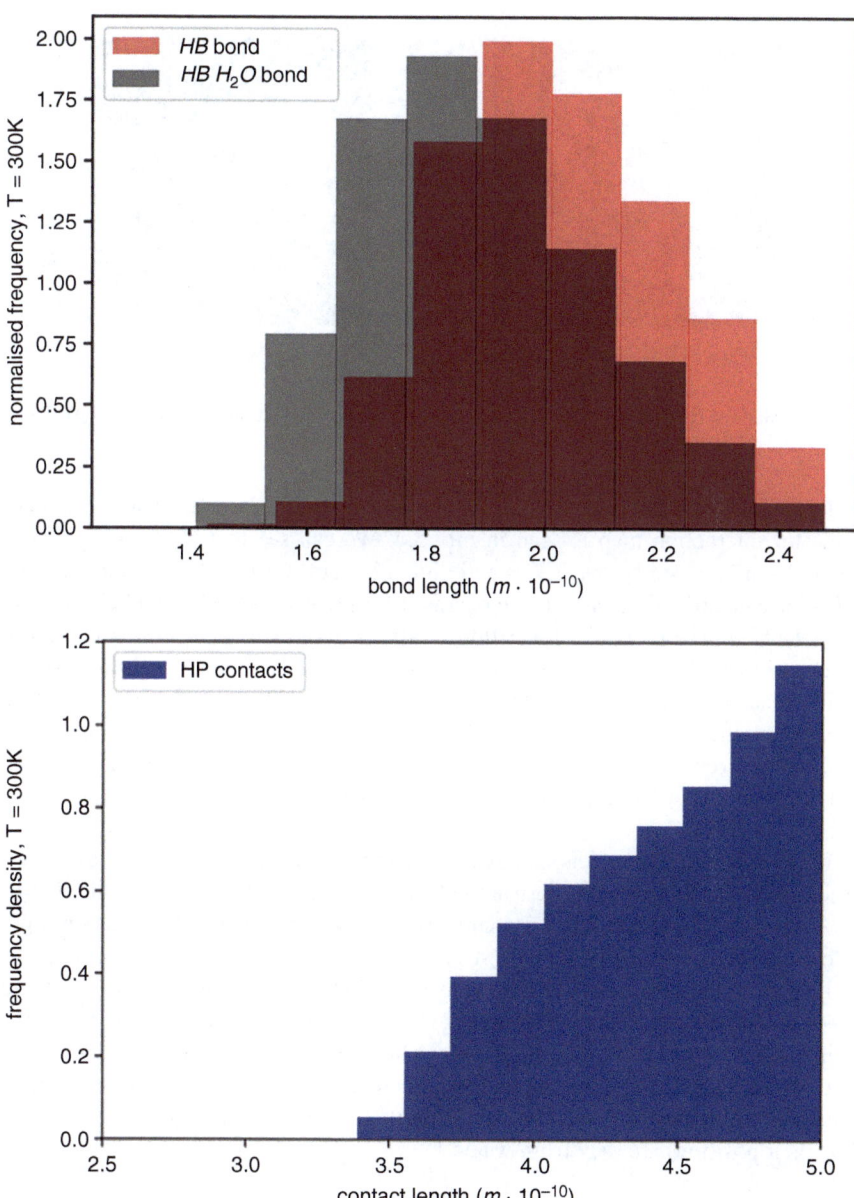

Fig. 5.3 Aggregate histograms of hydrogen bonds interactions lengths (top panel), and HP contacts lengths (bottom panel), exemplary temperature T = 300 K. Observe different mode length and skewness

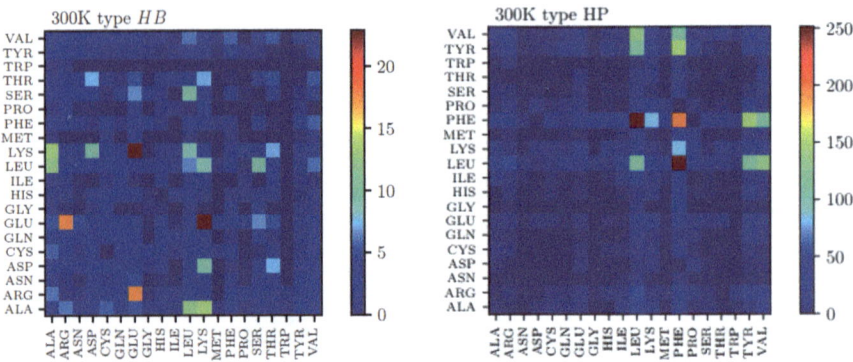

Fig. 5.4 Maps of HB hydrogen bonds (left panel) and HP contacts (right panel) between particular amino-acids, T = 300 K

In the case of HB bonds and HP contacts, the connection between two amino-acids for interaction map is seen in Fig. 5.4. We can see that the most hydrogen bonds form between lysine (LYS) and glutamic acid (GLU). Both of these amino acids have electrically charged side chains: LYS positive (-NH3+), and GLU negative (-COO-). Therefore the formation of strong HBs is evident in this case. The other amino acid with which GLU often forms hydrogen bonds is arginine (ARG), which also has a positively charged side chain (=NH2+). Furthermore, leucine (LEU), alanine (ALA), and serine (SER) play an essential role in the formation of HB. These three amino acids do not have charged side chains, but have groups that can be proton acceptor or donor (-NH2, =NH, -OH) and form HB: LYS-ALA, LEU-ALA, LYS-SER.

In the case of HP contacts, phenylalanine (PHE) and leucine (LEU) plays a significant role. That is because these amino acids have strong hydrophobic side chains: PHE has a benzyl group, LEU has an isobutyl group. Valine (VAL) with isopropyl side chain and tyrosine (TYR) with -CH2-C6H4-p-OH are slightly less important.

In case of the albumin-H_2O HB it is formed between the single amino acid and the H_2O molecule (see histogram in Fig. 5.5). It seems that glutamic acid (GLU), lysine (LYS) and aspartic acid (ASP) plays here an important role. It is noteworthy that GLU and ASP are proton acceptors, whereas LYS is a donor of proton. Radial distribution function for selected AA presented in Fig. 5.6 show how distant (and thus their position) are they from water.

3.2 Stability (Living Time) of Bonds Interactions and Contacts

Following analysis in (Voloshin and Naberukhin 2009), we present decay curves of particular bonds and interactions. At first specific bond, types interaction is recognized after one simulation step (to avoid any inconsistency with initial conditions).

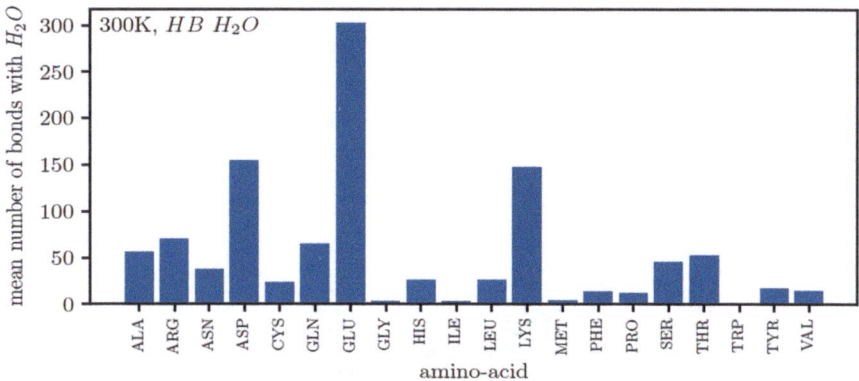

Fig. 5.5 Histograms of HB H_2O contacts with water molecules

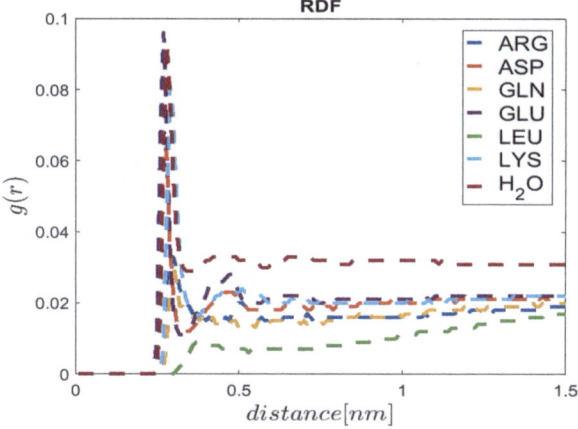

Fig. 5.6 Radial distribution function of water's oxygen from selected amino acids

Next, we compute how much of these survived after subsequent simulation steps and present its fraction in Fig. 5.7. Observe a significant difference in the decay rate between two types of hydrogen bonds.

Survival histograms of HP contacts and hydrogen bonds between chosen amino-acids and all amino-acids (or water particles in HB H_2O) are presented in Figs. 5.8, 5.9, and 5.10. For HP constants most stable in tyrosine (TYR), for HB bonds leucine (LEU) while for HB H_2O – arginine (ARG).

Presented results show in the molecular detail of a selected protein in its native, folded state. Properties of all amino acids from the albumin molecule reflect the nature of their interactions with other proteins and water. Charged AA, such as GLU, LYS, ASP form mainly hydrogen bonds, see Fig. 5.8. On the other hand, hydrophobic AA: PHE, LEU, TYR, and VAL create most hydrophobic contacts, as shown in Fig. 5.9. It is also worth mentioning that LEU shows up the high binding

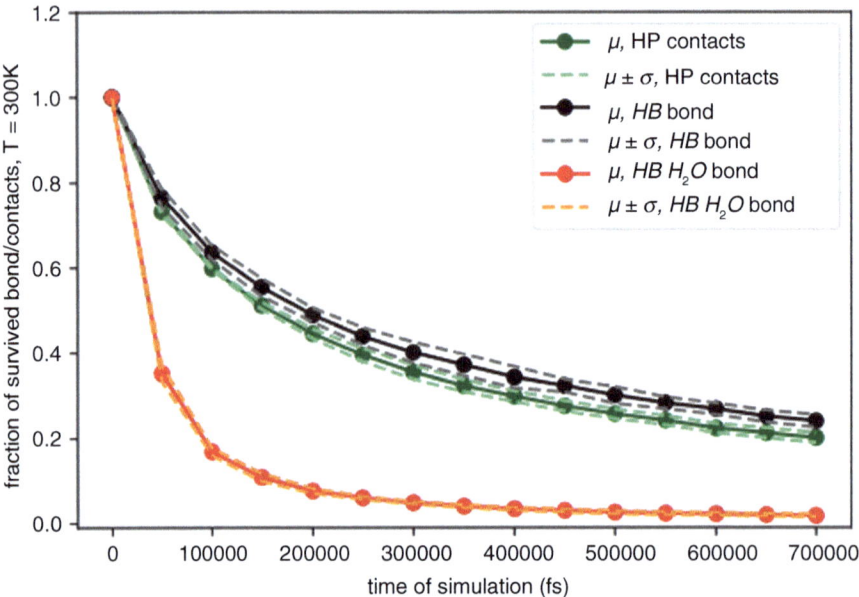

Fig. 5.7 Decay curve for various types of bonds interactions and contacts, statistics were taken over realizations. Observe that HB H₂O bonds decay in fastest rate, while HP contacts and HB bonds in slowest rate

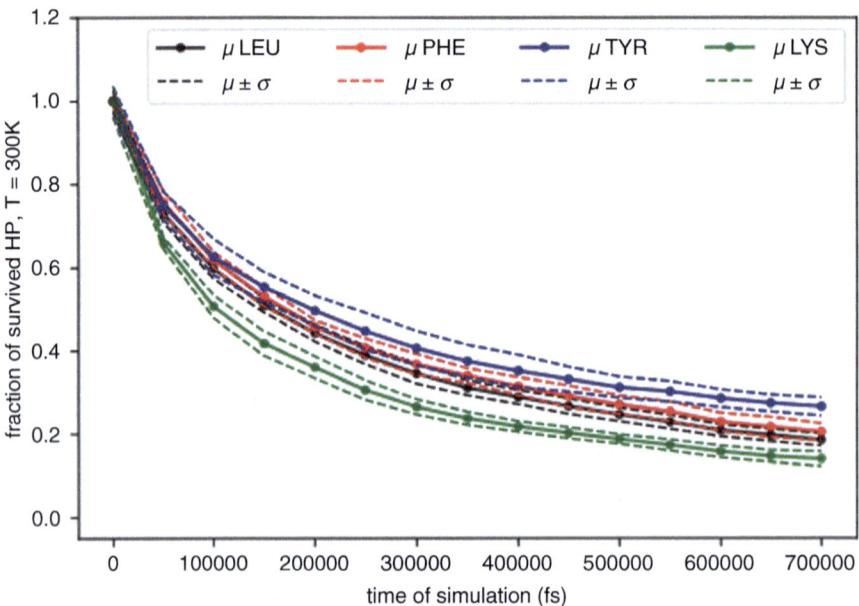

Fig. 5.8 Decay plot of HP contacts between chosen amino-acid and all amino-acids. tyrosine (TYR) produces most stable HP contacts

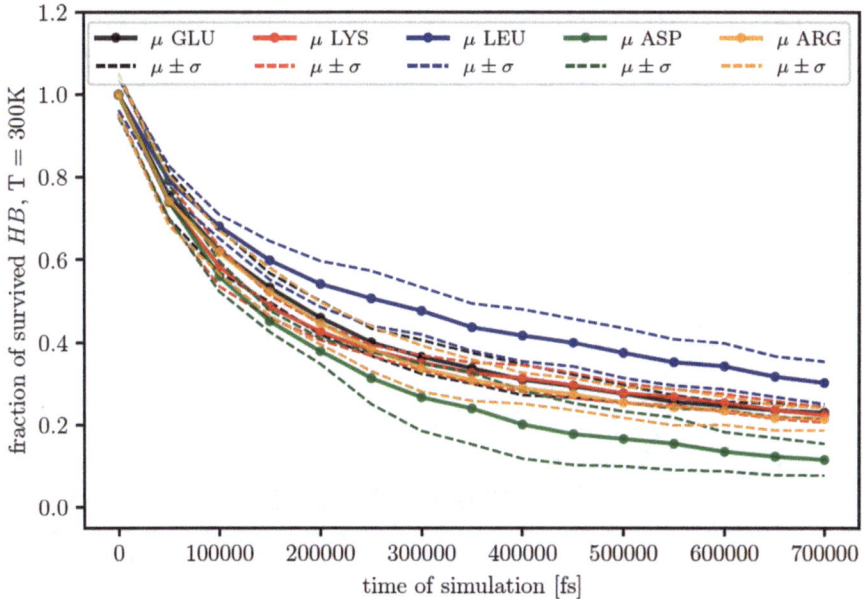

Fig. 5.9 Decay plot of HB bonds between chosen amino-acid and all amino-acids. leucine (LEU) produces most stable hydrogen bond

for HB, HP, and low with water. This indicates its position in the protein as it is buried in the core of albumin. Interestingly LEU forms the most stable HB with water, which also reflects above mention position, as water molecules can be trapped. Thus intramolecular bonding can be replaced with solvent HB. ARG creates less HB with water than other charged AA, but it tends to last longer, as presented in Fig. 5.10. Dynamical properties of water molecules near protein are determined by their proximity to surface. Thus, depending on position bonding between protein and water shows following pattern: (i) bulk water with a typical life time of the order of picoseconds; (ii) water present at the macromolecular surface with a bonding survival time of nanoseconds; (iii) buried water molecules boning can last of up to 50 nanoseconds (Martini et al. 2013).

4 Conclusions

In this chapter we have shown the importance of water in determining the protein structure, using the example of albumin. Discussed interactions of amino acids with water are the main factor that makes a protein take on a specific structure. The dynamics of these interactions are dependent on amino acid types, its position in the chain, and the type of interaction. The most stable are intramolecular hydrogen

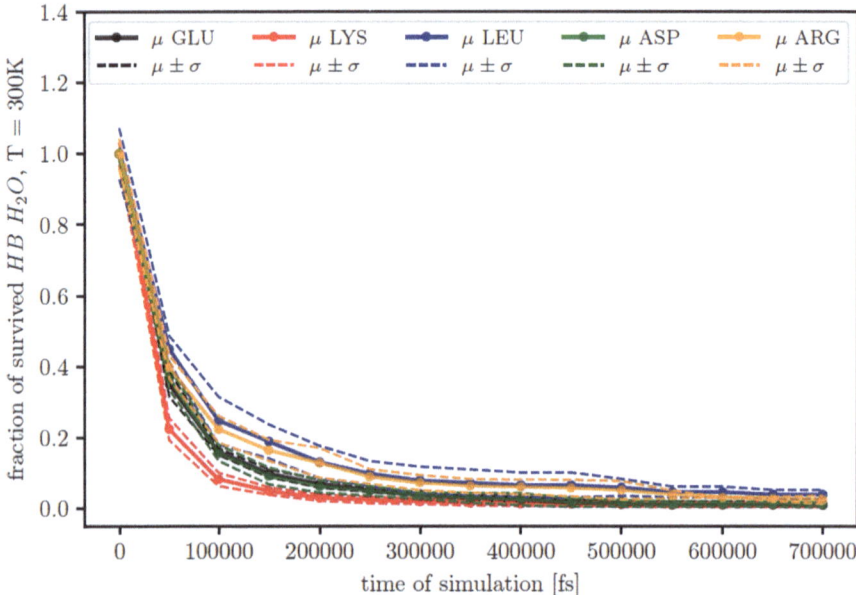

Fig. 5.10 Decay plot of HB H_2O bonds between chosen amino-acid and water particles. arginine (ARG) produces most stable hydrogen bond with water particles

bonds and hydrophobic contacts. Intermolecular bonding with water shows up a much less durable lifespan than intramolecular interactions.

Amino acids in albumin behave according to their properties. Polar and charged amino acids take part in the formation of HB, hydrophobic amino acids form HP contacts. The exception is leucine, which has higher stability for HB and HP, due to its position in the core of the protein.

References

Bełdowski, P., Weber, P., Dedinaite, A., Claesson, P. M., & Gadomski, A. (2018). Physical cross-linking of hyaluronic acid in the presence of phospholipids in an aqueous nano-environment. *Soft Matter, 14*, 8997–9004.

Bełdowski, P., Yuvan, S., Dedinaite, A., Claesson, P. M., & Pöschel, T. (2019). Interactions of a short hyaluronan chain with a phospholipid membrane. *Colloids and Surfaces, B: Biointerfaces, 184*, 110539.

Bhattacharya, A. A., Curry, S., & Franks, N. P. (2000). Binding of the general anesthetics propofol and halothane to human serum albumin. *The Journal of Biological Chemistry, 275*, 38731–38738.

Cerny, J., & Hobza, P. (2007). Non-covalent interactions in biomacromolecules. *Physical Chemistry Chemical Physics, 9*, 5291–5303.

Compiani, M., & Capriotti, E. (2013). Computational and theoretical methods for protein folding. *Biochemistry, 52*, 8601–8624.

Dedinaite, A., Wieland, D. C. F., Bełdowski, P., & Claesson, P. M. (2019). Biolubrication synergy: Hyaluronan–phospholipid interactionsat interfaces. *Advances in Colloid and Interface Science, 274*, 102050.

Evans, T. W. (2002). Review article: Albumin as a drug – Biological effects of albumin unrelated to oncotic pressure. *Alimentary Pharmacology & Therapeutics, 16*, 6–11.

IUPAC. (1997). *Compendium of chemical terminology*, 2nd ed. (the "Gold Book"). Compiled by McNaught, A. D and Wilkinson, A. Blackwell Scientific Publications, Oxford, p. 1123. Online version (2019-) created by Chalk, S.J.

Martini, S., Bonechi, C., Foletti, A., & Rossi, C. (2013). Water-protein interactions: The secret of protein dynamics. *The Scientific World Journal, 2013*, 138916.

Peters, T. (1996). *All about albumin: Biochemistry, genetics, and medical applications*. San Diego: Academic.

Quinlan, G. J., Martin, G. S., & Evans, T. W. (2005). Albumin: Biochemical properties and therapeutic potential. *Hepatology, 41*, 1211–1219.

Rose, G. D., Fleming, P. J., Banavar, J. R., & Maritan, A. (2006). A backbone-based theory of protein folding. *PNAS, 103*, 16623–16633.

Siódmiak, J., & Bełdowski, P. (2019). Hyaluronic acid dynamics and its interaction with synovial fluid components as a source of the color noise. *Fluctuations and Noise Letters, 18*, 1940013.

Subha Mahadevi, A., & Narahari Sastry, G. (2016). Cooperativity in noncovalent interactions. *Chemical Reviews, 116*, 2775–2825.

Voloshin, V. P., & Naberukhin, Y. I. (2009). Hydrogen bond lifetime distributions in computer-simulated water. *Journal of Structural Chemistry, 50*, 78–89.

Chapter 6
Water Behavior Near the Lipid Bilayer

Natalia Kruszewska, Krzysztof Domino, and Piotr Weber

Abstract In this chapter, we focus on the dynamics of water molecules situated in the vicinity of a phospholipid bilayer. Using a molecular dynamics simulation method, we studied interactions between water and the bilayer and tracked trajectories of the water molecules. Based on the hypothesis that molecules trapped inside the bilayer make different motions than the ones which are either attached to the surface or move freely in the water bulk, we divided the water molecules into three groups – the ones that exhibited subdiffusion (confined) motion, the ones that move diffusionally in the bulk and the ones that move superdiffusively due to interactions with the moving bilayer. In detail, the water behavior near the bilayer has been analyzed by mean squared displacement and entropy computed separately for the above mentioned three groups of molecules. To explain the subdiffusion motion of the water molecules, the number and the duration of hydrogen bonds created between water molecules and the bilayer have been investigated. In addition, we examined the mechanism of water molecule self-diffusion, by means of statistical tests. Our studies aim to present insight into the understanding of the lipid's role in water self-diffusion, which can be responsible for triggering different tribological responses of the system.

Keywords Water dynamics at lipid surface · Mechanism of water self-diffusion · Nanotribology with water's account · Molecular dynamics simulation · Viscoelasticity of lipid-water phases

N. Kruszewska
Institute of Mathematics and Physics, UTP University of Science and Technology, Bydgoszcz, Poland
e-mail: nkruszewska@utp.edu.pl

K. Domino (✉)
Institute of Theoretical and Applied Informatics, Polish Academy of Sciences, Gliwice, Poland
e-mail: kdomino@iitis.pl

P. Weber (✉)
Faculty of Applied Physics and Mathematics, Gdańsk University of Technology, Gdańsk, Poland
e-mail: Piotr.Weber@pg.edu.pl

1 Introduction

Phospholipids are macromolecules made of fatty acid monomers. These monomers have an amphiphilic nature: they are made of a hydrophilic head and two hydrophobic hydrocarbon tails. Such a structure allows phospholipids to spontaneously build bilayers due to hydrophobic-hydrophilic (HP) interactions between phospholipid molecules in the aqueous environment (Alberts et al. 2002). Hydrophobic tails of phospholipids attract other hydrophobic molecules and repulse water. Thus, water is moved towards hydrophilic heads. That way, tails connect back-to-back creating lamellar structures. Depending on the lipids' type and environmental properties, lipids can either form spherical micelles, with the tails directed inside, or they can form bimolecular sheets (Israelachvili et al. 1976). A sketch of a bilayer and exemplary snapshots of a computer model of a bilayer have been presented in Fig. 6.1.

The lipid bilayer is the core of most cell membranes and it is described as a rigid phase. The thermodynamics of transport processes are determined by the parameters of this phase. It is worth to mention, that such membranes play a very important role in living cells. They are selective barriers between separated structures and the environment, granting communication and protection (Meer et al. 2008).

The water surrounding the whole lipid matrix plays a crucial role – it is responsible for the matrix's functioning. The water determines the physical properties of the lipid matrix, its structure, and its stability. Depending on the cell function of the membrane, the lipid matrix's molecular composition varies even within the same organism. Further, looking at the mechanical properties of the lipid bilayers, they

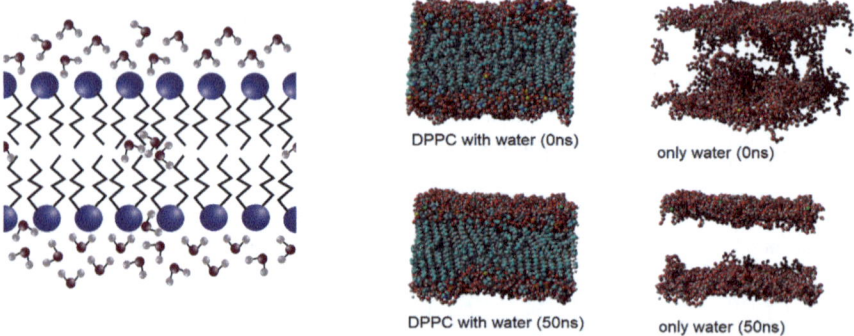

DPPC with water (0ns)

only water (0ns)

DPPC with water (50ns)

only water (50ns)

Fig. 6.1 Sketch of the lipid bilayer (build on hydrophilic heads – blue spheres and hydrophobic tails – black curves) with water semi-ordered on the surface and trapped inside the bilayer (left). Molecular dynamic simulation snapshots of the bilayer with water surrounding before (0 ns) and after (50 ns) simulation time (right). Additionally, pictures of water alone have been presented (bilayer set as invisible) to show the presence of water pockets at the start of simulations and their absence after 50 ns of simulations. In the beginning (0 ns) we can observe a more disordered structure of the bilayer than after 50 ns. Note the water atoms at the beginning and at the end of the simulation are not necessarily the same as the water detached from the bilayer and move freely in the simulation box (cf., Fig. 6.3)

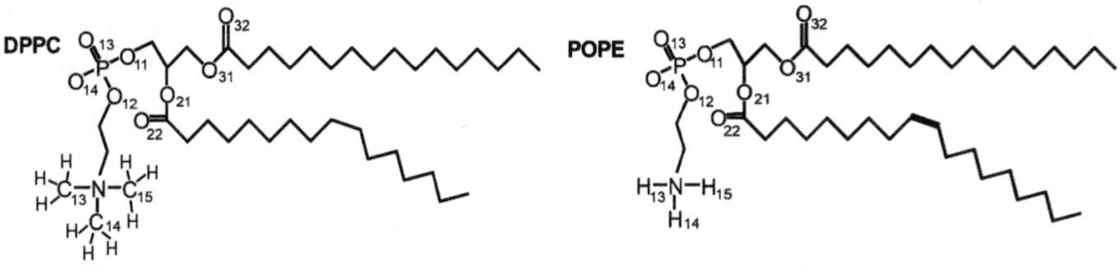

Fig. 6.2 Simplified chemical structures of dipalmitoylphosphatidylcholine (DPPC) and 1-palmito yl-2-oleoyl-sn-glycero-3-phosphoethanolamine (POPE) lipids

are durable and flexible. Hence in many body organs, lipid bilayers cover various structures. A very good example is the joint in which the phospholipid bilayers cover articular cartilage (AC). The lubrication mechanism of the joints is highly dependent on the properties of the bilayers (Pawlak et al. 2012). In more detail, the phenomena of the slippage of the bilayer and short-range repulsion between the negatively charged cartilage surfaces are causes of low frictional properties (Janicka et al. 2019). The bilayer changes its properties under load to facilitate lubrication, but it could be destroyed or even disappear with medical conditions such as most rheumatic diseases and osteoarthritis (Petelska et al. 2019; Dédinaité et al. 2019). From the molecular point of view, the healthy AC surface contains a mixture of phospholipids, mostly of phosphatidylcholine and phosphatidylethanolamine. Here types of the lipids differ in head groups, tail lengths, and having saturated or unsaturated hydrocarbon chains. The chemical structures of the two examples of lipids have been shown in Fig. 6.2.

The water located near hydrophilic parts of the lipids stabilizes the head-head interactions, creating hydrogen bonds or electrostatic interactions with water. The nature of the stabilization of the bilayer is not straight forward and it is still discussed by various groups of researchers. Usually, the hydrophobic effect of tails is indicated as the driving force of the association of amphiphilic molecules (Crowe and Bradshaw 2010). On contrary, in (Ben-Naim 2011) it is emphasized that the hydrophilic interactions, such as those between head groups, are stronger than the hydrophobic interactions due to the presence of strong water bridges built between hydrophilic molecules.

Thus, usually, water relocates and moves away from the bilayer to avoid contact with tails located within the bilayer. However, the water molecules sometimes could be trapped inside the bilayer structure and create water pockets. Their movements are then confined. This could happen in a time of the creation of the bilayer, or later, as a result of atom motions. One of the models explaining above mentioned water transportation through the bilayer is the Träuble model (Haines and Liebovitch 1995; Disalvo et al. 2015). In this model, water diffuses across the lipid membranes by occupying holes formed in the lipid matrix due to fluctuations of the lipids' chains, which can create kinks. Water trapped inside the bilayer can completely change the bilayer properties and, for example, allow ions or molecules to penetrate the bilayer, which normally could not occur. The formation of pockets filled with water can explain the insertion of polar peptides and amino acids into lipid membranes (Disalvo et al. 2015). Here the need to analyze trajectories of molecules arises.

Tracking of the trajectories of various molecules near the lipid bilayer, and even tracking the movement of lipids themselves inside the lipid matrix, demonstrate various types of motion (Saxton and Jacobson 1997; Kruszewska et al. 2020). Besides the observation of molecules performing Brownian motions, most of the particles undergo non-Brownian motion following anomalous diffusion regime because of their directed or confined motion and flow. This phenomenon significantly influences the kinetics of reactions among molecules in the close vicinity of the membrane. Such motions were first studied experimentally, e.g. by single-particle tracking computer-enhanced video microscopy raised by fluorescence recovery after photobleaching (FRAP) (Saxton and Jacobson 1997). Next, the NMR spectroscopy technique (combined with other methods, e.g. pulsed-field gradient, magic-angle spinning) started to be applied to examine lateral diffusion in the lipid bilayers (Wattraint and Sarazin 2005; Gaede and Gawrisch 2003). This method fits well for non-invasive studies, as neither a fluorophore probe nor a planar surface (as in the case of FRAP) is required. From the other perspective, the development of computer modeling techniques during the last decades provided new possibilities for investigating the properties of lipid bilayer systems, giving more detailed information about structures, kinetics, and dynamics of the molecules.

In this chapter, we have used molecular dynamic (MD) simulation techniques to investigate the interaction between water molecules and the bilayer. We also have described the motion of the water molecules near the lipid matrix. This work is an extension of (Kruszewska et al. 2020) where studies concerning the motion of the lipid bilayer, with Beta-2-glycoprotein-1 bound to it, have been shown. The presented analysis of water characteristics near the lipid bilayer system was designed to be useful in studies exploring the reasons for facilitated lubrication in the articular cartilage (Bełdowski et al. 2018). The purpose of the analysis was to explain the role of lipids in water diffusion, which could be responsible for triggering different tribological responses of the system.

2   Computer Experiment

The computer model of the system is built on a phospholipid bilayer immersed in a water solution with the addition of ions. The water behavior near this bilayer is a matter of great importance due to its influence on the properties of the membrane interface (Pasenkiewicz-Gierula et al. 2016). These properties have been studied since the 80s using various methods – at first experimental ones (Rand and Parsegian 1989; Nagle and Tristram-Nagle 2000). Such information was limited because of the complexity of the bilayers and of the limited spatial and temporal resolutions of the methods. The accessibility of computer modeling methods opened a large range of possibilities by providing structural and dynamic characteristics of molecules, averaged over a huge number of atoms. MD simulation may give information with an atomic resolution and may span the observation time by femtosecond steps. Such a resolution is necessary to understand the basic mechanisms of biomembrane functioning. Although MD simulation is a great tool to gain knowledge about molecular systems, it needs validation by comparing its results to experimental data. Most of the experimental and MD simulation results only match to a certain extent. However, in an overwhelming number of cases, they are similar or present similar trends, which allows us to hypothesize that both methods are accurate. The most important factor determining the correctness and reliability of the computer model is the parameterization of the atoms used in the model and their interactions (described by a force field). A more thorough overview of potential issues in MD simulations has been included in the Review (Pasenkiewicz-Gierula et al. 2016).

To analyze the water behavior near the phospholipid bilayer, we built an all-atom computer model of the system. To this end, we chose YASARA Structure Software (Vienna, Austria) to perform the simulations. Our all-atom MD modeling used the AMBER03 force field (Duan et al. 2003) to evaluate interactions between molecules in the model system. The lipids used for this study were dipalmitoylphosphatidyl-choline (DPPC – 130 atoms) and 1-palmitoyl-2-oleoyl-sn-glycero-3-phosphoetha-nolamine (POPE – 125 atoms). Both bilayers' structures have been taken from studies presented in (Yeghiazaryan et al. 2005) and consist of 288 lipids (144 lipids in each layer). To mimic the infinity of the system and ensure that edges do not influence the stability of the system, periodic boundary conditions have been applied. All-atom simulations were performed under the same conditions, namely: temperatures 310 K and pH = 7.0, which assures the "fluid" state of the bilayers similar to natural membranes (White and Wiener 1995). The system was immersed in a water-based solution containing 2% $CaCl_2$, TIP3P water model was used (Neria et al. 1996). A time-step of 2 fs was chosen. To maintain constant pressure and temperature (Berendsen et al. 1984), the Berendsen barostat and thermostat with a relaxation time of 1 ps were used. The YASARA simulation area was of size: X: 95 Å, Y: 86 Å, Z: 196 Å. After adding all components to the simulation environment, the system was equilibrated for 10 ns. After that, simulations were carried for 50 ns and repeated 3 times for each bilayer. To ensure reliability from the statistical point of view, the results were averaged across repetitions.

3   Results

Exemplary snapshots of simulation results have already been shown in Fig. 6.1, in which one can see the defects of the bilayers caused by thermal fluctuations of the atoms. After throwing out water molecules from the interior of the bilayer, the structure starts to be more organized. The three regions can be distinguished while looking at the bilayer. It can be studied by the electron density profile of the simulated system. They are as follows: I. the bulk water phase, II. the water-bilayer interface, and III. the hydrophobic core (see, Fig. 6.3).

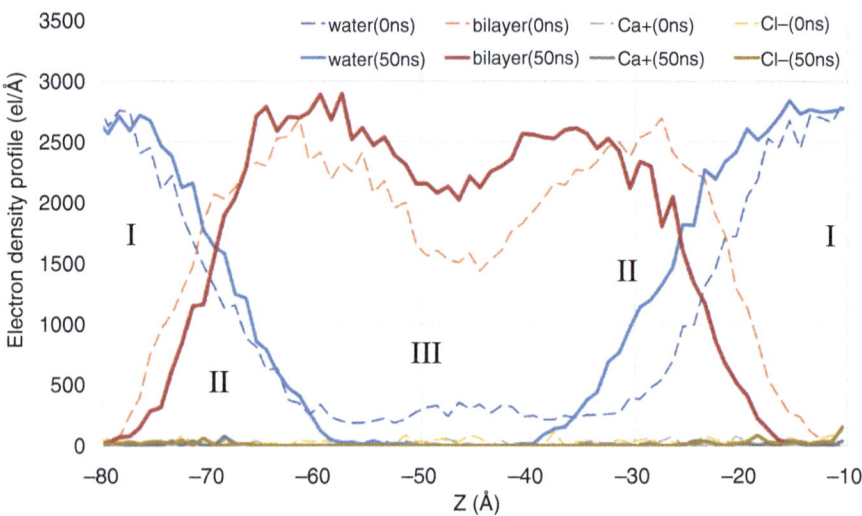

Fig. 6.3 Electron density profile across DPPC bilayer at the start of the simulation (dashed lines), and the end of the simulation (solid lines). The density of all phospholipids' atoms (red line), water atoms (blue line), and ions (grey: Ca^+ and yellow: Cl^- lines). Three regions have been marked: I. the bulk water phase, II. the water-bilayer interface, and III. the hydrophobic core

While the water behavior in the first case is not so interesting – the bulk contains only water and a few ions which just freely diffuse – the second and third regions contain a mix of water and phospholipid molecules, which thus influences the behavior of water. The number of water molecules in the second region decreases with the bilayer depth. The bilayer thickness, L_B, and the depth of water penetration into the bilayer, D_B, can be calculated from the water electron density profile and are as follows: $L_B \approx 65$ Å, $D_B \approx 20$ Å after 50 ns of simulation. The density profile at the start of the simulations (0 ns) has been presented in comparison to the one at 50 ns (cf., dashed lines in Fig. 6.3). It can be seen that during simulation time, when the water was thrown away from the bilayer, the bilayer shrunk for about 10 Å, and also the water depletion zone thinned from 30 Å to 20 Å. The water density lowered from about 300 el/Å to zero in the III region. Such a density profile can be obtained experimentally by the combined use of X-ray and neutron diffraction measurements (Wiener and White 1992) and provides comparable results.

The analysis of the simulation results has been divided into two groups: (i) we found the molecules at distance <4 Å from the bilayer and noted their locations in each simulation step up to 50 ns; (ii) we searched for molecules at distance <4 Å in each simulation step, looking for the molecules which built stable bonds with the bilayer. In the first case, trajectories of about $4.5 \cdot 10^3$ water molecules have been obtained, from which $2 \cdot 10^3$ have been chosen randomly to accelerate the computations. The first case allows us to study the dynamics of the water, and the second case allows us to investigate interactions between water and the bilayer.

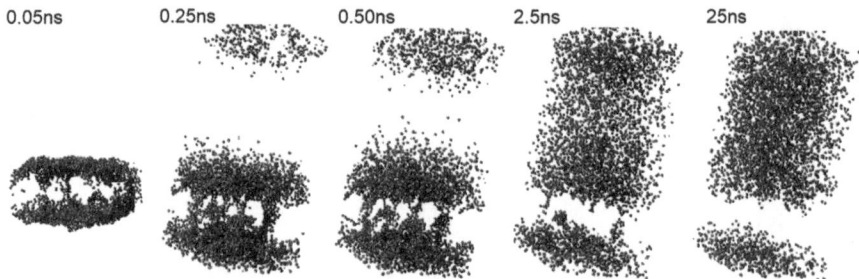

Fig. 6.4 Water molecules near the bilayer and the locations of the same water molecules in chosen time-steps, depiction of the case (ii), see text. Periodic boundary conditions can be seen by looking at the diffusion pattern of the water molecules (they go through the bottom wall of the simulation box and appear from the top). After 25 ns all molecules are evenly distributed in available space. The Z-axis is the vertical one (cf., Fig. 6.3)

Simulation snapshots of the case (i), showing the water in the vicinity of the bilayer, have been presented in Fig. 6.4.

It can be demonstrated that most of the water molecules located near the bilayer at the first time-step (0.05 ns) moved away from the bilayer vicinity during the simulation time and diffused over the whole simulation box (excluding the bilayer). In the beginning, water pockets inside the bilayer can be observed. This is a result of an even distribution of water molecules over the whole available space at the start of MD simulation. After equilibration of the system in the preparatory stage of the simulation, they were still present. However, the water was pushed out during the first few nanoseconds of the simulation process, and they became almost absent inside the bilayer after about 25 ns for DPPC and 15 ns for POPE. Note that natural watering of the lipids is also possible, e.g. as a result of kink moves of the lipid tails (as described in the previous section) or as applied pressure which causes shear moves of the two layers of the bilayer. This can result in the bilayer's disordering and e.g. rotating up-down (Briscoe 2017; Disalvo et al. 2015).

Results of the analysis of case (ii) have been used to investigate the duration time of interactions between single water molecules and the bilayer.

3.1   Interactions Between Water and the Bilayer

The number of H-bonds between water and the bilayer (per one lipid molecule) can be used to demonstrate how much the bilayer is hydrated. A hydrogen bond is regarded to be formed when the hydrogen bond energy is greater than 6.25 kJ/mol, which is 25% of the optimum value of 25 kJ/mol. The exact formula has been described in the YASARA manual (Krieger et al. 2012). The number of H-bonds, as a function of time, has been presented in Fig. 6.5, based on results from the case (ii) described in Sect. 3. Through close analysis of the curves, the results have been split

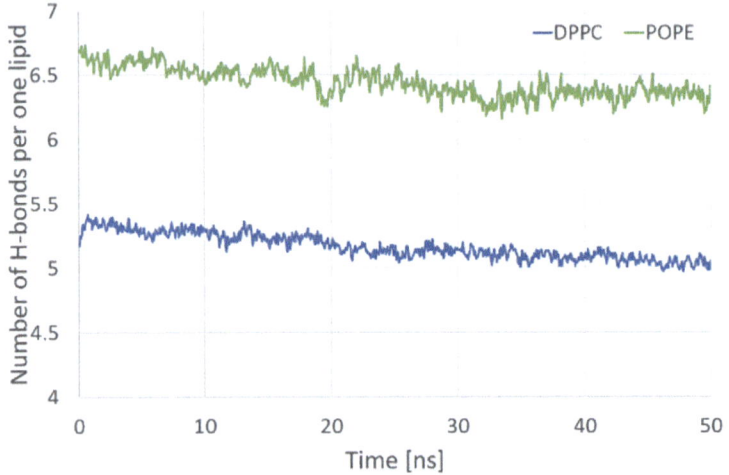

Fig. 6.5 Number of H-bonds as a function of time for DPPC and POPE bilayers

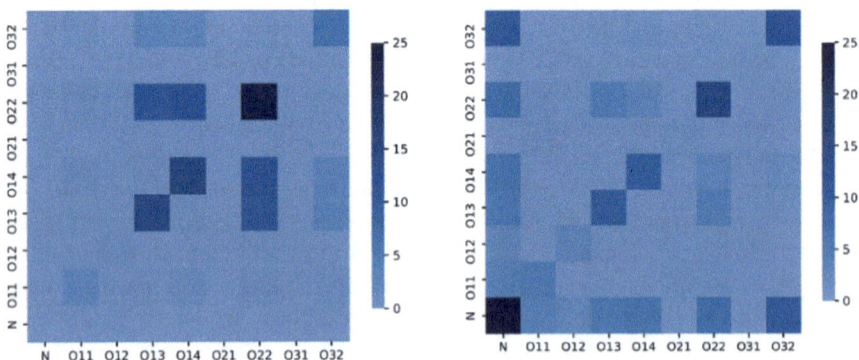

Fig. 6.6 Map of the water bridges for DPPC (left) and POPE (right) after 50 ns of simulation time

into two regimes: in the first one, the water has been removed from the interior of the bilayer (first 25 ns of simulations for DPPC and 20 ns for POPE), thus hydrogen bonds have been broken faster (slope of the trend line equals −0.0084/ns for DPPC and −0.0103/ns for POPE) than in the second regime, in which the number of H-bonds still decreased but about 2 times slower (slope of the trend line equals −0.0043/ns for DPPC and −0.0039/ns for POPE). In the last 5 ns of the simulation, the number of H-bonds stabilized near 5 for DPPC and 6.3 for POPE.

Thus, the POPE bilayer has disposed of the water trapped between the layers faster than in the case of DPPC. The number of H-bonds in the case of POPE is much higher due to the presence of nitrogen atoms in the head group of POPE.

A subset of H-bonds created water bridges. In Fig. 6.6, the water-bridge map has been presented. The map exhibits the differences between POPE and DPPC heads. In the case of DPPC, the bridges appear more often between oxygens O22, O13, and

O14 (cf., Fig. 6.2). The water molecules which take a part in this bonding are attached to more than one bilayer atom, which confines their movements and stabilizes the hydrophilic interactions (Ben-Naim 2011). In the case of the POPE bilayer, the water bridges are presented more often between nitrogen atoms – so on the top of the head – although bridges between O22-O22 and O13, O14, O32 have also been found. Thus, different parts of the hydrophilic head of the DPPC lipid have caused increased stabilization of the bilayer structure than for POPE.

In Fig. 6.7, the duration time of H-bonds between water and the lipids has been shown. Most of the bonds last 1 or 2 simulation steps (thus, 50–100 ps). A power-law relation has been noticed with fitting parameters as shown in Fig. 6.7.

The results differ between the two types of bilayer. The DPPC bilayer creates more durable bonds that cause the throwing out the water from inside the bilayer to occur more slowly. The water molecules near this bilayer better stabilize the surface of the bilayer than in the case of POPE, even though POPE creates more H-bonds with water than DPPC. This is important because by forming or breaking H-bonds at the interface, water can modulate ligand binding or recognition events of lipids or protein (Cheng et al. 2014; Roy and Bagchi 2012). The H-bond rearrangement of the solvent is implied by the diffusion dynamics of water molecules associated with the lipid bilayer surface. A faster diffusivity of water in the proximity of the bilayer causes a lower energy barrier for the formation and breaking of H-bonds between the water molecules (Song et al. 2014). The tempo of H-bond rearrangements and the change in water diffusivity near the protein-water or lipid-water interface are deemed crucial for protein folding, protein-protein, or protein-ligand interactions (Cheng et al. 2014; Roy and Bagchi 2012). They are also important for understanding, controlling, and predicting specific and nonspecific interactions between lipid bilayers and ions, small molecules, polycations, peptides, and (bio)nanomaterials (Doğangün et al. 2018). Additionally, in (Lopez et al. 2004) it was suggested that the water molecules "trapped" at the surface of the bilayer via H-bonds can aid

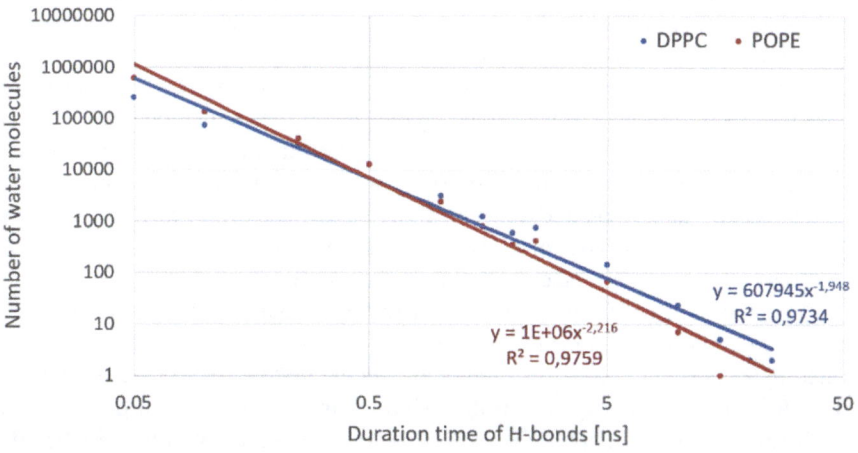

Fig. 6.7 Log-log plot of duration time of H-bonds for DPPC and POPE

proton transport via the Grotthuss mechanism. All phenomena mentioned above are very important when studying the synergistic role of the components of synovial fluid in articular cartilage (Dédinaité and Claesson 2017).

3.2 Mean Squared Displacement and Related Measures

A basic measure that can tell us how a water molecule moves near the lipid bilayer is the mean squared displacement (MSD). It measures how fast the molecules run from a starting location. The MSD of water molecules, $MSD(t) = <\Delta r^2(t)>$, at time step t, can be determined from its trajectories in the standard way:

$$\left\langle \Delta r^2 \left(t \right) \right\rangle = \frac{1}{N} \sum_{i=0}^{N} \left| \vec{r}_i \left(t \right) - \vec{r}_i \left(0 \right) \right|^2 , \qquad (6.1)$$

where N is the number of water molecules in the vicinity of the bilayer and $\vec{r}_i$ is the position vector of the geometrical center of a molecule i. The *MSD*, as a function of time of the water molecules, cf., Eq. (6.1), characterizes the viscoelastic properties of the system (Santamaría-Holek et al. 2007; Sarmiento-Gomez et al. 2014). It is related to the environment's mechanical response functions using the generalized Stokes-Einstein relation (Bellour et al. 2002) and in 3D Euclidean space, it follows the relationship (Metzler et al. 2014; Metzler and Klafter 2000):

$$\left\langle \Delta r^2 \left(t \right) \right\rangle = 6 D_\alpha t^\alpha , \qquad (6.2)$$

where D_α is a generalized self-diffusion coefficient, a constant that does not depend on time and is of the dimension of $[D_\alpha] = \mu m^2/s^\alpha$ (Dechant et al. 2014; Kneller et al. 2011). The exponent $\alpha = 1$ represents normal (Brownian) diffusion, and $\alpha \neq 1$ represents anomalous diffusion (when $\alpha > 1$, a superdiffusion process appears and when $\alpha < 1$, we can talk about subdiffusion).

Using the information about changes in time of the coordinates of water molecules (located closer than 4 Å from the bilayer at the first simulation step – case (i) described in Sect. 3), we have tested whether the motion of the molecules fits the scaling relation $<\Delta r^2(t)> \sim t^\alpha$ (cf., Eq. (6.2)), where α is the scaling exponent fitted from the data. The Python MDAnalysis Toolkit (Michaud-Agrawal et al. 2011) has been used in the fitting. The values of the fitting parameters, obtained separately for the atoms of each water molecule, are shown in a histogram in Fig. 6.8.

To compute the *MSD*, we have eliminated the periodic boundary conditions by changing the proper coordinates each time the particle crossed the wall. For all the water molecules which were near the bilayer in the first simulation step (0 ns) (cf., Fig. 6.8.1), the averaged results show almost normal diffusion with $\alpha = 0.8975 \pm 0.0008$ and $D_\alpha = (3970 \pm 11) \cdot \mu m^2/s^\alpha$. However, we are aware that the water molecules trapped inside the bilayer have different dynamics than the water

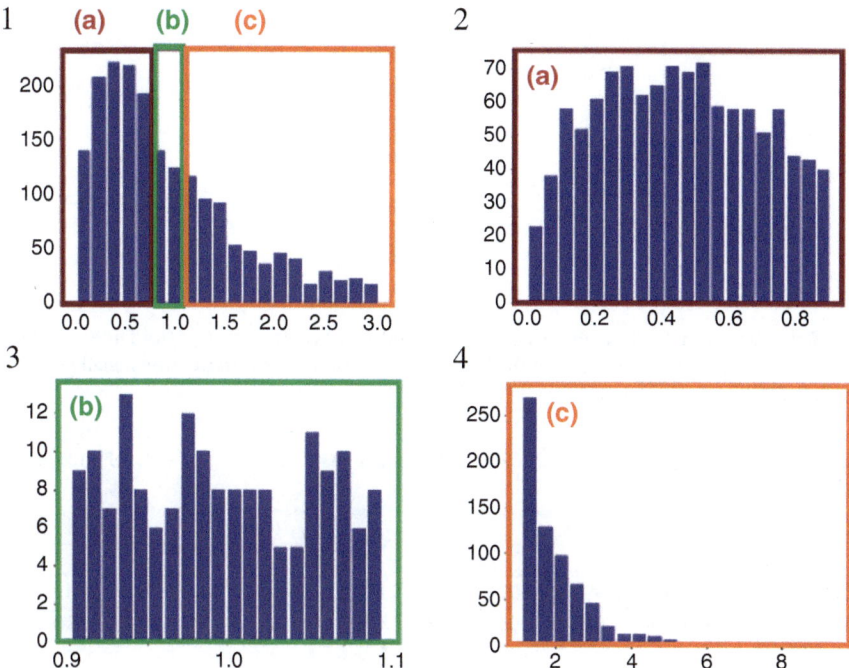

Fig. 6.8 Probability density (not normalized) of α for POPE. The first picture presents a histogram of α for 2000 water molecules picked randomly from molecules closer than 4 Å from the bilayer (with a cutoff value of 3 for better clarity). The exponent divides particles into three cases: (**a**) the ones which move subdiffusionally (α < 1), (**b**) the ones which move diffusionally (α ≈ 1), (**c**) and the ones which make a superdiffusive motion (α > 1). The histograms for the three groups are shown separately in Pictures 2–4

molecules at the surface. The ones located near the surface should detach easier compared to those trapped inside. But the surface can attract water which is hydrogen bound to lipid oxygen atoms and diffuses slowly (Yamamoto et al. 2014; Roy and Bagchi 2012; Lopez et al. 2004). On the other hand, the bilayer still moves and thus has a possibility to "launch" the water to speed it up (introducing flow) and cause its superdiffusive (accelerated) motion (Cheng et al. 2014). That's why it is worth looking at the behavior of water molecules from at least three points of view by separating the water molecule collection into: (a) those with α < 0.9 - subdiffusion motion, (b) the ones with α ∈ (0.9,1.1) - moves diffusionally, (c) and the ones with α > 1.1 - moves superdiffusionally. About 55% of water molecules are in subgroup (a), 9% in subgroup (b), and 36% in subgroup (c) (cf., Fig. 6.8.2-4). The *MSD* of water molecules as a function of time, located at the first simulation step near DPPC and POPE bilayers (placed closer than 4 Å from the bilayer), divided into the three groups (a), (b), and (c) has been presented in Fig. 6.9 (left). As most of the experimental results provide an apparent diffusion coefficient, D_{app}, we have also computed it based on equation $D_{app} = D_{a}t^{\alpha - 1}$ (Banks and Fradin 2005) and the

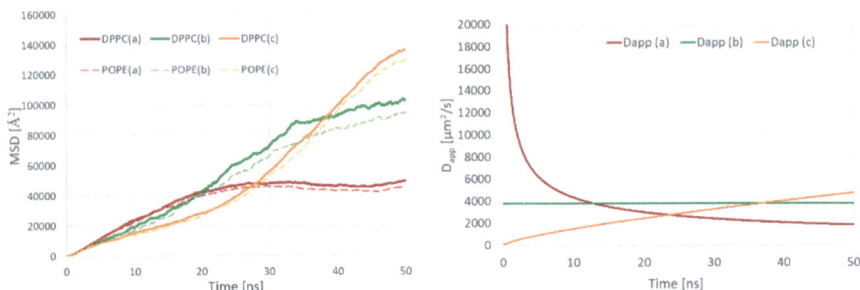

Fig. 6.9 *MSD* (for DPPC and POPE) and apparent diffusion coefficient, D_{app} (for DPPC) for three cases: (**a**) the ones which move subdiffusionally ($\alpha < 1$), (**b**) the ones move diffusionally ($\alpha \approx 1$), (**c**) and the ones move superdiffusively ($\alpha > 1$)

Table 6.1 α and D_α parameters based on the fitting function presented in Eq. (6.2) for three groups of water molecules

	α			$D_\alpha\,[\cdot 10^2 \mu m^2/s^a]$		
	(a)	(b)	(c)	(a)	(b)	(c)
DPPC	0.473 ± 0.008	0.999 ± 0.007	1.726 ± 0.007	146.1 ± 4.1	38.1 ± 0.9	2.78 ± 0.08
POPE	0.460 ± 0.008	0.996 ± 0.006	1.755 ± 0.008	143.6 ± 4.1	34.8 ± 0.7	2.38 ± 0.07

results have been presented in Fig. 6.9 (right). A mesoscopic study of similar type based on quite the same lipid membrane involving hydration rationale has been presented in the past by means of a phenomenological argumentation line, disclosing an algebraic time dependence of the principal kinetic coefficient coming out from decisive water vs. lipid interactions, facilitated by a certain adjustable pressure factor (Gadomski 1996).

The *MSD* results for DPPC and POPE bilayers are very similar. In all cases, the *MSD* curves nearly overlap being slightly lower for POPE than DPPC. The fitting parameters have been collected in Table 6.1.

The experimental results of the lateral apparent diffusion coefficients for water molecules near a mixed bilayer, coming from NMR, match the order of magnitude but exhibit about 3 times lower values (Wattraint and Sarazin 2005). In (Gaede and Gawrisch 2003) the D_{app} results for lateral diffusion of water near hydrated and dehydrated POPC lipids have been described and equal about 450 μm²/s. The discrepancies are a result of differences in dimensionality (we compute not 2D lateral diffusion but diffusion in 3D), differences in studied types of the bilayer, and environment properties (e.g., temperature). In Lopez et al. 2004, the authors studied the 3D MSD of water molecules H-bonded to oxygen atoms, and the provided value for the diffusion coefficient equals D = 20 μm²/s, which is similar to the diffusion coefficient for lipids in the bilayer. The difference between this approach and the one presented is the time range. The aforementioned approach was performed on a small-time range (up to 0.7 ns) in T = 298 K and regarded only the part of the MSD curve when the increase was stable (a linear function of time). In our case, water molecules that were H-bonded to the bilayer or were in close proximity to it (distance smaller than 4 Å) were tracked during a longer simulation time (50 ns). Some remained nearby, or H-bonded, while others managed to break away from the vicinity of the bilayer to the bulk. Thus, our approach looks at the process more generally, ie. closer to the real biomolecular systems.

3.2.1   Viscoelastic Moduli

From the MSD, the viscoelastic complex shear moduli of the water can be obtained. Using the generalized Stokes-Einstein relation in the Fourier domain we can form it as

$$G^* (\omega) = \frac{k_B T}{\pi R_g i \omega F \{MSD(t)\}}, \tag{6.3}$$

where k_B is the Boltzmann constant, T is temperature, ω is frequency, $F\{MSD(t)\}$ is the Fourier transform of the MSD, and R_g is the radius of gyration. It can be estimated as (Mason 2000)

$$G^* (\omega) \approx \frac{k_B T}{\pi R_g MSD\left(\dfrac{1}{\omega}\right)\Gamma\left[1+\alpha(\omega)\right]}, \tag{6.4}$$

where

$$\alpha(\omega) = \frac{dlnMSD(t)}{dlnt}\bigg|_{t=1/\omega} \tag{6.5}$$

The real and imaginary part of the complex viscoelastic modulus can be obtained from relations: $G'(\omega) = G^*(\omega) \cos(\pi\alpha(\omega)/2)$ and $G''(\omega) = G^*(\omega) \sin(\pi\alpha(\omega)/2)$, where $G^*(\omega) = G'(\omega) + iG''(\omega)$. The real part of the shear modulus, $G'(\omega)$, is the storage modulus and is the elastic susceptibility – the elastic component of the stress. The imaginary part of the shear modulus, $G''(\omega)$, is the loss modulus and is the viscous component of the stress.

Taking α from the fitting described in Sect. 3.2, the complex viscoelastic moduli, G' and G", have been computed using Eq. (6.4). Their values for DPPC have been presented in Fig. 6.10 (for POPE, the results were similar due to similarity in MSD). When the molecules move diffusively, G" dominates over G'; this scenario is recognizable in case (b), where viscosity prevails over elasticity. When the molecule is confined by the elastic structures of the complex system, G' dominates; this is true for case (a) but the domination of G' is small. For case (c), G' decreases in

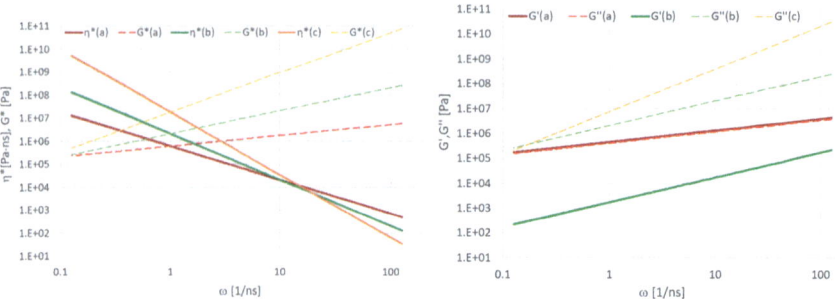

Fig. 6.10 Complex viscoelastic modulus, G^*, its components G' and G''; and viscosity, η^*, of water near the DPPC bilayer for three groups of water molecules (**a**), (**b**), and (**c**) as described in Sect. 3.2

frequency and becomes negative, thus it cannot be drawn on a log-log plot. A negative modulus means that the system is unstable, which sometimes happens in systems under high pressure (Lee et al. 2008; Gadomski 1996).

The complex viscosity, η^*, computed as $\eta^* = G^*/\omega$ has been presented in Fig. 6.10 (left). The viscosity lowers with frequency in each case. The slope of the viscosity changes in each case and is the smallest for (a) and highest for (c). System responses, presented in Fig. 6.10, are typical for viscoelastic liquids in the regime of high frequencies (Mason 2000; Guigas et al. 2007; Huang et al. 2017).

3.2.2 Schlitter Entropy

Another parameter that describes structural changes observed in a molecular system is configurational entropy. Based on a quantum-mechanical harmonic approximation, Schlitter introduced an expression for such entropy (Schlitter 1993). In this case, the entropy of the conformational ensemble of the single water molecule can be computed with formula:

$$S = k_B . \ln \det \left(1 + \frac{k_B . T . e^2}{\hbar^2} . M . \sigma \right),$$

$$(6.6)$$

where **1** is the unit matrix, **M** is the mass matrix (containing atomic masses in the diagonal), e is Euler's number, and $\boldsymbol{\sigma}$ is the symmetrical covariance matrix whose elements are given by:

$$\sigma_{ij} = \left\langle \left(x_i - \langle x_i \rangle \right) . \left(x_j - \langle x_j \rangle \right) \right\rangle.$$

$$(6.7)$$

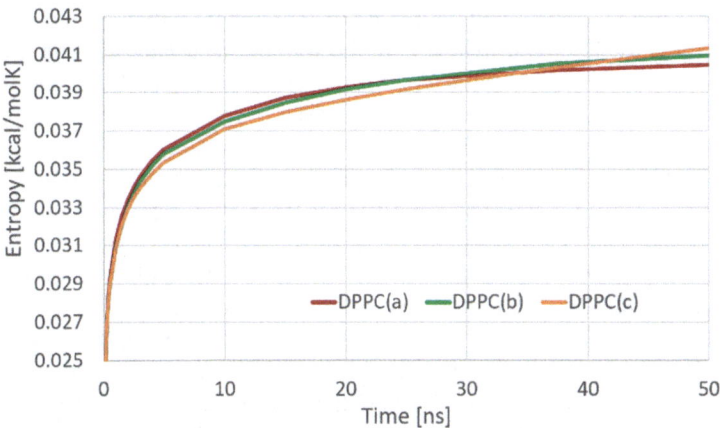

Fig. 6.11 Entropy as a function of time for three cases: (**a**) the ones which move subdiffusionally ($\alpha < 1$), (**b**) the ones which move diffusionally ($\alpha \approx 1$), and (**c**) the ones which move superdiffusionally ($\alpha > 1$)

The mean value in angular brackets is an average of all positions of atoms of water i and j in Cartesian space (separately for x, y, and z coordinates) calculated for a given time.

The entropy calculated from the mass-weighted covariances, Eq. (6.6), of single water molecules and averages over all molecules in each subset (a), (b), and (c), has been presented in Fig. 6.11.

Observe, that Schlitter's entropy approach has been successfully used to analyze the dynamics of molecules in the system with the bilayer. For example, it has been used in the analysis of drug molecular dynamics (Zapata-Morin et al. 2014). For the discussion on the entropy of the solvent, refer to (Fengler 2011) where the author suggests that although Schlitter's formula may be successfully applied to the analysis of the lipid's entropy, it may fail when applied to the solvent's entropy. This observation can be deduced from the fact that the free energy of the solvent (and the entropy) does not have the sharp minimum to which it can easily converge. Nevertheless, if there is a subset of water particles that "mimic" the behavior of the molecule, such a subset can be analyzed using the method suitable for the molecule.

The problem of the entropy of the solvent was investigated by Reinhard and Grubmüller (2007), where trajectories of the solvent were reduced by recognizing the permutation symmetry of the solvent. The confrontational space explored by the diffusive motion of the solvent particles was reduced as well, and the convergence of the Schlitter entropy was improved. We propose a different approach based on dividing the set of water molecules into subsets yielding various diffusive regimes: (a), (b) and (c) as described in Sect. 3.2.1 (see, Fig. 6.8). Given such a transformation, we expect the confrontational space to be smaller for particular subsets. Further, as the super-diffusion spreads faster, the super-diffusive regime could result in a faster exploration of the corresponding confrontational space. To find such a subset we look at the diffusive dynamics of particular water particles.

From Fig. 6.11, one can observe that the water particle dynamics obey various regimes such as the sub-diffusive one, the normal diffusive one, and the super-diffusive one. It is consistent with *MSD* investigations (cf., Figs. 6.8 and 6.9). We expect that the sub-diffusive regime is caused by the mechanism of trapping the particles inside the bilayer. However, a detailed investigation shows that most trapped particles leave the bilayer region in the simulation time, hence we cannot differentiate between trapped and non-trapped ones. The super-diffusive regime in the tail of the histogram is probably caused by the mechanism of pushing away water particles by moving elements of the bilayer. The parts of lipid heads forming the bilayer are subjects of rapid movement (Kruszewska et al. 2020). We hope these super-diffusive regimes could have a high contribution to transnational entropy, making it dominant among entropies of the water particles. Such transnational entropy is rather well-estimated by Schlitter's formula, which is contrary to the rotational (and vibrational) ones (Carlsson and Åqvist 2005).

Schlitter's entropy computed for particles in the sub-diffusive regime (cf., Fig. 6.11) converges to the constant value as expected. On the other hand, the super-diffusive particles converge to the positively sloped straight line. This is an interesting finding, most probably tied to the entropy production phenomenon apparent for the super-diffusive regime (Li et al. 2003; Prehl et al. 2010).

3.2.3 Testing Data on the Anomalous Diffusion Mechanism

Anomalous diffusion motions of the water molecules, presented above, can be realized based on several mechanisms. Three popular mechanisms usually are mentioned in the literature that may cause this phenomenon (Metzler and Klafter 2000; Metzler et al. 2014): continuous-time random walk (CTRW), fractional Brownian motion (FBM), and fractional Lévy stable motion (FLSM).

One of the most important physical quantities used in the statistical description of a random particle is MSD. It can be calculated using averaging over an ensemble of trajectories (cf., Eq. (6.1)). Alternatively, when abstracting for a while from the virtually present nonergodicity problem, we can use the temporal average over a single trajectory and calculate MSD as a time average over the interval [0,*T*] according to the formula (Neusius et al. 2009)

$$\delta^2\left(\Delta,T\right) = \int_0^{T-\Delta} \frac{\left[x\left(\Delta+\tau\right)-x\left(\tau\right)\right]^2}{T-\Delta}\,d\tau, \tag{6.8}$$

where Δ is the lag time. This quantity becomes the basis for characterizing the underlying stochastic process.

In the CTRW mechanism, a particle waits some time between two consecutive jumps. If the waiting time is taken from long-tailed distribution, i.e. waiting time probability density with the inverse power asymptotic, and a jump-length variance is finite, then the ensemble-averaged MSD becomes subdiffusive. In another case, when the waiting probability density is described by Poisson's distribution and the jump length probability density is the Lévy distribution with the stability parameter from the interval (1;2), then the described process is superdiffusive (Metzler and Klafter 2000).

The FBM is a generalization of the Brownian motion. Statistical properties of this process are mainly expressed by the Hurst exponent $0 < H < 1$. In this case, ensemble-averaged MSD is proportional to t^{2H}. So we can see, that for $0 < H < 1/2$ we can observe subdiffusion, for $1/2 < H < 1$ we observe superdiffusion, whereas for $H = 1/2$ we have the normal diffusion (Burnecki et al. 2011). The FBM of index H is the mean-zero Gaussian process $B_H(t)$, which has the following integral representation:

$$B_H(t) = \int_{-\infty}^{+\infty} \left[(t-u)_+^{H-\frac{1}{2}} - (-u)_+^{H-\frac{1}{2}} \right] dB(u), \tag{6.9}$$

where the symbol $B(t)$ represents a standard Brownian motion and $(y)_+ = \max(y,0)$. Increments of the FBM process are stationary, and their properties depend on the value of H. If $1/2 < H < 1$ then increments are positively correlated also exhibit long memory (long-range dependent), whereas if $0 < H < 1/2$, the increments are negatively correlated and exhibit short memory (Burnecki et al. 2011).

The FMB process can be generalized to the FLSM, which is defined by the following integral representation:

$$L_H^{\alpha}(t) = \int_{-\infty}^{+\infty} \left[(t-u)_+^d - (-u)_+^d \right] dL_{\alpha}(u), \tag{6.10}$$

where the symbol $L_{\alpha}(t)$ represents Lévy α-stable motion, $0 < \alpha \leq 2$, $0 < H < 1$ and $d = H - 1/\alpha$ (memory parameter). The FLMS process is α-stable and has stationary increments. The increments are positively correlated (has long-range dependence) for $d > 0$ and negatively correlated for $d < 0$. The MSD calculated using the time average for FLSM has different properties than the corresponding MSD calculated as an ensemble average (Burnecki et al. 2011).

These three above cases (CTRW, FBM, FLMS) do not exhaust all possible sources of anomalous diffusion (Burnecki and Weron 2010). Here we want to answer the question of which of the above three mechanisms best fits the water molecule trajectories obtained from molecular dynamics. To obtain this aim we study properties of single trajectories of the randomly chosen water molecules, which are considered in various situations described in the previous section, and then we use the p-variation method and sample MSD method to identify the type of anomalous diffusion (Ślęzak and Weron 2012).

Our data (single trajectory for each case (a)–(c)) follow from the molecular dynamics, which are performer by assumed time interval [0,T]. They are, from the stochastic processes point of view, a realization of particular stochastic process $X(t)$. In Fig. 6.12, we present trajectory components of the center of mass of the water

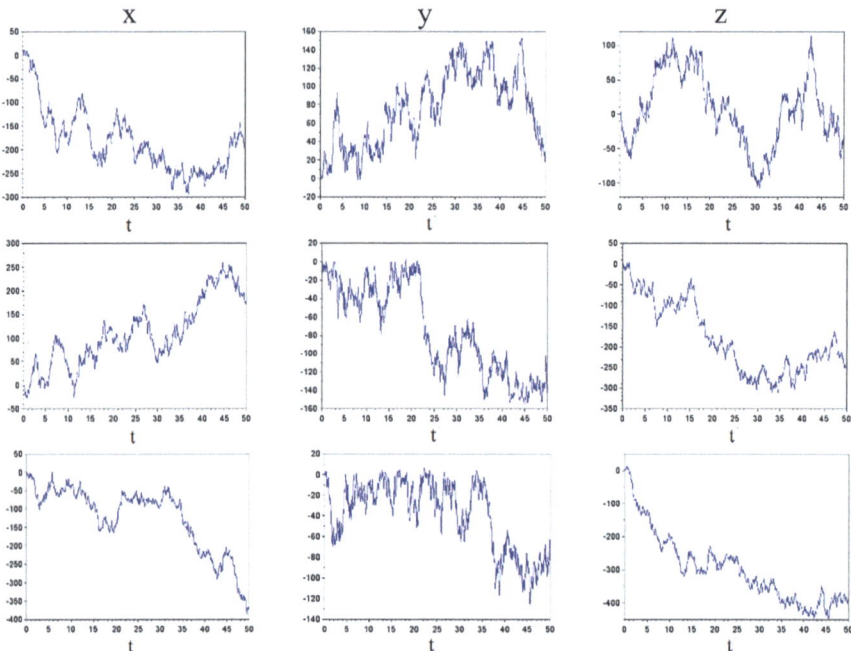

Fig. 6.12 Three components (x,y,z) of the center of the mass (as a function of time) of randomly chosen water molecule for POPE: case (a) (first row) – single water molecule for the subdiffusive regime, case (b) (second row) single water molecule near pure diffusion regime and case (c) (third row) single water molecule for the superdiffusive regime

molecule near POPE in the three cases described in the previous sections (cases (a)–(c)).

Next, we would like to know which of the above three processes best fit the data presented in Fig. 6.12. Here we use the p-variational method (Burnecki and Weron 2010). The p-variation is defined by the following formula:

$$V^{(p)}(t) = \lim_{n \to \infty} V_n^{(p)}(t), p > 0, \qquad (6.11)$$

where $V_n^{(p)}(t)$ for stochastic process $X(t)$ is given by the following expression:

$$V_n^{(p)}(t) = \sum_{j=0}^{2^n-1} \left| X\left(\frac{(j+1)T}{2^n} \wedge t\right) - X\left(\frac{jT}{2^n} \wedge t\right) \right|^p, \qquad (6.12)$$

and $(a \wedge b) = \min\{a,b\}$.

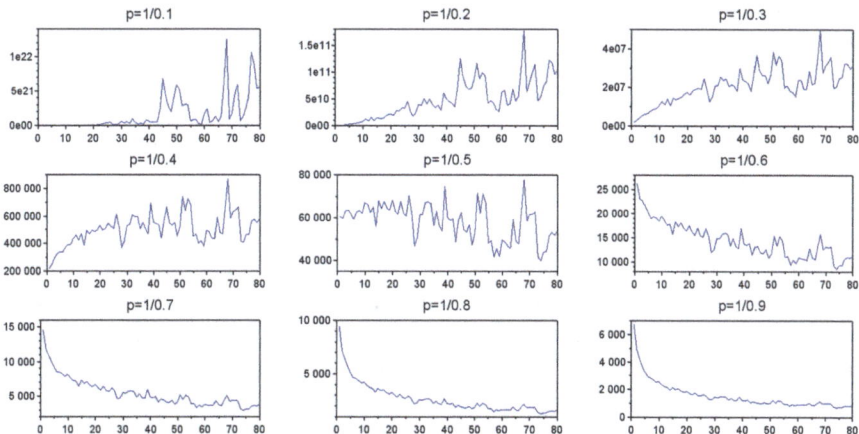

Fig. 6.13 The p-variation test's result for the randomly picked water molecule near POPE for case (a). One can see that tendency of the p-variation, as a function of n, changes

The p-variational method for underlying model FLSM when $d < 0$, gives the sample p-variation $V_n^{(p)}(t)$ which increases with n for any parameter $p > 0$. In this case, the sample p-variation for FBM, as a function of n, changes its trend from a decreasing to an increasing for $p = 1/H$, $0 < H < 1$. The p-variation for FBM additionally has the property, that it is a linear function of time for $p = 1/H$. These properties can be used to distinguish which of the two mechanisms, FMB or FLSM, best describes presented (anomalous) diffusion processes.

The example result of the p-variational method for a stochastic trajectory of randomly picked water molecule near POPE bilayer for case (a) is presented in Fig. 6.13. One can see that p-variation as a function of n, changes its trend between $H = 0.4$–0.5 from increasing to decreasing one. The possibility of the CTRW mechanism can be omitted as numerically obtained time sample MSD is not linear with time. Numerically obtained memory parameter d (cf., Eq. (6.10)) is smaller than zero ($d \sim -0.03$). As it was mentioned above these properties are characteristic of the FBM mechanism (Magdziarz et al. 2009). For a sufficiently large n, one can observe that numerical p-variational as a function of time is almost linear.

The same analysis has been performed for other cases (case (b) and case (c)). In case (b), which is similar to pure diffusion, the numerically obtained memory parameter $d \sim -0.028$, thus, it has a small absolute value and is negative. It means that the diffusion process can be interpreted as the process of a normal character, but some properties of anomalous diffusion can be slightly visible. We numerically establish that p-variational as a function of n for some parameter p changes its trend, like in the previous case, but this time, the change is near $H = 0.5$. Thus, we can say that the mechanism of the process is also FBM.

In case (c), the parameter $d \sim 0.0015$, thus, it has a small absolute value and is positive. It indicates that a superdiffusive process has been observed. The

p-variational method suggests that for the case (c) the FLSM mechanism governs this process.

4 Summary

The diffusion dynamics of water molecules situated in the proximity of one of the two types of phospholipid bilayers (DPPC and POPE) has been studied. The goal of the studies was to give insight into the understanding of the role of lipid membranes in tribological phenomena (Pawlak et al. 2016). For example, the water molecules, which shield the bilayer from the components of synovial fluid in articular cartilage, could be responsible for triggering different tribological responses of the system by changing the energy barrier for the formation of various interactions. A faster diffusive movement of water in the proximity of the bilayer lowers the energy barrier for the formation and causes breakages of H-bonds between the water molecules. On contrary, the presence of the "slow" water molecules at the surface of the bilayer indicates they are strongly H-bonded and separated from the bulk water by a large free energy barrier. "Slow" here means, that the motion of the water molecules has been characterized by long residence times, of the order of 100 ps or more (Roy and Bagchi 2012; Yamamoto et al. 2014). Moreover, it is proposed in (Roy and Bagchi 2012) that the stabilizing effect of quasi-bound water molecules might be a general phenomenon in various water-soluble proteins and enzymes.

Our interaction and motion investigations have shown that the water molecules undergo various types of dynamics and can be split into, at least, three groups: the ones which made subdiffusional (so confined) motions, the ones which move diffusionally in the bulk, and the ones which move superdiffusively due to interactions with the moving bilayer. The water molecules' diffusivity has been studied in more detail by looking at MSD of the water molecules and also by looking at the hydration of DPPC and POPE-based membranes. Hydration was measured in terms of the number of hydrogen bonds formed.

Although the bilayer type (DPPC and POPE) has a small influence on the water motion near the bilayer, the differences in head groups can provide smaller or greater H-bonding of the bilayer with the water. Importantly, differences in H-bonding causes differences in the stability of the bilayer. It was found that the DPPC bilayer creates more durable bonds than POPE, a fact that better stabilizes the membrane compared with the larger number of H-bonds formed by the POPE. It has been also demonstrated, that all water molecule groups exhibit slightly different viscoelastic responses.

To make a more detailed investigation of the water particle dynamics the statistical testing was carried out. The p-variation and sample MSD tests performed on the trajectories of random water molecules near the bilayers suggest that the CTRW model is not a proper one to explain the evolution of the trajectory. In most cases of waters' trajectories near DPPC and POPE, the FBM process was recorded. This is attributed to the viscoelasticity of lipid bilayers, what has been also reported in (Yamamoto et al. 2014).

Further, our studies have shown that most of the H-bonds last up to 100 ps and most of the water molecules from the proximity of the bilayer have diffused from the bilayer. This indicates that for a better description of the water motion, a greater resolution computer simulation is needed. We plan to perform such simulations as the next step in our studies.

Our research has also encompassed the analysis of the ordering inside the water system based on the computation of Schlitter's configurational entropy. The differences in entropy behavior for the three groups of water molecules have been presented and were in accordance with the analysis results of MSD. In practice, Schlitter's entropy is treated rather as the upper bound of the real entropy of the system. However, as such bound rises monotonically with time, we should expect a similar increase in the real entropy of the system. Nevertheless, to confirm the findings, further research is necessary. First, in spite of the fact that overall the entropy is most reliably approachable at thermodynamic equilibrium (Ben-Naim 2019), it should here be calculated in the moving observation window of constant length to eliminate the effect of the different data sizes. Next, other more accurate measures of entropy can be used.

Acknowledgements This work is supported by the internal grant BN-10/2019 of the Institute of Mathematics and Physics, UTP (NK) and by the grant of National Science Centre in Poland (Miniatura Grant) 2019/03/X/ST3/01482 (PW). KD acknowledges financial support by the Foundation for Polish Science through TEAM-NET project (contract no. POIR.04.04.00-00-17C1/18-00). Calculations were carried out at the Academic Computer Centre in Gdańsk. Authors also want to thank Maryellen Zbrozek (University of Michigan), Fulbright ETA at UTP, for her impact in obtaining the final form of the paper and proofreading.

References

Alberts, B., Johnson, A., Lewis, J., Raff, M., Roberts, K., & Walter, P. (2002). The lipid bilayer. In *Molecular biology of the cell* (4th ed.). New York: Garland Science.

Banks, D. S., & Fradin, C. (2005). Anomalous diffusion of proteins due to molecular crowding. *Biophysical Journal, 89*, 2960–2971.

Bełdowski, P., Weber, P., Dèdinaitè, A., Claesson, P. M., & Gadomski, A. (2018). Physical cross-linking of hyaluronic acid in the presence of phospholipids in an aqueous nano-environment. *Soft Matter, 14*, 8997–9004.

Bellour, M., Skouri, M., Munch, J.-P., & Hébraud, P. (2002). Brownian motion of particles embedded in a solution of giant micelles. *The European Physical Journal E, 8*, 431–436.

Ben-Naim, A. (2011). The role of water in protein folding, self-assembly and molecular recognition. In *Molecular theory of water and aqueous solutions: Part II*. Hackensack: World Scientific Publishing Company.

Ben-Naim, A. (2019). Entropy and information theory: Uses and misuses. *Entropy, 21*, 1170.

Berendsen, H. J. C., Postma, J. P. M., van Gunsteren, W. F., DiNola, A., & Haak, J. R. (1984). Molecular dynamics with coupling to an external bath. *The Journal of Chemical Physics, 81*, 3684–3690.

Briscoe, W. H. (2017). Aqueous boundary lubrication: Molecular mechanisms, design strategy, and terra incognita. *Current Opinion in Colloid & Interface Science, 27*, 1–8.

Burnecki, K., & Weron, A. (2010). Fractional Lévy stable motion can model subdiffusive dynamics. *Physical Review E, 82*, 021130.

Burnecki, K., Magdziarz, M., & Weron, A. (2011). Identification and validation of fractional subdiffusion dynamics. In *Fractional Dynamics: Recent Advances* (pp. 331–351). Singapore: World Scientific.

Carlsson, J., & Åqvist, J. (2005). Absolute and relative entropies from computer simulation with applications to ligand binding. *The Journal of Physical Chemistry. B, 109*, 6448–6456.

Cheng, C.-Y., Olijve, L. L. C., Kausik, R., & Han, S. (2014). Cholesterol enhances surface water diffusion of phospholipid bilayers. *The Journal of Chemical Physics, 141*, 22D513.

Crowe, J., & Bradshaw, T. (2010). *Chemistry for the biosciences: The essential concepts* (2nd ed.). New York: OUP Oxford.

Dechant, A., Lutz, E., Kessler, D. A., & Barkai, E. (2014). Scaling Green-Kubo relation and application to three aging systems. *Physical Review X, 4*, 011022.

Dédinaité, A., & Claesson, P. M. (2017). Synergies in lubrication. *Physical Chemistry Chemical Physics, 19*, 23677–23689.

Dédinaité, A., Wieland, D. C. F., Bełdowski, P., & Claesson, P. M. (2019). Biolubrication synergy: Hyaluronan – Phospholipid interactions at interfaces. *Advances in Colloid and Interface Science, 274*, 102050.

Disalvo, E. A., Pinto, O. A., Martini, M. F., Bouchet, A. M., Hollmann, A., & Frías, M. A. (2015). Functional role of water in membranes updated: A tribute to Träuble. *Biochimica et Biophysica Acta (BBA) – Biomembranes, 1848*, 1552–1562.

Doğangün, M., Ohno, P. E., Liang, D., McGeachy, A. C., Bé, A. G., Dalchand, N., Li, T., Cui, Q., & Geiger, F. M. (2018). Hydrogen-bond networks near supported lipid bilayers from vibrational sum frequency generation experiments and atomistic simulations. *The Journal of Physical Chemistry. B, 122*, 4870–4879.

Duan, Y., Wu, C., Chowdhury, S., Lee, M. C., Xiong, G., Zhang, W., Yang, R., et al. (2003). A point-charge force field for molecular mechanics simulations of proteins based on condensed-phase quantum mechanical calculations. *Journal of Computational Chemistry, 24*, 1999–2012.

Fengler, S. (2011). *Estimating orientational water entropy at protein interfaces*. Diss. Georg-August-Universität Göttingen, Göttingen.

Gadomski, A. (1996). Stretched exponential kinetics of the pressure induced hydration of model lipid membranes. A possible scenario. *Journal de Physique II France, 6*, 1537–1546.

Gaede, H. C., & Gawrisch, K. (2003). Lateral diffusion rates of lipid, water, and a hydrophobic drug in a multilamellar liposome. *Biophysical Journal, 85*, 1734–1740.

Guigas, G., Kalla, C., & Weiss, M. (2007). Probing the nanoscale viscoelasticity of intracellular fluids in living cells. *Biophysical Journal, 93*, 316–323.

Haines, T. H., & Liebovitch, L. S. (1995). Chapter 6: A molecular mechanism for the transport of water across phospholipid bilayers. In E. A. Disalvo & S. A. Simon (Eds.), *Permeability and stability of lipid bilayers*. Boca Raton: CRC Press.

Huang, J., Zhong, C., & Wu, X. (2017). Shear behavior at high pressures and viscoelastic properties in water and in brine solutions with high salinities for a tetra-polymer containing poly(ethylene oxide) side chains. *RSC Advances, 7*, 47624–47635.

Israelachvili, J. N., Mitchell, D. J., & Ninham, B. W. (1976). Theory of self-assembly of hydrocarbon amphiphiles into micelles and bilayers. *Journal of the Chemical Society, Faraday Transactions 2: Molecular and Chemical Physics, 72*, 1525–1568.

Janicka, K., Bełdowski, P., Majewski, T., Urbaniak, W., & Petelska, A. D. (2019). The amphoteric and hydrophilic properties of cartilage surface in mammalian joints: Interfacial tension and molecular dynamics simulation studies. *Molecules, 24*, 2248.

Kneller, G. R., Baczyński, K., & Pasenkiewicz-Gierula, M. (2011). Communication: Consistent picture of lateral subdiffusion in lipid bilayers: Molecular dynamics simulation and exact results. *The Journal of Chemical Physics, 135*, 141105.

Krieger, E., Dunbrack, R. L., Hooft, R. W. W., & Krieger, B. (2012). Assignment of protonation states in proteins and ligands: Combining PKa prediction with hydrogen bonding network optimization. *Methods in Molecular Biology, 819*, 405–421.

Kruszewska, N., Domino, K., Drelich, R., Urbaniak, W., & Petelska, A. D. (2020). Interactions between Beta-2-glycoprotein-1 and phospholipid bilayer – Molecular dynamic study. *Membranes, 10*, 396.

Lee, B., Rudd, R. E., Klepeis, J. E., & Becker, R. (2008). Elastic constants and volume changes associated with two high-pressure rhombohedral phase transformations in vanadium. *Physical Review B, 77*, 134105.

Li, X., Essex, C., Davison, M., Hoffmann, K. H., & Schulzky, C. (2003). Fractional diffusion, irreversibility and entropy. *Journal of Non-Equilibrium Thermodynamics, 28*, 279–291.

Lopez, C., Nielsen, S., Klein, M., & Moore, P. (2004). Hydrogen bonding structure and dynamics of water at the dimyristoylphosphatidylcholine lipid bilayer surface from a molecular dynamics simulation. *The Journal of Physical Chemistry. B, 108*, 6603–6610.

Magdziarz, M., Weron, A., Burnecki, K., & Klafter, J. (2009). Fractional Brownian motion versus the continuous-time random walk: A simple test for subdiffusive dynamics. *Physical Review Letters, 103*, 180602.

Mason, T. G. (2000). Estimating the viscoelastic moduli of complex fluids using the generalized stokes–Einstein equation. *Rheologica Acta, 39*, 371–378.

Metzler, R., & Klafter, J. (2000). The random walk's guide to anomalous diffusion: A fractional dynamics approach. *Physics Reports, 339*, 1–77.

Metzler, R., Jeon, J.-H., Cherstvy, A. G., & Barkai, E. (2014). Anomalous diffusion models and their properties: Non-stationarity, non-ergodicity, and ageing at the centenary of single particle tracking. *Physical Chemistry Chemical Physics, 16*, 24128–24164.

Michaud-Agrawal, N., Denning, E. J., Woolf, T. B., & Beckstein, O. (2011). MDAnalysis: A toolkit for the analysis of molecular dynamics simulations. *Journal of Computational Chemistry, 32*, 2319–2327.

Nagle, J. F., & Tristram-Nagle, S. (2000). Structure of lipid bilayers. *Biochimica et Biophysica Acta (BBA) – Reviews on Biomembranes, 1469*, 159–195.

Neria, E., Fischer, S., & Karplus, M. (1996). Simulation of activation free energies in molecular systems. *The Journal of Chemical Physics, 105*, 1902–1921.

Neusius, T., Sokolov, I. M., & Smith, J. C. (2009). Subdiffusion in time-averaged, confined random walks. *Physical Review E, 80*, 011109.

Pasenkiewicz-Gierula, M., Baczyński, K., Markiewicz, M., & Murzyn, K. (2016). Computer modelling studies of the bilayer/water interface. *Biochimica et Biophysica Acta (BBA) – Biomembranes, 1858*, 2305–2321.

Pawlak, Z., Urbaniak, W., Gadomski, A., Yusuf, K. Q., Afara, I. O., & Oloyede, A. (2012). The role of lamellate phospholipid bilayers in lubrication of joints. *Acta of Bioengineering and Biomechanics, 14*, 101–106.

Pawlak, Z., Gadomski, A., Sojka, M., Urbaniak, W., & Bełdowski, P. (2016). The amphoteric effect on friction between the bovine cartilage/cartilage surfaces under slightly sheared hydration lubrication mode. *Colloids and Surfaces B: Biointerfaces, 146*, 452–458.

Petelska, A. D., Kazimierska-Drobny, K., Janicka, K., Majewski, T., & Urbaniak, W. (2019). Understanding the unique role of phospholipids in the lubrication of natural joints: An interfacial tension study. *Coatings, 9*, 264.

Prehl, J., Essex, C., & Hoffmann, K. H. (2010). The superdiffusion entropy production paradox in the space-fractional case for extended entropies. *Physica A, 389*, 215–224.

Rand, R. P., & Parsegian, V. A. (1989). Hydration forces between phospholipid bilayers. *Biochimica et Biophysica Acta (BBA) – Reviews on Biomembranes, 988*, 351–376.

Reinhard, F., & Grubmüller, H. (2007). Estimation of absolute solvent and solvation shell entropies via permutation reduction. *The Journal of Chemical Physics, 126*, 014102.

Roy, S., & Bagchi, B. (2012). Free energy barriers for escape of water molecules from protein hydration layer. *The Journal of Physical Chemistry B, 116*, 2958–2968.

Santamaría-Holek, I., Rubí, J. M., & Gadomski, A. (2007). Thermokinetic approach of single particles and clusters involving anomalous diffusion under viscoelastic response. *The Journal of Physical Chemistry B, 111*, 2293–2298.

Sarmiento-Gomez, E., Santamaría-Holek, I., & Castillo, R. (2014). Mean-square displacement of particles in slightly interconnected polymer networks. *The Journal of Physical Chemistry B, 118*, 1146–1158.

Saxton, M. J., & Jacobson, K. (1997). Single-particle tracking: Applications to membrane dynamics. *Annual Review of Biophysics and Biomolecular Structure, 26*, 373–399.

Schlitter, J. (1993). Estimation of absolute and relative entropies of macromolecules using the covariance matrix. *Chemical Physics Letters, 215*, 617–621.

Ślęzak, J., & Weron, K. (2012). Revisited approach to statistical analysis of ionic current fluctuations. *Acta Physica Polonica B, 43*, 1215–1226.

Song, J., Franck, J., Pincus, P., Won Kim, M., & Han, S. (2014). Specific ions modulate diffusion dynamics of hydration water on lipid membrane surfaces. *Journal of the American Chemical Society, 136*, 2642–2649.

van Meer, G., Voelker, D. R., & Feigenson, G. W. (2008). Membrane lipids: Where they are and how they behave. *Nature Reviews Molecular Cell Biology, 9*, 112–124.

Wattraint, O., & Sarazin, C. (2005). Diffusion measurements of water, ubiquinone and lipid bilayer inside a cylindrical nanoporous support: A stimulated echo pulsed-field gradient MAS-NMR investigation. *Biochimica et Biophysica Acta (BBA) – Biomembranes, 1713*, 65–72.

White, S. H., & Wiener, M. C. (1995). Chapter 1: Determination of the structure of fluid lipid bilayer membranes. In E. A. Disalvo & S. A. Simon (Eds.), *Permeability and stability of lipid bilayers*. Boca Raton: CRC Press.

Wiener, M. C., & White, S. H. (1992). Structure of a fluid dioleoylphosphatidylcholine bilayer determined by joint refinement of X-ray and neutron diffraction data. III. Complete structure. *Biophysical Journal, 61*, 434–447.

Yamamoto, E., Akimoto, T., Yasui, M., & Yasuoka, K. (2014). Origin of subdiffusion of water molecules on cell membrane surfaces. *Scientific Reports, 4*, 4720.

Yeghiazaryan, G., Poghosyan, A., & Shahinyan, A. (2005). Structural and dynamical features of hydrocarbon chains of dipalmitoylphosphatidylcholine (DPPC) molecules in phospholipid bilayers: A molecular dynamics study. *Electronic Journal of Natural Sciences, 4*, 44.

Zapata-Morin, P. A., Sierra-Valdez, F. J., & Ruiz-Suárez, J. C. (2014). The interaction of local anesthetics with lipid membranes. *Journal of Molecular Graphics and Modelling, 53*, 200–205.

Chapter 7
Water Molecules Organization Surrounding Ions, Amphiphilic Protein Residues, and Hyaluronan

Jacek Siódmiak

Abstract Water, that is so common in our environment and that is an essential component of our life, is an object of interest for researchers for centuries. Some of its physicochemical properties are extraordinary. Although water is a liquid, many experimental studies indicate that under certain conditions water has the characteristics of an ordered structure. We show here the results of molecular dynamics simulations of bulk water as well as water interacting with different ions, amino acids, and hyaluronic acid. Using a tool called radial distribution function to analyze simulation results we try to answer whether the water in the considered systems shows features of an ordered structure.

Keywords Water clusters · Ions assisted water clusters · Amphiphiles · Radial distribution function · Organized water matrices · Hyaluronic acid with water assistance · Computer simulation

1 Introduction

Water is indispensable for of all life. This is the most important liquid in our ecosystem. All plants, insects, and animals, including us, cannot survive without it. Its unique properties make it the object of continuous research. It is estimated that the total amount of water on Earth is approximately 1,338,000,000 km³ (Eakins and Sharman 2010). Of which 1120 cubic kilometers populate the surfaces of proteins, inhabit the grooves of DNA, reside at the surfaces of lipid bilayers, or in tissues and cells can exhibit properties that are quite distinct from those found in the bulk. Due to its unique characteristics and certain common features observed in different

J. Siódmiak (✉)
Institute of Mathematics and Physics, UTP University of Science and Technology,
Bydgoszcz, Poland
e-mail: siedem@utp.edu.pl

A. Gadomski (ed.), *Water in Biomechanical and Related Systems*,
Biologically-Inspired Systems 17, https://doi.org/10.1007/978-3-030-67227-0_7

biological systems, this water has been termed biological water to distinguish it from bulk water (Bagchi 2013). Biological water is about 0.000081% of the total amount of water. Water is also indispensable in industry. It is often used to cool industrial installations and engines. Many chemical processes take place with the participation of water. It is called the "universal solvent" because, thanks to its polar properties, it can dissolve many substances.

From the point of view of life processes taking place in the living cells of all organisms, water seems to be even more important than it is in the industry. Water makes up 70–85% of the weight of an adult human. Up to 90% of the cell weight is water. It is the environment of biochemical reactions, and sometimes it is their substrate (in the photosynthesis process) or product (in the cell respiration process). Our cells need to be filled with water to work properly. This is because enzymes inside cells work only in solution. Water also has a transport function. Plasma makes up more than half of total blood volume consists of 90% water and has hormones and gasses dis-solved in it as well as toxins such as urea, which are removed from the body with yet more water. Water prevents excessive temperature fluctuations and is involved in thermoregulation in warm-blooded organisms.

By comparing the properties of water with other substances, it can be said that water is the only natural substance found in all three physical states: liquid, solid and as a gas, at the temperatures that naturally occur on Earth. Moreover, there is a unique combination of temperature and pressure at which solid, liquid, and gaseous phases can coexist in thermodynamic equilibrium. This unique combination of temperature and pressure is called the triple point. A stable equilibrium occurs at approximately 273.1575 K (0.0075 °C) and a partial vapor pressure of 611.657 Pa (6.11657 mbar) (Wagner et al. 1994).

In liquid or solid water, atoms constituting water molecule engage in hydrogen bonding interactions with surrounding water molecules. Most molecules in liquid water are in two hydrogen-bonded configurations with one strong donor and one strong acceptor hydrogen bond in contrast to the four hydrogen-bonded tetrahedral structure in ice. Upon heating from 25 to 90 °C, insignificant amount of the molecules (5–10%) change from tetrahedral layout to two hydrogen-bonded configurations. Many experiments show that water at ambient conditions is classified as a uniform, continuous tetrahedral liquid (Morgan and Warren 1938; Wernet et al. 2004). The wide-angle X-ray and neutron scattering experiments and analysis confirm that coordination number for water is between 4 and 5, with well-defined peaks in the radial distribution functions that implies a tetrahedral layout similar to that of ice (Clark et al. 2010; Head-Gordon and Hura 2002).

Water is a very good solvent. Due to their polarity, water molecules can attach to other molecules (ions, macromolecules, proteins). Under certain conditions, the so-called solvation shell may occur. A solvation shell is the solvent interface of any chemical compound or biomolecule that constitutes the solute. When the solvent is water, it is often referred to as a hydration shell or hydration sphere. A classic example is water molecules arranging around a metal ion. For example, if the latter is a cation, the positive charge on the metal ion electrostatically attract the electronegative oxygen atom of the water molecule. The result is a solvation shell of water molecules that surround the ion. This shell can be several molecules thick, depending on the charge of the ion, its distribution, and spatial dimensions.

For molecules, which exhibit amphiphilic properties, hydration is the process of attaching dipolar water molecules through polar, or hydrophilic, groups. The groups of a hydrophilic nature include the following groups: -COOH (carboxylic), -COO⁻ (ester), -NH₂ (amine), -NH₃⁺ (amine), -OH (hydroxyl), -O- (oxide), -SH (thiol) and H2N-(C=NH)-NH- (guanidine). The degree of protein hydration depends on its type, environmental pH, the presence of other water-binding substances and temperature. The hydration process also takes place for single amino acids. The amphiphilicity of amino acids depends on the side chain type. The hydration shell that forms around proteins is of importance in biochemistry. This interaction of the protein surface with the surrounding water is fundamental to the activity of the protein (Zhang et al. 2007).

Here we present several issues related to the organization of water molecules in a bulk as well as around the different types of molecule. In Sect. 2 the research methodology is described. In Sect. 3 the radial distribution function and how this tool is used to discover an ordered structure in the studied medium is presented. In Sect. 4 the results of water simulation in a very confined (almost 2D) space are shown. Next, in Sect. 5 it is shown how single water molecules are arranged around the Na⁺ and Cl⁻ ions. In Sect. 6, we analyze the arrangement of water around selected amino acids. In Sect. 7 we show the results of simulation of water interacting with hyaluronic acid. The last Section summarizes the presented issues.

2 Methods

In all the examples presented here, full-atom molecular dynamics simulations (Frenkel and Smit 2002) were performed in program YASARA (Yet Another Scientific Artificial Reality Application) (Krieger and Vriend 2015). NOVA force field was used (Krieger et al. 2002). The NOVA force field looks like common molecular dynamics force fields, with the total energy being expressed as a sum of individual contributions: bonds, angles, planarity, Van der Waals, and electrostatic terms. Most negative point charges are placed outside the nuclei (off-center charges). Van der Waals interactions are modeled by Born-Mayer potential ($A\ exp\ (-BR)$) instead of the familiar Lennard-Jones (A/R^{12}) potentials (Van Vleet et al. 2016). Planarity is treated by least-squares plane fitting instead of improper torsions. The YASARA NOVA force field has the following form:

$$E_a = \sum_{j<a} 0.5 \cdot k_j \cdot \left(R_{j0} - R_j\right)^2 + 0.5 \cdot l_p \cdot R_p^2 + \sum_{i<a}\left(A_i \cdot e^{-B_i \cdot R_i} - \frac{C_i}{R_i^6}\right)$$

$$+ \sum_{k<a}\sum_m\sum_n \frac{q_m \cdot q_n}{4\pi\varepsilon_0 \cdot R_{mn}}$$

Atom distances are named R, equilibrium values R_0. The energy contribution of atom a: (E_a) is the sum over all chemical bonds to atom j, with bond-stretching force constant k_j, $j < a$, plus a planarity term, where l_p is a plane stretching force constant and R_p is distance from the plane, plus the sum over all non-bonded Van der Waals interactions with atom i, using *EXP6*-potential parameters A_i, B_i and C_i, and for $i < a$, plus the electrostatic Coulomb interactions between all m point charges on atom a and n point charges on atom k (where $k < a$), ε_0 is a vacuum permittivity.

By default, YASARA uses TIP3 (Transferable Intermolecular Potential with 3 points) water model (Jorgensen 1981; Jorgensen et al. 1983; Harrach and Drossel 2014). Three-site models have three interaction points corresponding to the three atoms of the water molecule. Each site has a point charge, and the site corresponding to the oxygen atom also has the Lennard-Jones parameters (Allen and Tildesley 2017). The potential for a TIP3P model is represented by:

$$E_{mn} = \sum_i^{\text{on } m} \sum_j^{\text{on } n} \frac{q_i q_j e^2}{r_{ij}} + \frac{A}{r_{OO}^{12}} - \frac{B}{r_{OO}^6}$$

where e is elementary charge, q_i and q_j are the partial charges relative to the charge of the electron ($q_O = -0.834$ and $q_H = 0.417$). The original TIPS 3 site model has positive charges on the hydrogens and a negative charge on oxygen and $q_O = -2q_H$ (Jorgensen 1981). r_{ij} is the distance between two atoms or charged sites (r_{OO} distance between two oxygen atoms), $A = 582.0 \cdot 10^{-3} \, \text{kcal} \, \text{Å}^{12} / \text{mol}$ and $B = 595.0 \, \text{kcal} \, \text{Å}^6 / \text{mol}$ are the Lennard-Jones parameters (Jorgensen 1981; Jorgensen et al. 1983). The charged sites may be on the atoms or on dummy sites (such as lone pairs). In most water models, the Lennard-Jones term applies only to the interaction between the oxygen atoms.

3 Radial Distribution Function – Ordered Structures Discovery Tool

The radial distribution function RDF (or pair distribution function) $g(r)$, gives the probability of finding a particle in the distance r from another particle (Allen and Tildesley 2017; Levine et al. 2011; Chandler 1987). If we count the appearance of two atoms/molecules at separation r, from $r = 0$ to $r = \infty$, we can get the radial distribution function $g(r)$. The radial distribution function is a useful tool to describe the structure of a system, particularly of liquids.

Let us consider a spherical shell of thickness dr at a distance r from a chosen atom/molecule (Fig. 7.1). The volume of the shell is given by:

$$V_{shell} = \frac{4}{3}\pi(r + dr)^3 - \frac{4}{3}\pi r^3 \approx 4\pi r^2 \cdot dr$$

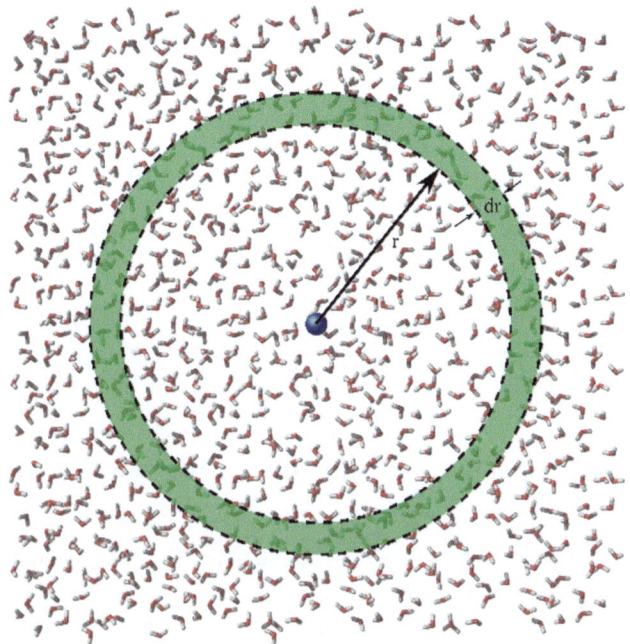

Fig. 7.1 Space discretization for the evaluation of the radial distribution function

If the number of particles per unit volume is ρ, then the total number of atoms/molecules in the shell $dn(r)$, can be calculated as follows:

$$dn(r) = g(r) \cdot 4\pi r^2 \cdot \rho \cdot dr.$$

After conversion of the above equation, RDF gets the following form:

$$g(r) = \frac{dn(r)}{4\pi r^2 \cdot \rho \cdot dr}$$

In solids, the radial distribution function has an infinite number of sharp peaks whose separations and heights are characteristic of the lattice structure (Franchetti 1975). The radial distribution function of a liquid is something between RDF of solid and the gas (Yoon et al. 1981; Chandler 1987), with a small number of peaks at short distances, superimposed on a steady decay to a constant value at longer distances (Fig. 7.2). Radial distribution functions are usually equal zero at small distances because of interatomic repulsion at these distances. There are often sharp peaks at distances that correspond to structure in the system: for example, the first solvation shell of a molecule in solution (Henao et al. 2016). The shape and location of peaks are related to the temperature and the intermolecular potential energy between molecules (Fig. 7.2) (Costa et al. 2011; Ding et al. 2014). The location of

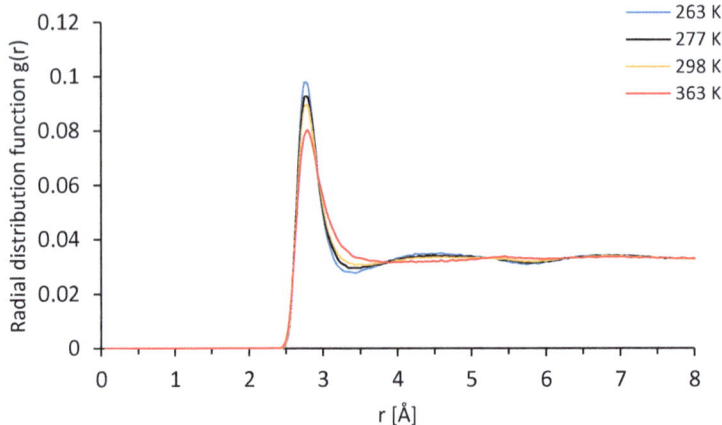

Fig. 7.2 The oxygen-oxygen radial distribution function of bulk water at: 263 K, 277 K, 298 K and 363 K

the first peak in the radial distribution function is close to the minimum of the Lennard-Jones potential. The integral of radial distribution function can provide information about average coordination numbers (number of the nearest neighbors) (Chandler 1987).

The radial distribution function curves from Fig. 7.2 were obtained from molecular dynamics simulation of water. The TIP3P water model was used. The simulation box was $50 \times 50 \times 50$ Å, the number of water molecules was 4130 and pressure 1 bar. From Fig. 7.2 alongside increasing temperature, the high of the first peak decreases. This is due to the fact, that water molecules move at higher velocities and are not able to stay in close contact with other water molecules for a long time.

4 Self-Organization of Water in Tightly Confined Space

A characteristic feature of gases and liquids is that they adopt the shape of the vessel in which they are located. In addition, the gases, unlike liquids, occupy the entire available volume. If the liquid is confined in a very limited space, it will behave slightly different from the bulk liquid (Thompson 2018; Chaplin 2009). Confined water is found widespread in nature in granular and porous materials, both around and within cells, macromolecules, supramolecular structures, and gels. When water is confined, there is a conflict between the energetic minimization of the hydrogen-bonded network and the fit within the space available. The properties of the confined water are difficult to predict and may be very different from those of bulk water. Molecular dynamics simulations of water enclosed in a $30 \times 30 \times 5$ Å and $30 \times 30 \times 10$ Å box show that the more available space is limited (in this case close to two-dimensional space), the more water molecules are arranged in regular geometric shapes. The simulation box thickness of 5 Å, after considering reflective boundary conditions (walls exert a repulsive force to reflect atoms when they bump into the walls), corresponds to a single layer of water. This is because the height of the tetrahedron formed by the water molecule with its closest neighbors, to which it is connected by hydrogen bonds, is about 4 Å. Similarly, a simulation box thickness of 10 Å corresponds to two layers of water particles. The two parameters: the width and length of the simulation box equal to 30 Å and the cut-off distance equal to 10 Å are sufficient to ignore the influence of the box walls on the bulk water.

Figure 7.3 shows how water molecules are arranged in a space practically limited to two dimensions. For both sizes of the simulation box, two snapshots were recorded. One is a dynamic picture, i.e. from the course of the simulation. The second shows the arrangement of water molecules after energy minimization

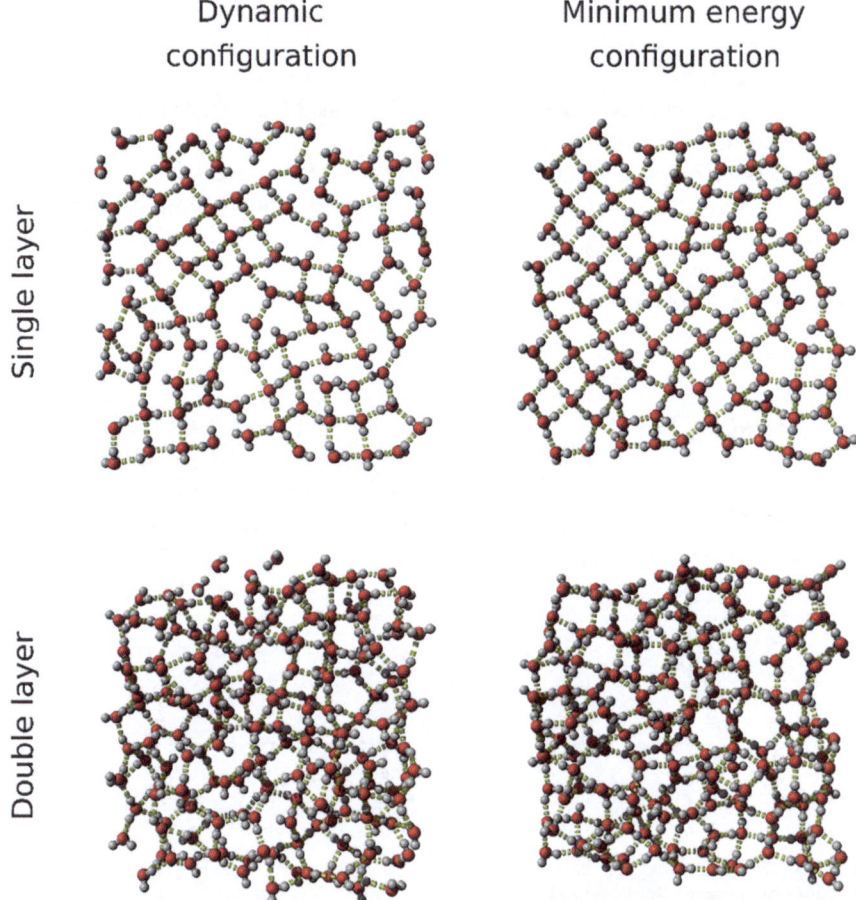

Fig. 7.3 Simulation snapshots showing single and double layer of water in a dynamic situation and after minimizing energy

procedure. The dynamic image shows that even though the water molecules are in constant motion, certain polygonal structures are visible. This is especially noticeable for a single layer of water. Quadrangles dominate but pentagons are also visible. In the case of a double layer, the distribution of the particles is practically chaotic. In both cases not all possible hydrogen bonds have been produced.

When looking at the arrangement of molecules after the energy minimization the order is clearly seen. For a single layer almost all possible hydrogen bonds have been produced (water molecules have reached the maximum coordination number for this system). A lot of quadrangles and a few pentagons are seen. The distances between the particles are almost identical. In turn, hexagons appear in the case of a double layer. Hexagonal structures are characteristic of the structure of ice (Satarifard et al. 2017; Liu et al. 1996). This is evidenced by the split of the second peak in radial distribution function, Fig. 7.4.

Looking at the RDF for both dynamic and static cases, Fig. 7.4, for both sizes of the simulation box the first peaks for the static systems overlap. The second peaks, and further, are separated. Between them, there is a clear minimum that is not visible for the dynamic system and for bulk water, cf. Figure 7.2. Clearly separated peaks indicate an orderly structure. It is also clear that the first peaks for the dynamic case are much lower than for the static case.

The presented system can be identified with the surface layer of biological membranes. Spectroscopic data indicate that water has distinct ordered behavior near membrane surfaces (Cheng et al. 2003; Fukuma et al. 2007). Molecular dynamics simulations show that water confined between membrane interfaces differs substantially from bulk solvent (Kasson et al. 2011).

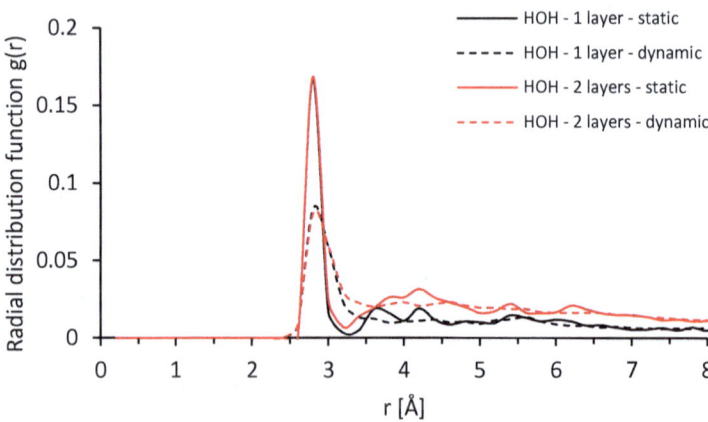

Fig. 7.4 The oxygen-oxygen radial distribution function of single and double layer of water in its dynamic and energy minimized case

5   Organisation of Water around Ions

As it is well known, water is a polar molecule. Water can dissolve many organic and inorganic substances that have polar regions. The strong polarity of water can interrupt ionic bonds and dissociate salts (Atkins et al. 2018). In water molecule the bond angle between the hydrogen atoms (H-O-H angle) is 104.5°. The hydrogen atoms are close to two corners of a tetrahedron centered on the oxygen. At the other two corners are lone pairs of valence electrons that do not participate in the bonding. As a result, there is slightly more negative charge near the oxygen atom in a covalent bond between hydrogen and oxygen. When electrons are not shared equally the result is a polar covalent bond (Atkins et al. 2018). On the water molecule, the electrons on the oxygen atom bend the shape of the molecule. As a result, there is a slightly more positive pole and a slightly more negative pole. The molecule has a net dipole, meaning that it is polar overall (Eisenberg and Kauzmann 2005).

Because of its polarity, a molecule of water in the liquid or solid state can form up to four hydrogen bonds with neighboring molecules (Fig. 7.5). Hydrogen bonds are about ten times as strong as the Van der Waals force that attracts molecules to each other in most liquids (Berg et al. 2002).

Water is the most important component of life on the Earth. It is necessary for both plants and animals. Water usually does not appear in its purest form. Because it is the basic component of body fluids, there are many ingredients dissolved in it. They are nutrients and minerals. Sodium and chlorine ions are the most common microelements found in body fluids (Waterhouse and Farmery 2012). The result of their occurrence is, among others, the salty taste of sweat or tears. Normal serum

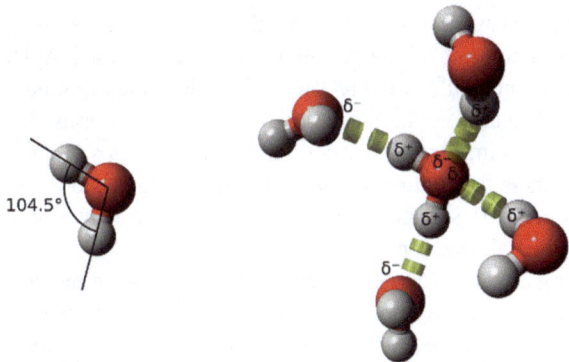

Fig. 7.5 Molecular geometry of water molecule (left) and hydrogen bonds between molecules of water (right). As a result of two covalent bonds (between oxygen and hydrogen atoms of the same molecule) and two hydrogen bonds (between neighboring molecules), the geometry around each oxygen atom is approximately tetrahedral

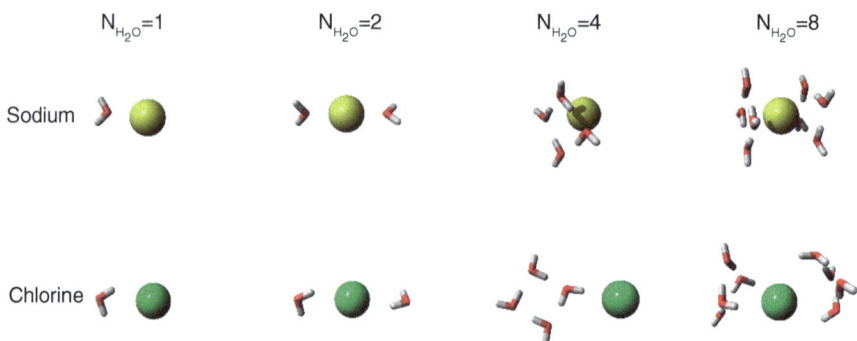

Fig. 7.6 Rearrangement of water molecules around Na^+ and Cl^- ions

sodium concentration is 135–145 mmol/l and chlorine concentration is 95–105 mmol/l. To check how water behaves in the presence of ions, whether it tends to form some structures with itself, or if with ion participation, molecular dynamics simulations were performed. To avoid interactions between electrically charged particles, single Na^+ and Cl^- ions were placed in a $50 \times 50 \times 50$ Å simulation box filled with water molecules at 298 K and 1 bar pressure. To avoid the interaction of ions, both cases were simulated separately, i.e. the first system contained water plus sodium ions and the second system contained water plus chlorine ions.

In Fig. 7.6 it is shown how water molecules are organized around Na^+ and Cl^- ions. The structures presented here are the result of a process of energy minimization. The results for different amounts of N_{H_2O} water molecules are shown in Fig. 7.6.

For a single water molecule, the results are obvious. As might be expected for a sodium ion that has a positive charge, water molecule rotates its oxygen atom towards the ion. However, in the case of a chlorine ion, which has a negative charge, water molecule rotates one of its hydrogen atoms towards the ion. This one example shows that a water molecule is an electric dipole with a clearly separated charge. For $N_{H_2O} = 2$ water molecules, which are initially located on opposite sides of the ion, after the process of minimizing energy remain in their places. As with a single water molecule case, their orientation is the same, i.e. oxygen atoms are turned to the sodium ion, and in the second case, water molecule rotates one of its hydrogen atoms towards the chlorine ion.

An interesting effect have been observed for $N_{H_2O} = 4$. In the case of the sodium ion, the water molecules have positioned themselves in the corners of a square (with improper dihedral angle equal zero, i.e. the square is flat) with a side length of 2.695 Å (distance measured between oxygen atoms), see Fig. 7.7 (left). In addition, the formed structure was located at the side of the ion but symmetrically to it (the vector from the ion to the center of the square is perpendicular to the tetramer plane). This is indicated by the distances of oxygen atoms to the ion, which are equal, see Fig. 7.3 (central).

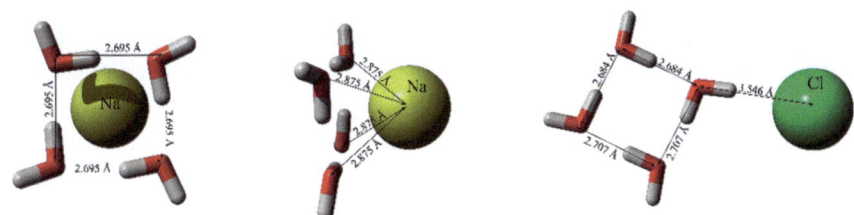

Fig. 7.7 Structures formed by four water molecules near the sodium ion (left – front view and central – side view) and chlorine ion. The distances between oxygen atoms in water molecules and between oxygen atoms and ions are shown

The situation is different for the chlorine ion. Here, water molecules also form a tetramer with an improper dihedral angel equal zero, but this time it is set sideways to the ion (the vector connecting the ion and the center of the tetramer is parallel to it). The closest to the chlorine ion is one of the hydrogen atoms. For the nearest to ion water molecule the vector connecting the oxygen atom with the hydrogen atom rotated toward ion is parallel to the vector connecting the ion and the oxygen atom of this water molecule, see Fig. 7.7 (right). The distance between the ion and the nearest water molecule (exactly to the oxygen atom) is greater than for the sodium ion.

For $N_{H_2O} = 8$ the differences between the structures formed by water molecules vary considerably. For the sodium ion, the water molecules formed two flat tetrameters located symmetrically on both sides of the ion. However, around the chlorine ion, water molecules also formed two tetrameters, except that they were not flat and the distances between the molecules forming them were not equal. In addition, it was observed that water molecules tended to push the chlorine ion out of the water shell.

To ensure that water molecules around the ions form orderly structures radial distribution function was generated, see Fig. 7.8. As was shown earlier, the water molecules surrounding the chlorine ion are situated in the way that hydrogen atoms are turned towards the ion (Hribar et al. 2002). In addition, RDF analysis indicates that water molecules are at a greater distance from the ion than is the case with the sodium ion. As mentioned before, the area under the curve indicates the number of the nearest neighbors (coordination number). As can be seen, in both cases the height and width of the first peak is identical. So, the first-order hydration shell for both types of ion are the same, i.e. they are made of the same number of water molecules. In the case of chlorine, a major peak at 6.2 Å may suggest the formation of a second-order hydration shell. The remaining fluctuations of the RDF curve are not significant.

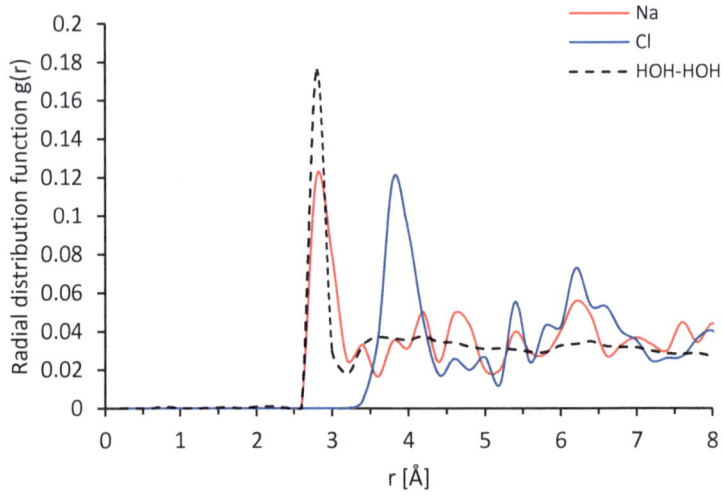

Fig. 7.8 Radial distribution function of pair ion-water (in particular oxygen atoms) obtained for Na⁺ and Cl⁻ ions. RDF of oxygen-oxygen of water is shown for comparison

6   Hydration of Amino Acids

Amino acids are commonly found in animal organisms, plants, and microorganisms. They are components of proteins and peptides. Amino acids consist of a basic amino group (-NH₂), an acidic carboxyl group (-COOH), and an organic R group (or side chain) that is unique to each amino acid (Nelson and Cox 2005). Each molecule contains a central carbon (C_α) atom, called the α-carbon, to which both an amino and a carboxyl group are attached. The remaining two bonds of the α-carbon atom are used by a hydrogen (H) atom and the R group. The formula of a general amino acid is shown of Fig. 7.9. The amino acids differ from each other in the chemical structure of the R group.

One of the most useful manners by which to classify the standard (or common) amino acids is based on the polarity (that is, the distribution of electric charge) of the R group (side chain). Four groups are distinguished (Reddy 2019).

Group I: Nonpolar amino acids:

Glycine, alanine, valine, leucine, isoleucine, proline, phenylalanine, methionine, and tryptophan. The R groups of these amino acids have either aliphatic or aromatic groups. This makes them hydrophobic.

Group II: Polar, uncharged amino acids:

Serine, cysteine, threonine, tyrosine, asparagine, and glutamine. The side chains in this group possess a spectrum of functional groups. However, most have at least one atom (nitrogen, oxygen, or sulfur) with electron pairs available for hydrogen bonding to water and other molecules.

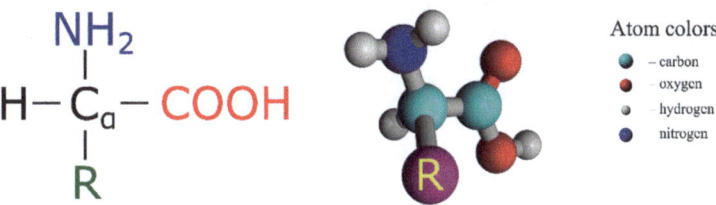

NH$_2$

H—C$_\alpha$— COOH

R

Atom colors
- – carbon
- – oxygen
- – hydrogen
- – nitrogen

Fig. 7.9 A template structure of an amino acids. Amino group (-NH$_2$), carboxyl group (-COOH), side chain (R) and hydrogen atom are bonded to the central carbon atom (C$_\alpha$)

Group III: Acidic amino acids:

Aspartic acid and glutamic acid. Each has a carboxylic acid on its side chain that gives it acidic (proton-donating) properties. In an aqueous solution at physiological pH, all three functional groups on these amino acids will ionize, thus giving an overall charge of −1.

Group IV: Basic amino acids:

Arginine, histidine, and lysine. Each side chain is basic (i.e., can accept a proton). Lysine and arginine both exist with an overall charge of +1 at physiological pH.

For the purpose of this work, the amino acids were divided into three groups according to their hydrophobic properties. To do this the Kyte-Doolittle scale was used (Kyte and Doolittle 1982). This scale is widely used for detecting hydrophobic regions in proteins. Regions with a positive value are hydrophobic. List of amino acids and their abbreviations (3-letter and 1-letter notation) and the value of their hydrophobicity on the Kyte-Doolittle scale is presented in Table 7.1.

Eight amino acids were selected for further consideration: isoleucine, leucine, phenylalanine, glycine, serine, tyrosine, arginine, lysine. The selection criterion was hydrophobicity (extremely hydrophobic, neutral and extremely hydrophilic) and type of side group (simple and aromatic ring), see Fig. 7.10.

Amino acids are the building blocks of proteins (Kessel and Ben-Tal 2010). They differ in size and shape. Many experimental studies and computer simulations indicate the existence of hydration shell which surrounds proteins (Fogarty and Laage 2014; Brovchenko and Oleinikova 2008; Svergun et al. 1998; Ebbinghaus et al. 2007; Frauenfelder et al. 2009). It may be a single layer of ordered molecules of water, but there are cases of shells with a much more complex structure. The source of this phenomenon are amino acids that form protein. It is an effect of amphiphilic properties of individual amino acids.

To investigate the phenomenon of formation of the water shell around amino acids, molecular dynamics simulations of systems containing the selected amino acids in water were performed. The following setup was used for this purpose. The 50 × 50 × 50 Å simulation box was filled with water at a temperature of 298 K and then the selected amino acid was placed in it. The pressure was set at 1 bar. During the simulation lasting 1 ns, one hundred of radial distribution functions of water molecules were recorded. In contrast to the RDF shown in Fig. 7.2, where the pair

Table 7.1 List of amino acids and their abbreviations (3-letter and 1-letter notation) and the value of their hydrophobicity on the Kyte-Doolittle scale (Kyte and Doolittle 1982)

Amino acid	Three-letter symbol	One-letter symbol	Hydrophobicity on the Kyte-Doolittle scale	Hydropathy class
Isoleucine	ILE	I	4.5	Hydrophobic
Valine	VAL	V	4.2	
Leucine	LEU	L	3.8	
Phenylalanine	PHE	F	2.8	
Cysteine	CYS	C	2.5	
Methionine	MET	M	1.9	
Alanine	ALA	A	1.8	
Tryptophan	TRP	W	−0.9	Neutral
Glycine	GLY	G	−0.4	
Threonine	THR	T	−0.7	
Serine	SER	S	−0.8	
Tyrosine	TYR	Y	−1.3	
Proline	PRO	P	−1.6	
Histidine	HIS	H	−3.2	
Asparagine	ASN	N	−3.5	Hydrophilic
Aspartic acid	ASP	D	−3.5	
Glutamine	GLN	Q	−3.5	
Glutamic acid	GLU	E	−3.5	
Lysine	LYS	K	−3.9	
Arginine	ARG	R	−4.5	

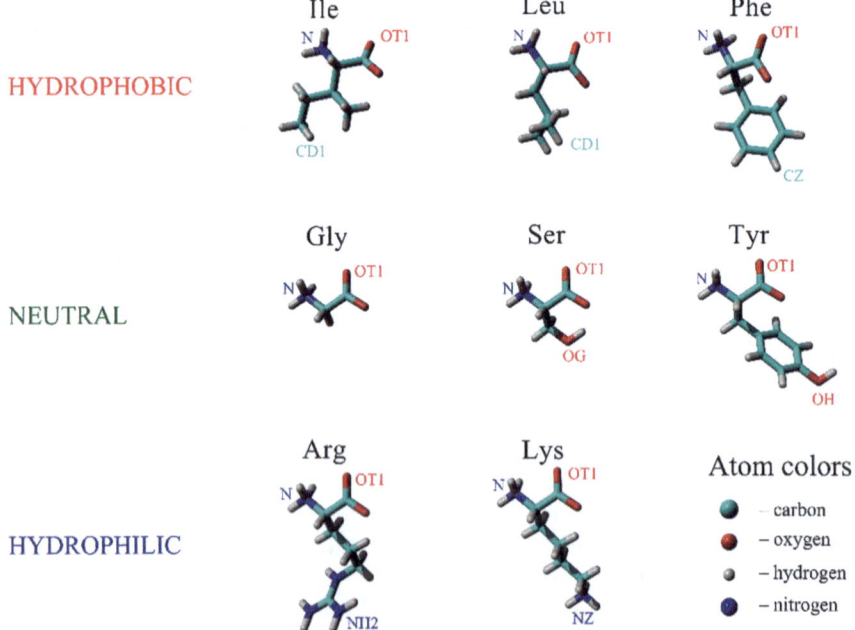

Fig. 7.10 Selected amino acids after consideration of hydrophobicity and type of side chain. The most exposed to external environment atoms are marked with color letters

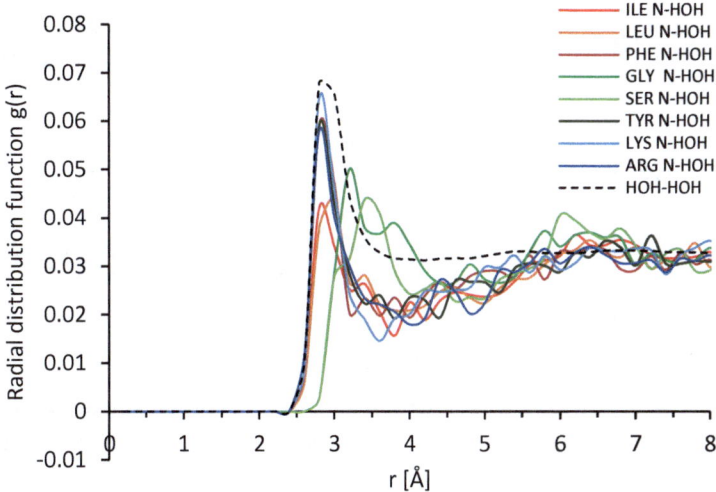

Fig. 7.11 Radial distribution function of water for a water-nitrogen pair (N-HOH) for all selected amino acids. To locate the nitrogen atom (N) see Fig. 7.10

consisted of water molecules, here the pair consisted of selected atom of amino acid (as indicated in Fig. 7.10) and water molecules. For all amino acids RDF was generated for the following pairs: N-HON (Fig. 7.11) and OT1-HOH (Fig. 7.12), and depending on the amino acid type, the following pairs: CD1-HOH, CZ-HOH, OG-HOH, OH-HOH, NH2-HOH, NZ-HOH (Fig. 7.13). Here HOH means water molecule. By the water molecule, we mean only the oxygen atom in the water molecule.

Figure 7.11 shows the radial distribution function of water for a water-nitrogen pair (to locate the nitrogen atom – N see Fig. 7.10) for all selected amino acids. The position of the first peak for hydrophobic and hydrophilic amino acids is the same. Its position coincides with the position of the first water-water pair peak. Only for neutral amino acids, GLY and SER, the first peak is slightly shifted towards greater distances. The height of the first peak for hydrophilic amino acids is like the height of the first peak for water-water pair. For hydrophobic amino acids, except for PHE, the height of the first peak is much lower than for a water-only system. The same is true for neutral amino acids, except for TYR. Both PHE and TYR have a side chain in the form of an aromatic ring. In both cases, the height of the first peak is like that of the characteristic hydrophilic amino acids.

Figure 7.12 shows the radial distribution function of water for a water-oxygen pair (to locate the oxygen atom – OT1 see Fig. 7.10). The position of the first peak for amino acids is the same for all types of amino acids. This time the position of the first peak is shifted comparing to the first peak for water in the direction of greater distances. In addition, the peak is not very high and clearly visible, which indicates that the water molecules do not form any structures.

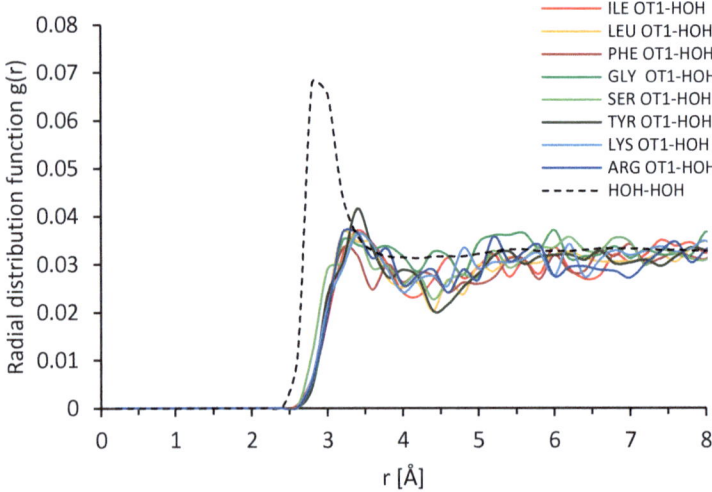

Fig. 7.12 Radial distribution function of water for a water-oxygen pair (OT1-HOH) for all selected amino acids. To locate the oxygen atom (OT1) see Fig. 7.10

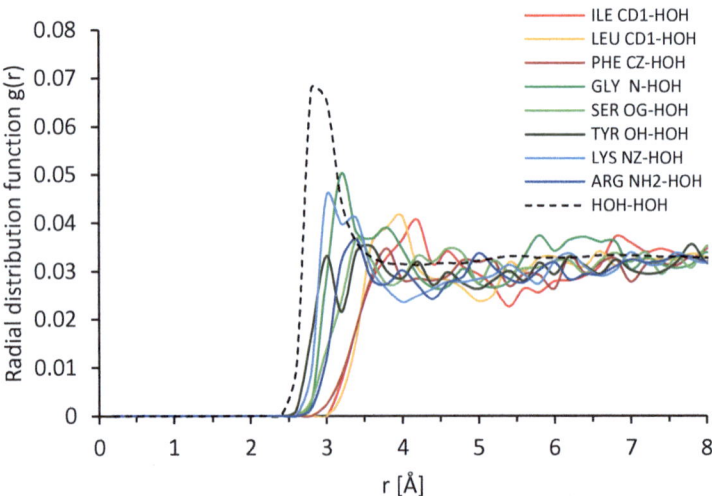

Fig. 7.13 Radial distribution function of water for a water-side chain atom (R-HOH) for all selected amino acids. To locate the appropriate side chain atom (R) see Fig. 7.10

Figure 7.13 shows the radial distribution function of water for the pair: water and side-chain atom (R) which is most exposed towards the water. The shift of the first peak towards greater distances for hydrophobic amino acids: ILE, LEU and PHE, is clearly visible. This proves that regardless of the type of side chain (straight or aromatic ring), R-group determine the hydrophobicity of the amino acid.

From the above it can be concluded that in the case of the isolated and neutral amino acid the side chain determines the hydrophobicity more strongly than the amino and carboxylic groups.

An interesting effect can be additionally noted. Analyzing RDF for the water-nitrogen pair (Fig. 7.11) it is seen that the first minimum that appears after the first peak is quite deep. It is much deeper than in the case of other analyzed pairs of atoms. It can be concluded that the first-degree hydration shell is clearly separated. Its thickness corresponds to a single layer of water molecules, whose diameter is about 2.75 Å. This minimum indicates the existence of a depleted zone (the average water density in the depleted zone is less than for bulk water). This could be caused by the fact that water at a hydrophobic surface loses a hydrogen bond, therefore has increased enthalpy. Water molecules compensate for this by doing pressure-volume work, that is, the network expands to form low-density water with lower entropy (Chalikian 2001). Then bulk water spreads out at a distance approximately of 6 Å. The above results are consistent with experimental observations based on crystallographic data (Biedermannová and Schneider 2015).

A single amino acid cannot generate a large/multilayer hydration shell. Protein, which is often made up of several hundred or even several thousand amino acids, has sufficient potential to create such a shell around itself for which first hydration shell with an average density ~ 10% larger than that of the bulk solvent (Svergun et al. 1998).

7   Hydration of Hyaluronic Acid

Hyaluronic acid (HA), also called hyaluronan, is a polyanionic natural polymer occurring as linear polysaccharide composed of D-glucuronic acid and N-acetyl-D-glucosamine, see Fig. 7.14. Hyaluronic acid can be 25,000 disaccharide repeats in length. Polymers of hyaluronic acid can range in size from 5 kDa to 20 MDa *in vivo* (Andrysiak et al. 2018). The average molecular weight in human synovial fluid is 3–4 MDa, and hyaluronic acid purified from human umbilical cord is 3.14 MDa. Hyaluronic acid binds cells together and helps to lubricate joints (Siódmiak et al. 2017). As one of the chief components of the extracellular matrix (in aqueous solutions HA forms specific stable tertiary structures), hyaluronan contributes significantly to cell proliferation and migration, and may also be involved in the progression of some malignant tumors (Atkins et al. 1980; Hari and Hales 2008). Moreover, HA has unique capacity in retaining water. HA can bind up to 1000 times more water than the weight of the macromolecule itself. The water binding and hydration properties of HA provide water to the skin preventing skin aging. This moisturizing effect is widely used in the cosmetic industry (Papakonstantinou et al. 2012). Skin HA accounts for most of 50% of total body HA. The average 70 kg person has roughly 15 grams of hyaluronan in the body, one-third of which is turned over (degraded and synthesized) every day (Stern 2004).

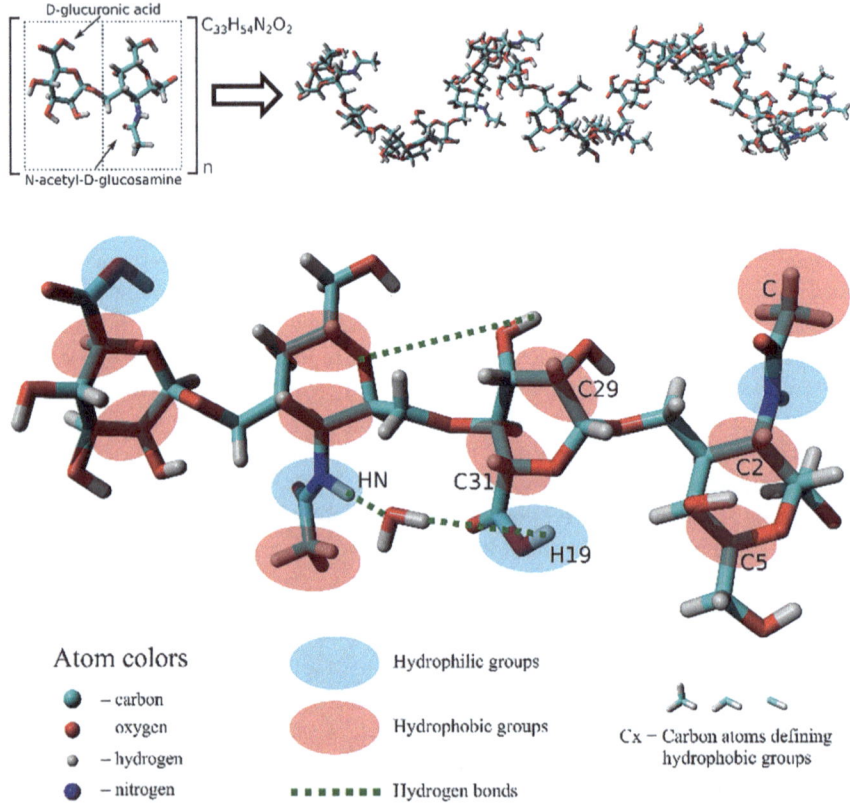

Fig. 7.14 Chemical structures of HA disaccharide unit and the polymeric structure (built of ten mers) used in the simulation (top). Hyaluronan tetrasaccharide unit (middle) where the hydrophilic and the hydrophobic functional groups, and hydrogen bonds leading to conformational rigidity are depicted. Three groups of non-polar atoms (bottom). The atoms for which RDF was gained are marked: carbon – C, C2, C5, C29, C31, hydrogen – H19, HN

In Fig. 7.14, regions that have amphiphilic properties are marked with blue ellipses. These areas are important from the point of view of the presented issue. As can be seen, the hydrophobic areas containing Cx-labeled carbon atoms are mainly located along the hyaluronan chain. The exception is the highly hydrophobic CH_3 group. The Cx atoms are "hidden" and are not exposed to the outside, usually to water. In turn, the most hydrophilic fragments of the macromolecule are located on its outskirts and are strongly exposed to water. The green dotted lines represent intra-molecular hydrogen bonds resulting in a rigid conformation. Water molecule shown on Fig. 7.14 links HA carboxyl and acetamido groups with hydrogen bonds that stabilize the secondary structure of the biopolymer, described as a single-strand left-handed helix with two disaccharide residues per turn (twofold helix) (Fallacara et al. 2018). Three groups of non-polar atoms, labeled Cx, are also shown and are marked with red ellipses. Groups of atoms containing Cx carbon atoms are divided

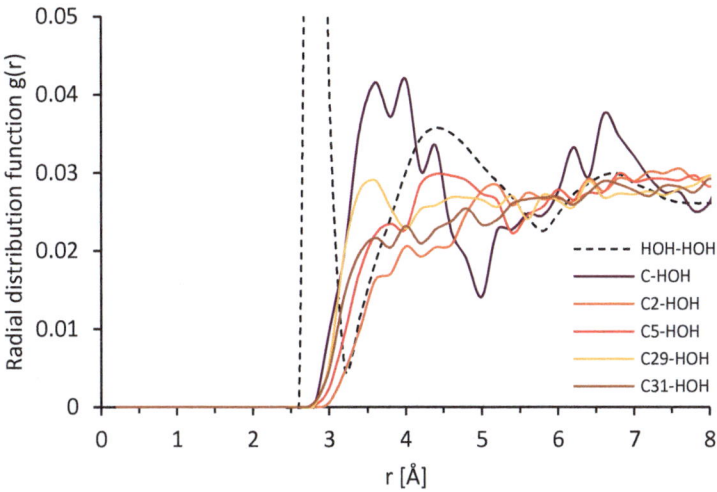

Fig. 7.15 Radial distribution function obtained for the following pairs of hydrophobic atoms: (C, C2, C5, C29, C31) carbon-water and bulk water for comparison. For the abbreviations of atomic names see Fig. 7.10. The peak corresponding to bulk water has been cut off for better visualization of the remaining curves

into three types. In group one, the carbon atom is bonded with three hydrogen atoms, in group two with two hydrogen atoms and in group three only with one. The first group has the highest hydrophobicity, while the third group has the smallest (Tanford 1973; Adamson and Gast 1997). Carbonaceous groups (CH, CH_2, CH_3) are hydrophobic because they are non-polar and thus do not attract water strongly. Moreover, they are polarizable and thus damp nearby water fluctuations (Scott and Ridgway 2017; Stirnemann et al. 2013; Zhang et al. 2017).

Figure 7.15 shows the radial distribution function obtained for the following pairs of hydrophobic atoms: (C, C2, C5, C29, C31) carbon-water and bulk water for comparison. It is clearly seen that for all curves corresponding to hydrophobic atoms the first maximum is shifted towards greater distances compared to clean water. This is particularly visible for the carbon atom C, or more precisely for the CH_3 group, which has strong hydrophobic properties. The triple structure of the first peak is because there are three hydrogen atoms in the CH_3 group. In addition, there is a second strong maximum, followed by a deep minimum (depletion zone). This may indicate the formation of a two-layer hydration shell around CH_3 group. For other hydrophobic groups, this effect is not very pronounced because these groups are slightly hidden and consequently less exposed towards the water.

In Fig. 7.16 it is shown the radial distribution function obtained for the following pairs of hydrophilic atoms: (H19, HN) hydrogen-water and bulk water for comparison.

The significant shift of the first peaks towards smaller distances in comparison to clean water confirms the hydrophilicity of the areas marked with blue ellipses in Fig. 7.14. In the case of hydrogen H19, a deep minimum behind the first peak is

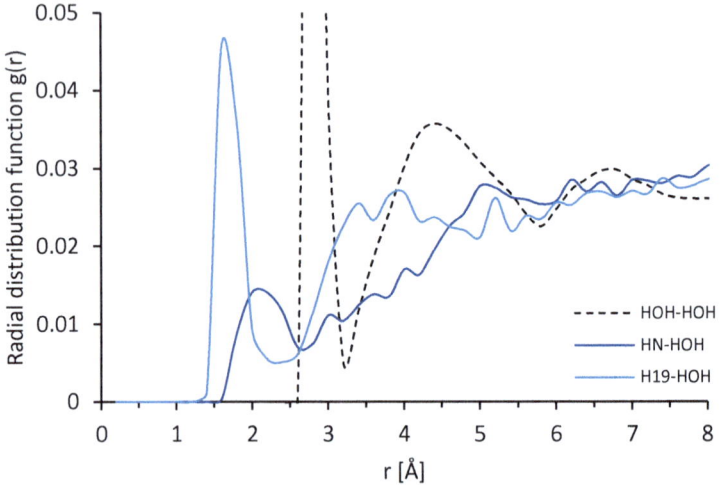

Fig. 7.16 Radial distribution function obtained for the following hydrophilic pairs of atoms: (HN, H19) hydrogen-water and bulk water for comparison. For the abbreviations of atomic names see Fig. 7.10. The peak corresponding to bulk water has been cut off for better visualization of the remaining curves

visible. Hydrogen H19 is bonded to water and together with hydrogen HN contributes to the stiffening of the hyaluronic acid structure. Kaufmann et al. carried out similar studies but using the GROMOS force field (Kaufmann et al. 1998). They observed similar effects. In addition, they studied chain dynamics by analyzing torsional angles in the β(1,4) dimmer of HA.

8 Summary

According to the results of experimental studies as well as the results of computer simulations, it is obvious that water is not an unordered set of particles. Under favorable conditions, bulk water is a uniform, continuous, tetrahedral liquid. The situation is much more complex when different compounds are dissolved in water. Whether they are simple salts or complex biomolecules (e.g. proteins), water interacts with them and form complex structures called hydration shells. Depending on the size of the dissolved molecules, these shells can be either simple two-layer, like in case of ions (Lee et al. 2017), but they can also be complex and multilayer as probably in the case of proteins (Persson et al. 2018). Both in experimental work and in computer simulations, radial distribution function is used to determine whether we are dealing with an ordered structure. In the case of data from the experiment the radial distribution function does not pose any problems of interpretation. The problems are with results obtained from computer simulations. If we want to make radial distribution function corresponding to reality, the orientation radial distribution function must be applied (Li et al. 2014). The reason for this is that if we

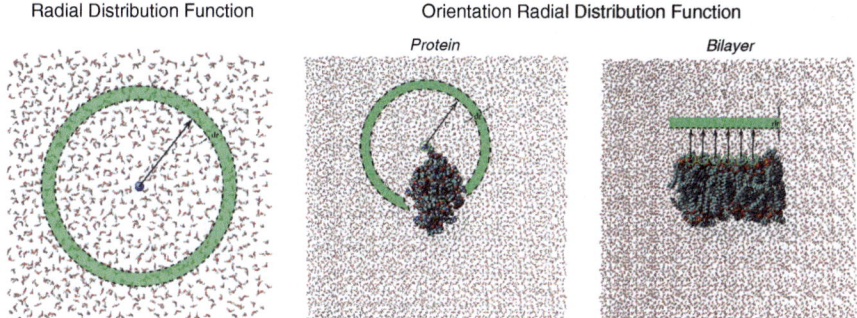

Fig. 7.17 Construction of the radial distribution function appropriate for small molecules (e.g. solvent) vs. orientation radial distribution function appropriate for the solvent molecules in the vicinity of macromolecule, bilayer and/or membrane

choose the atom or chemical group around which we want to check the distribution of solvent molecules, the remaining atoms of the molecule obscure the solvent molecules on the opposite side of the macromolecule, see Fig. 7.17.

Structural fluctuations and molecular excitations of hydrating water molecules cover a broad range in space and time. These include the scope from individual water molecules to larger pools and from femtosecond (OH stretching and bending vibrations) to microsecond (mainly governed by the dynamics of the biomolecule, inducing structural changes and energy exchange with the water shell) time scales. Recent progress in theory and molecular dynamics simulations as well as in ultrafast vibrational spectroscopy has led to new and detailed insight into fluctuations of water structure, elementary water motions, electric fields at hydrated biointerfaces, and processes of vibrational relaxation and energy dissipation (Laage et al. 2017). The development of these experimental techniques allows us to observe these subtle effects.

Water is generally an incompressible liquid. Due to the formation of hydration shells around a given biomolecule, the distribution of water density is not uniform. Nevertheless, the system remains in a state of equilibrium. The non-uniform density distribution is indicated by the peaks and minima visible in radial distribution function. From the point of view of fluid mechanics, under the influence of external forces, an exchange of matter can take place between regions of different density. The regions of reduced density will behave like a compressible liquid under the influence of external pressure, and then the hydration shell, especially multilayered, will act as a micro shock absorber.

The number of amphiphilic molecules in the synovial fluid is very large (Andrysiak et al. 2018). As already mentioned, hyaluronic acid can bind a large amount of water. In addition, the synovial fluid contains a large amount of proteins, approx. 25 mg/ml (Bennike et al. 2014). If a hydration shell is formed around each of the macromolecules (the larger molecule, then more complex the shell is), then the ability of such liquid to absorb mechanical vibrations caused by changing joint loads (e.g. while walking or running) is significant.

References

Adamson, A. W., & Gast, A. P. (1997). *Physical chemistry of surfaces*. New York: Wiley.
Allen, M. P., & Tildesley, D. J. (2017). *Computer simulation of liquids*. Oxford: Oxford University Press.
Andrysiak, T., Bełdowski, P., Siódmiak, J., Weber, P., & Ledziński, D. (2018). Hyaluronan-chondroitin sulfate anomalous crosslinking due to temperature changes. *Polymers, 10*(5), 560–511.
Atkins, E. D. T., Meader, D., & Scott, J. E. (1980). Model for hyaluronic acid incorporating four intramolecular hydrogen bonds. *International Journal of Biological Macromolecules, 2*(5), 318–319.
Atkins, P., de Paula, J., & Keeler, J. (2018). *Physical chemistry*. Oxford: Oxford University Press.
Bagchi, B. (2013). *Water in biological and chemical processes: From structure and dynamics to function*. Cambridge: Cambridge University Press.
Bennike, T., Ayturk, U., Haslauer, C. M., Froehlich, J. W., Proffen, B. L., Barnaby, O., Birkelund, S., Murray, M. M., Warman, M. L., Stensballe, A., & Steen, H. (2014). A normative study of the synovial fluid proteome from healthy porcine knee joints. *Journal of Proteome Research, 13*(10), 4377–4387.
Berg, J. M., Tymoczko, J. L., & Stryer, L. (2002). *Biochemistry*. New York: W.H. Freeman.
Biedermannová, L., & Schneider, B. (2015). Structure of the ordered hydration of amino acids in proteins: Analysis of crystal structures. *Acta Crystallographica Section D: Biological Crystallography, 71*(11), 2192–2202.
Brovchenko, I., & Oleinikova, A. (2008). Which properties of a spanning network of hydration water enable biological functions? *ChemPhysChem, 9*, 2695–2702.
Chalikian, T. V. (2001). Structural thermodynamics of hydration. *The Journal of Physical Chemistry. B, 105*, 12566–12578.
Chandler, D. (1987). *Introduction to modern statistical mechanics*. New York: Oxford University Press.
Chaplin, M. F. (2009). Structuring and behaviour of water in nanochannels and confined spaces. In L. in Dunne & G. Manos (Eds.), *Adsorption and phase behaviour in nanochannels and nanotubes* (pp. 241–255). Dordrecht: Springer.
Cheng, J. X., Pautot, S., Weitz, D. A., & Xie, X. S. (2003). Ordering of water molecules between phospholipid bilayers visualized by coherent anti-stokes Raman scattering microscopy. *Proceedings of the National Academy of Sciences, 100*(17), 9826–9830.
Clark, G. N. I., Cappa, C. D., Smith, J. D., Saykally, R. J., & Head-Gordon, T. (2010). The structure of ambient water. *Molecular Physics, 108*(11), 1415–1433.
Costa, L., Thaciana, M., Eudes, F., & Mauro, R. (2011). Molecular dynamics simulation of liquid trimethylphosphine. *The Journal of Chemical Physics, 135*, 064506–064507.
Ding, J., Xu, M., Guan, P. F., Deng, S. W., Cheng, Y. Q., & Ma, E. (2014). Temperature effects on atomic pair distribution functions of melts. *The Journal of Chemical Physics, 140*, 064501–064508.
Eakins, B. W., & Sharman, G. F. (2010). Volumes of the world's oceans from ETOPO1, NOAA. *National Geophysical Data Center.* https://www.ngdc.noaa.gov. Accessed 20 July 2020.
Ebbinghaus, S., Kim, S. J., Heyden, M., Yu, X., Heugen, U., Gruebele, M., Leitner, D. M., & Havenith, M. (2007). An extended dynamical hydration shell around proteins. *Proceedings of the National Academy of Sciences, 104*(52), 20749–20752.
Eisenberg, D., & Kauzmann, W. (2005). *The structure and properties of water*. Oxford/New York: Oxford University Press.
Fallacara, A., Baldini, E., Manfredini, S., & Silvia, V. (2018). Hyaluronic acid in the third millennium. *Polymers, 10*, 701–736.
Fogarty, A. C., & Laage, D. (2014). Water dynamics in protein hydration shells: The molecular origins of the dynamical perturbation. *The Journal of Physical Chemistry. B, 118*(28), 7715–7729.

Franchetti, S. (1975). Radial distribution functions in solid and liquid argon. *Nuovo Cimento B Serie, 26*, 507–521.

Frauenfelder, H., Chen, G., Berendzen, J., Fenimore, P., Jansson, H., McMahon, B., Stroe, I., Swenson, J., & Young, R. (2009). A unified model of protein dynamics. *Proceedings of the National Academy of Sciences, 106*(13), 5129–5134.

Frenkel, D., & Smit, B. (2002). *Understanding molecular simulation: From algorithms to applications*. San Diego: Academic.

Fukuma, T., Higgins, M. J., & Jarvis, S. P. (2007). Direct imaging of individual intrinsic hydration layers on lipid bilayers at Ångstrom resolution. *Biophysical Journal, 92*(10), 3603–3609.

Hari, G. G., & Hales, C. A. (2008). *Chemistry and biology of Hyaluronan*. Amsterdam: Elsevier Science.

Harrach, M., & Drossel, B. (2014). Structure and dynamics of TIP3P, TIP4P, and TIP5P water near smooth and atomistic walls of different hydroaffinity. *The Journal of Chemical Physics, 140*, 174501–174514.

Head-Gordon, T., & Hura, G. (2002). Water structure from scattering experiments and simulation. *Chemical Reviews, 102*(8), 2651–2670.

Henao, A., Busch, S., Guàrdia, E., Tamarit, J. L., & Pardo, L. C. (2016). The structure of liquid water beyond the first hydration shell. *Physical Chemistry Chemical Physics, 18*, 19420–19425.

Hribar, B., Southall, N. T., Vlachy, V., & Dill, K. A. (2002). How ions affect the structure of water. *Journal of the American Chemical Society, 124*(41), 12302–12311.

Jorgensen, W. L. (1981). Transferable intermolecular potential functions for water, alcohols, and ethers. Application to liquid water. *Journal of the American Chemical Society, 103*(2), 335–340.

Jorgensen, W. L., Chandrasekhar, J., Madura, J. D., Impey, R. W., & Klein, M. L. (1983). Comparison of simple potential functions for simulating liquid water. *The Journal of Chemical Physics, 79*(2), 926–935.

Kasson, P. M., Lindahl, E., & Pande, V. S. (2011). Water ordering at membrane interfaces controls fusion dynamics. *Journal of the American Chemical Society, 133*(11), 3812–3815.

Kaufmann, J., Möhle, K., Hofmann, H.-J., & Arnold, K. (1998). Molecular dynamics study of hyaluronic acid in water. *Journal of Molecular Structure, 422*(1–3), 109–121.

Kessel, A., & Ben-Tal, N. (2010). *Introduction to proteins: Structure, function, and motion*. Boca Raton: CRC Press.

Krieger, E., & Vriend, G. (2015). New ways to boost molecular dynamics simulations. *Journal of Computational Chemistry, 36*, 996–1007.

Krieger, E., Koraimann, G., & Vriend, G. (2002). Increasing the precision of comparative models with YASARA NOVA – A self-parameterizing force field. *Proteins, 47*(3), 393–402.

Kyte, J., & Doolittle, R. F. (1982). A simple method for displaying the hydropathic character of a protein. *Journal of Molecular Biology, 157*(1), 105–132.

Laage, D., Elsaesser, T., & Hynes, J. T. (2017). Water dynamics in the hydration shells of biomolecules. *Chemical Reviews, 117*(16), 10694–10725.

Lee, Y., Thirumalai, D., & Hyeon, C. (2017). Ultrasensitivity of water exchange kinetics to the size of metal ion. *Journal of the American Chemical Society, 139*(36), 12334–12337.

Levine, B. G., Stone, J. E., & Kohlmeyer, A. (2011). Fast analysis of molecular dynamics trajectories with graphics processing units-radial distribution function histogramming. *Journal of Computational Physics, 230*(9), 3556–3569.

Li, C., Burney, K., Bergler, K., & Wang, X. (2014). Structural evolution of nanoparticles under picosecond stress wave consolidation. *Computational Materials Science, 95*, 74–83.

Liu, K., Cruzan, J. D., & Saykally, R. J. (1996). Water clusters. *Science, 271*(5251), 929–933.

Morgan, J., & Warren, B. E. (1938). X-ray analysis of the structure of water. *The Journal of Chemical Physics, 6*, 666–673.

Nelson, D. L., & Cox, M. M. (2005). *Lehninger principles of biochemistry*. New York: W.H. Freeman and Company.

Papakonstantinou, E., Roth, M., & Karakiulakis, G. (2012). Hyaluronic acid: A key molecule in skin aging. *Dermato-Endocrinology, 4*(3), 253–258.

Persson, F., Söderhjelm, P., & Halle, B. (2018). The geometry of protein hydration. *The Journal of Chemical Physics, 148*, 215101-21.

Reddy, M. K. (2019). Amino acid. *Encyclopedia Britannica.* https://www.britannica.com/science/amino-acid. Accessed 20 July 2020.

Satarifard, V., Mousaei, M., Haddadi, F., Dix, J., Fernandez, M., Carbone, P., Beheshtian, J., Peeters, F., & Neek-Amal, M. (2017). Reversible structural transition in nanoconfined ice. *Physical Review B, 95*, 064105–064108.

Scott, L., & Ridgway, F. A. (2017). *A mathematical approach to protein biophysics.* Cham: Springer.

Siódmiak, J., Bełdowski, P., Augé, W. K., II, Ledziński, D., Śmigiel, S., & Gadomski, A. (2017). Molecular dynamic analysis of hyaluronic acid and phospholipid interaction in tribological surgical adjuvant design for osteoarthritis. *Molecules, 22*(9), 1436–1457.

Stern, R. (2004). Hyaluronan catabolism: A new metabolic pathway. *European Journal of Cell Biology, 83*(7), 317–325.

Stirnemann, G., Wernersson, E., Jungwirth, P., & Laage, D. (2013). Mechanisms of acceleration and retardation of water dynamics by ions. *Journal of the American Chemical Society, 135*, 11824–11831.

Svergun, D. I., Richard, S., Koch, M. H. J., Sayers, Z., Kuprin, S., & Zaccai, G. (1998). Protein hydration in solution: Experimental observation by x-ray and neutron scattering. *Proceedings of the National Academy of Sciences, 95*(5), 2267–2272.

Tanford, C. (1973). *The hydrophobic effect.* New York: Wiley.

Thompson, W. H. (2018). Perspective: Dynamics of confined liquids. *The Journal of Chemical Physics, 149*, 170901–170914.

Van Vleet, M. J., Misquitta, A. J., Stone, A. J., & Schmidt, J. R. (2016). Beyond Born-Mayer: Improved models for short-range repulsion in ab initio force fields. *Journal of Chemical Theory and Computation, 12*(8), 3851–3870.

Wagner, W., Saul, A., & Pruss, A. (1994). International equations for the pressure along the melting and along the sublimation curve of ordinary water substance. *Journal of Physical and Chemical Reference Data, 23*(3), 515–527.

Waterhouse, B. R., & Farmery, A. D. (2012). The organization and composition of body fluids. *Anaesthesia and Intensive Care Medicine, 13*(12), 603–608.

Wernet, P., Nordlund, D., Bergmann, U., Cavalleri, M., Odelius, M., Ogasawara, H., Näslund, L. Å., Hirsch, T. K., Ojamäe, L., Glatzel, P., Pettersson, L. G. M., & Nilsson, A. (2004). The structure of the first coordination shell in liquid water. *Science, 304*(5673), 995–999.

Yoon, B. J., Jhon, M. S., & Eyring, H. (1981). Radial distribution function of liquid argon according to significant structure theory. *Proceedings of the National Academy of Sciences, 78*(11), 6588–6591.

Zhang, L., Wang, L., Kao, Y. T., Qiu, W., Yang, Y., Okobiah, O., & Zhong, D. (2007). Mapping hydration dynamics around a protein surface. *Proceedings of the National Academy of Sciences, 104*(47), 18461–18466.

Zhang, Q., Wu, T. M., Chen, C., Mukamel, S., & Zhuang, W. (2017). Molecular mechanism of water reorientational slowing down in concentrated ionic solutions. *Proceedings of the National Academy of Sciences, 114*, 10023–10028.

Chapter 8
Pathological Water Science – Four Examples and What They Have in Common

Daniel C. Elton and Peter D. Spencer

Abstract Pathological science occurs when well-intentioned scientists spend extended time and resources studying a phenomena that isn't real. Researchers who get caught up in pathological science are usually following the scientific method and performing careful experiments, but they get tricked by nature. The study of water has had several protracted episodes of pathological science, a few of which are still ongoing. We discuss four areas of pathological water science – polywater, the Mpemba effect, Pollack's "fourth phase" of water, and the effects of static magnetic fields on water. Some common water-specific issues emerge such as the contamination and confounding of experiments with dissolved solutes and nanobubbles. General issues also emerge such as imprecision in defining what is being studied, bias towards confirmation rather than falsification, and poor standards for reproducibility. We hope this work helps researchers avoid wasting valuable time and resources pursuing pathological science.

Keywords Basics of water · Pathological water science · Polywater · "Fourth" phase · Static magnetic field effect on water

D. C. Elton
Radiology and Imaging Sciences, National Institutes of Health Clinical Center, Bethesda, MD, USA
e-mail: daniel.elton@nih.gov

P. D. Spencer (✉)
School of Biomedical Sciences, Faculty of Health, Institute of Health and Biomedical Innovation, Queensland University of Technology (QUT), Brisbane, Australia
e-mail: p.spencer@hdr.qut.edu.au

1   Introduction to Pathological Science

In 1953 the Nobel prize winning chemist Irving Langmuir gave a talk on pathological science, which he referred to as the "the science of things that aren't so" (Langmuir and Hall 1989). Langmuir had observed several cases (often firsthand) where scientists were tricked into believing in a phenomena, often for years or decades. Eventually it was found the purported phenomena was actually caused by confounding factors in an experiment or faulty methods of data analysis. Some of the examples that Langmuir discussed are N-rays, mitogenic rays, and extrasensory perception. Some prominent examples since 1954 are polywater, cold fusion, and magnet therapy. Many other mini-episodes of pathological science can be found in the psychological and social sciences, which are currently undergoing a major reproducibility crisis. We believe the scientific community needs to get better at detecting pathological science. The first reason for this view is the obvious one – pathological science wastes scientist's time and (usually) taxpayer money. The second reason is that properly sorting out the "wheat from the chaff" can be very hard both for other scientists and the public when great volumes of pathological science are being published. Intense competition for funding has led to rushed work and exaggerated or sensationalized findings. Increased pressure to publish has led to a proliferation of low tier journals with weaker standards of peer review. Together with the rise of preprint servers, scientists and the public now have to deal with a deluge of low-quality papers. Finally, we note that pathological science is often used to promote products which actually have no utility to the end user. This is especially a problem in the area of health-treatments because resources are sometimes misallocated away from treatments that would have actually helped the patient.

We wish to emphasize that pathological science is distinct from pseudoscience. While some pseudoscience may also be called pathological science, not all pathological science is pseudoscience. The reason not all pathological science is pseudoscience is that most researchers working on pathological science are trained career scientists who use the scientific method well. They simply are tricked! We also want to emphasize that those who have fallen prey to pathological science are generally well intentioned and often very bright and talented researchers. Even Nobel prize winners have fallen for pathological science – Brian Josephson (1973, Physics) and Luc Montagnier (2008, Physiology or Medicine) have both endorsed water memory as a real phenomenon.

The features of pathological science that Langmuir identified in his talk are:

1. "The maximum effect that is observed is produced by a causative agent of barely detectable intensity, and the magnitude of the effect is substantially independent of the intensity of the cause."
2. "The effect is of a magnitude that remains close to the limit of detectability; or, many measurements are necessary because of the very low statistical significance of the results."
3. "Claims of great accuracy."
4. "Fantastic theories contrary to experience."
5. "Criticisms are met by ad hoc excuses."

6. "Ratio of supporters to critics rises up to somewhere near 50% and then falls gradually to oblivion"

In this paper, we take a very broad view of what pathological science is. So, not every example of pathological science we discuss exhibits features 1–6. To us, pathological science is simply any area of science where nature tricks researchers into believing in a phenomenon for an extended period of time. It is our observation that research on water is particularly prone to pathological science and we explore why this might be. Liquid water, the substrate in which all known life operates, holds a privileged position in human culture and science. Phillip Ball explores this in his book H_2O: A Biography of Water and argues that the idea that "water is special" is a bias instilled in us by thousands of years of human culture (Ball 1999). This is undoubtably true, but scientifically such a bias is not entirely off-the-mark – water does have many anomalous properties and is special in many ways amongst liquids. Issues only occur when people latch onto the idea that water is more special than it really is and then do not properly criticize their ideas and only seek confirmation of them rather than falsification. Humans are subject to many cognitive biases (Kahneman 2011), and some of these, such as the confirmation bias and extension neglect (neglect of magnitude) undoubtably play a role in pathological science. Our eye in this work however is less on human psychology and more on the specific properties of water which make it difficult to study and thus prone to pathological science. We hope this work helps researchers studying water develop a more critical attitude and avoid wasting time pursuing pathological science.

2   Polywater

Perhaps the most famous example of pathological science is polywater. The polywater saga has been explicated in many places, so we keep a summary here in brief. Polymeric water ("polywater") was purported to be a special phase of water which formed when water was condensed into tiny capillary tubes with diameters smaller than 100 micrometers. The earliest papers on polywater originated from the group of Boris Deryaguin at the Institute of Surface Chemistry in Moscow, USSR in the early 1960s. In 1962 Fedayakin proposed that polywater had a honeycomb like structure with each oxygen bonded to 3 hydrogens. Lectures by Deryaguin in England and the United States in 1966, 1967 and 1968 drew the attention of western researchers. Research interest peaked after a 1969 a "a" should be removed.paper by Lippincott et al. in Science which reported spectroscopic results which were said to provide conclusive evidence of a stable polymeric structure" (Lippincott et al. 1969). Over 160 papers on polywater were published in 1970 alone (Eisenberg 1981). In 1971 Hasted noted problems with hexagonal water structures in general, noting that high energy cost of placing hydrogens between oxygens was enough to make such structures explode if they were ever created (Hasted 1971). By 1972 it became apparent that the observed phenomena were due to trace amounts of impurities (Rousseau and Porto 1970), some of which likely came from human sweat

(Rousseau 1971). In some cases, it was found that the sample tubes contained very little water at all. Altogether, over 500 publications were authored on polywater between 1963 and 1974 (Eisenberg 1981; Bennion and Neuton 1976).

3   Exclusion Zone Phenomena and Pollack's "Fourth Phase"

Recently we reviewed the literature on exclusion zone (EZ) phenomena in water (Elton et al. 2020). The EZ occurs when plastic microspheres are repelled away from the surface of some material leaving a region of microsphere-free water near the surface. An EZ near first observed in the laboratory of Gerald Pollack in 2003 near polyvinyl alcohol gels (Zheng and Pollack 2003). Later, in 2006 Pollack reported larger EZs near the surface of Nafion (Zheng et al. 2006). As we review in our article, the existence of an EZ near Nafion has been replicated many times by at least 10 different laboratories and constitutes a real phenomenon in search of an explanation. The finding of an EZ near metals has only been found by two independent groups and in our own experiments we were not able to replicate it in. Full stop should be removed.either zinc, copper, or aluminum with neutral latex microspheres (Fig. 8.1) (Spencer et al. 2018). It is unclear how big EZs are near hydrophilic materials other than Nafion. In his 2003 work, Pollack observed EZs near several hydrophilic gels such as polyacrylic acid, poly acrylamide, and agarose, but none of these results have been replicated. More specifically, Pollack reported that positively charged functionalized spheres were repelled by agarose gels (which is weakly

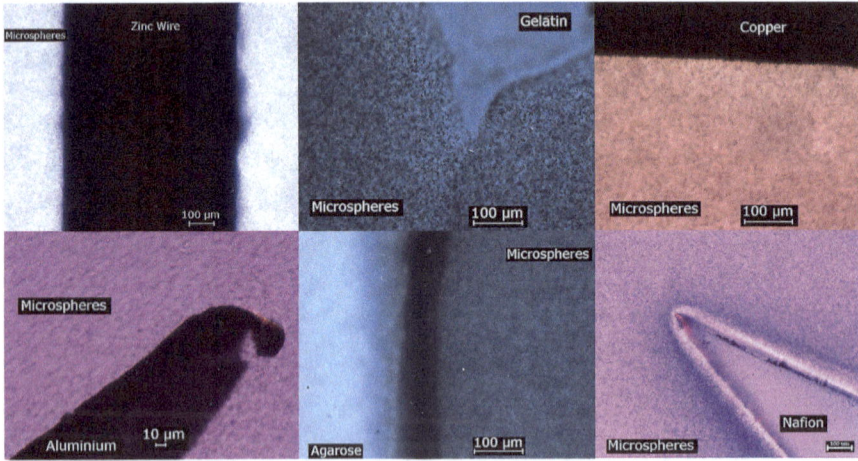

Fig. 8.1 Top row from left to right – zinc, gelatin, copper. Bottom row – aluminum, agarose, and Nafion. All EZs were visualized using a 1:500 suspension of 1.0 μm carboxylated solid-latex microspheres. The last image (Nafion) is the only one which shows an exclusion zone, evidenced by the much lower density of microspheres near the surface. The image with Nafion was taken with a polarized light microscope where birefringent materials appear brightly colored

negatively charged), but he does not report any results for other types of microspheres (Zheng and Pollack 2003). In our own study we did not find an EZ near either agar, agarose, or gelatin using neutral latex microspheres (Fig. 8.1). The pH of the agarose started very close to 7 and decreased to 6 after about 20 h and no changes in pH were observed for agar. Despite the lack of replication of the EZ phenomena beyond Nafion, Pollack often claims in interviews that an EZ is generated near all hydrophilic materials and plays an important role in biological processes.

It appears the core phenomenon of an EZ near Nafion is real and is likely caused by diffusiophoresis (also called chemotaxis) due to a long-lived pH gradient generated by the negatively charged sulfonic groups which are particular to the surface of Nafion (Elton et al. 2020; Schurr 2013; Florea et al. 2014b; Musa et al. 2013). Functionalized microspheres contain surface charges which lead to counterions near their surface. The surface charge is distributed uniformly, but a non-uniform pH gradient causes the counter ion distribution to become non-uniform. This sets up different electrostatic forces on both sides of the particle, leading to a net force on the particle (Fig. 8.2 illustrates this). As shown by Florea, the theory of diffusiophoresis precisely explains the kinetics (growth) of the EZ over time. The existence of a large pH gradient near Nafion has been shown using indicator dye in several of Pollack's works (for instance (Chai et al. 2009)).

Indeed, it seems the EZ phenomenon isn't really specific to water – research from Pollack's own lab showed that it occurs in a variety of liquids – methanol, ethanol, isopropanol, acetic acid, and dimethyl sulfoxide (Chai and Pollack 2010). Further investigations into EZ water, however, have generated much work we regard as verging on pathological because it meets several of Langmuir's criteria (in particular principles 1, 3, and 4). While Pollack is usually careful about what he says in his journal articles, typically sticking to the observed experimental facts, Pollack's book contains many wild conjectures which fly in the face of basic science, such as

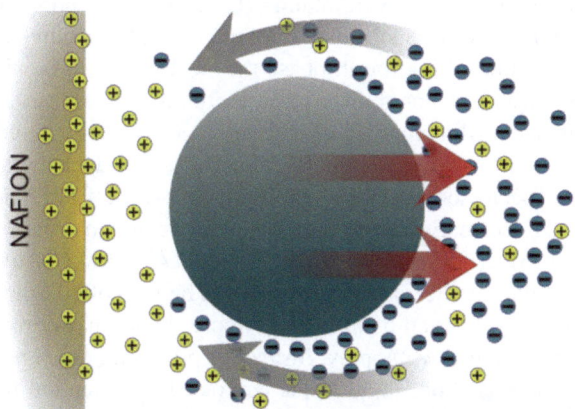

Fig. 8.2 Illustration of the mechanism by which diffusiophoresis generates a force on plastic microspheres which leads to the exclusion zone. (Schurr 2013; Florea et al. 2014a)

the idea that bloodflow is powered by sunlight (Pollack 2013). The most famous of these is Pollack's proposal that the EZ contains a "fourth phase" of water – a claim which is explored and discussed in several of his peer reviewed papers as well (Chai et al. 2009). Pollack hypothesizes that EZ water is structured in hexagonal sheets, with the hydrogens lying directly between oxygens, a structure which is very similar to polywater. He further proposes that when these sheets are stacked, hydrogen atoms bond to the oxygens in neighboring layers such that each hydrogen forms three bonds. Oehr and LeMay present a similar theory that the observed EZ water may comprise tetrahedral oxy-subhydride structures (Oehr and LeMay 2014). Pollack also hypothesizes that when light is shined on EZ water it causes positive and negative charges to separate, and the EZ water region to grow (Chai et al. 2009). This is obviously problematic since water is a good conductor and charge separation would be difficult to sustain.

Pollack points to enhanced absorption at 270 nm as evidence for a possible phase change in the EZ (Zheng et al. 2006; So et al. 2012). This absorption peak was not found in quantum chemistry simulations (Segarra-Martí et al. 2014). Strikingly, results from Pollack's own lab show that a similar absorption peak is seen in pure salt solutions (LiCl, NaCl, KCl) (Chai et al. 2008), so the source of this enhanced absorption appears to be related to dissolved solutes. A study of Arrowhead Spring water found absorption at 270 nm, so even trace dissolved solutes can create it (Dibble et al. 2014). Hypothesizing that EZ water would be a transitionary form between ice and liquid water, Pollack performed UV absorption measurements of melting ice (So et al. 2012). During the course of these experiments the 270 nm peak sometimes (but not always) appeared transiently (i.e., for a few seconds) while the ice was melting (Langmuir's criteria #1 and #2). In the same work they also report that degassing the water reduced of the appearance of the peak (So et al. 2012). Thus, it is also possible that the peak is related to tiny bubbles trapped in the ice which migrate to the surface while the ice is melting. As we discuss in our review (Elton et al. 2020), a possible mechanism for the absorption near 270 nm would absorption from superoxide anions (O_2^-) and their protonated form, the hydroperoxyl radical (HO_2). Such absorption may be enhanced by nanobubbles.

Pollack's promotion of his fourth phase theory deserves to be vigorously criticized not only because it contradicts basic thermodynamics, but also because it lends support to a sprawling number of enterprises selling "structured" or "hexagonal" water for health purposes. Tests of some of these products with nuclear magnetic resonance spectroscopy (NMR) show no difference from pure water (Shin 2006). Companies currently selling EZ water products who cite Pollack's work include Divinia Water, Structured Water Unit LLC, Flaska, Advanced Health Technologies (vibrancywater.ca), and Adya Inc. The idea of utilizing EZ water for health has also been promoted by influential figures in alternative medicine such as Dr. Joseph Mercola and Dave Asprey. Instead of providing much needed words of scientific skepticism caution, Pollack has embraced the attention he has received from alternative medicine community by participating in podcasts with Mercola, Asprey, and many others where he has promoted the idea that EZ water is important for health.

4   The Mpemba Effect

The idea that hot water can freeze faster than cold water has a long history (Jeng 2006). Brief mentions of this phenomena can be found in the writings of several famous thinkers including Aristotle, Thomas Bacon, and Descartes (Jeng 2006). In 1969 a Tanzanian high school student named Erasto Mpemba co-authored an article on the subject in the journal *Physics Education* (Mpemba and Osborne 1969). Erasto was actually studying sugared milk while making ice cream, but his finding spurred research searching for the effect in pure liquid water. In the period of 1970–1990, dozens of papers were published which purported to find such an effect. The literature is confusing due to the lack of experimental standards leading to many variables coming into play (some used distilled water, some used tap, some studied the effects of dissolved salts, authors used different cooling schedules/methods, etc). It appears none of the studies attempted to explicitly replicate a prior study, which resulted in all the studies having different types of experimental setup. Researchers also used slightly different definitions regarding the precise circumstances that would constitute confirmation or falsification of the Mpemba effect. Katz (2008) has analyzed this perplexing literature and postulates that most of the experiments were contaminated by solutes, either gaseous or solid (Katz 2008). He proposes that dissolved solutes (either gaseous or solid) are removed during heating, and that solutes accumulate along the freezing front and reduce the heat flux. Later, Linden & Burridge also reviewed the prior literature (Burridge and Linden 2016). They also performed their own study which showed that the height at which the temperature is measured determines what relative freezing times are observed. Since most prior work did not report this variable, it is hard to compare literature results. The conclusion from their own experiments was that the effect does not exist (Burridge and Linden 2016).

Assuring that containers with hot and cold water are cooled in identical and measured in an identical fashion which accurately determines the freezing time requires a careful experimental setup. To give a simple example of a pitfall that students might encounter at home, freezers have a thin layer of ice crystals coating their interior surfaces. If you place a container of hot water in such a freezer, the ice crystals will melt, allowing for better thermal contact between the container and the freezer. Thus, it's not surprising the container with the hot water freezes faster in such a case.

It has only been recently that very careful experiments have been performed which attempt to cool hot and cold water under perfectly identical conditions. One such series of experiments was published by Dr. James D. Brownridge in 2010 (Brownridge 2010). Brownridge took ultra-pure samples of distilled water, sealed them in small glass vials and suspended the vials by threads in a vacuum. The vials were then cooled using radiative cooling. This completely removed the possibility of a difference in the thermal contact between the hot & cold vials and ensured they were cooled in exactly the same fashion. Brownridge found that in some cases that the hot water vial would freeze first. This only occurred though when the cold water would supercool further than the hot before freezing. At all times, the hot water was always warmer than the cold water, and both vials were cooling at the same rate – it was just the cold water supercooled more. Brownridge found that each glass vial has a highest temperature nucleation site (HTNS) which determines the temperature water will freeze in that vial (Brownridge 2010). Comparing what appear to be identical vials, the HTNS are random and they can be between anywhere from 0 °C to −45 °C. Brownridge showed that the HTNS is a constant of the container by rerunning the freezing many times. So, in the end, the two containers (hot and cold) were not actually identical because they had different nucleation sites! The basic idea behind this -- that supercooling to different degrees as a result of unpredictable nucleation factors was responsible for the Mpemba effect being observed, had been proposed earlier by Auerbach in 1995. (Auerbach 1995).

Brownridge and Auerbach's work showing that the Mpemba effect is just due to unpredictable supercooling seems to have been completely lost on the Royal Society of Chemistry, which in 2012 held a much-publicized competition to explain the effect. The winner of that competition proposed that the effect is due to some or all of the following: (a) evaporation, (b) dissolved gases, (c) mixing by convective currents, and (d) supercooling. Brownridge, by carefully removing the confounds of (a)–(c), showed that supercooling is enough to generate the effect if containers with different nucleation sites are used. Still, (a)–(c) are quite possibly confounding factors which were responsible for observations of the Mpemba effects in previous works.

5   Magnetic Fields and Water

The number of different effects that magnets have been claimed to have on water is truly mind boggling, and there are too many to properly analyze in this small chapter. Magnetic fields have been reported to change the physicochemical properties of liquid water, including viscosity (Ghauri and Ansari 2006; Cai et al. 2009), refractive index (Hosoda et al. 2004), melting temperature (Inaba et al. 2004), rate of vaporization (Nakagawa et al. 1999), adsorption (Ozeki et al. 1991; Higashitani et al. 1993), electrolyte conductivity (Holysz et al. 2007), and conductivity (Szcześ et al. 2011). Some authors report that the property changes remain for many hours even after the magnetic fields are turned off (Mahmoud et al. 2016; Silva et al. 2015; Coey and Cass 2000; Szcześ et al. 2011). Magnetic fields have been claimed to inhibit the formation of ice crystals in both pure water and biological products (Otero et al. 2016). There is also research that purports that magnetic fields can be used to "treat" water in some way – either to purify (Ambashta and Sillanpää 2010), de-scale (Coey and Cass 2000), or disinfect water (Biryukov et al. 2005). Authors who have attempted to review this massive and perplexing literature have lamented the lack of independent reproduction of most results (Knez and Pohar 2005; Smothers et al. 2001). Some of these experimental findings are said to be supported by molecular dynamics simulations that show that magnetic fields enhance hydrogen bonding (Chang and Weng 2006). However, the effect size is extremely small – a 10 Telsa magnetic field caused a size increase of only 0.34% in water clusters in one such study (Toledo et al. 2008). This is not surprising because the magnetic susceptibility of water molecules is very small – about -9.0×10^{-6} (Otero et al. 2016).

Research on the effect of magnetic fields on the freezing of water is very mixed. For instance, work in 2000 on pure water droplets found that magnetic fields *reduced* the degree of supercooling before freezing in contrast to many works which suggest that magnetic fields inhibit freezing (Aleksandrov et al. 2000). That is not to say that magnetic fields have no effect – water is weakly diamagnetic so strong enough magnetic fields induce a magnetic dipole in the opposite direction. One group of researchers levitated water droplets in a 15 T magnetic field and found that the droplets supercooled to $-10\,°C$ (Tagami et al. 1999). This is not a remarkable degree of supercooling, especially for droplets of pure water not in contact with any nucleating agents. There are also papers on the effects of "oscillating magnetic fields" on water but, as any physicist knows, oscillating magnetic fields are always accompanied by oscillating electric fields, so speaking of them in isolation doesn't make much sense. To the degree that a weak oscillating magnetic fields inhibit freezing in food, as has been claimed by the Japanese company ABI Corporation with their "Cells Alive" freezer, which is maybe due to heating of trace metals (iron etc.) in the food, or to non-magnetic sources altogether such as acoustic vibrations in the freezer system (Wowk 2012).

An analysis of the massive literature on magnetic fields and water could easily fill several review articles so in the remainder of this section we focus on a specific subfield – magnetic treatment for preventing scaling/corrosion. There is no universally agreed upon mechanism by which magnetic fields inhibit scaling but a common theory is that they work by changing the morphology of the precipitates to prevent them from depositing in flat sheets (Barrett and Parsons 1998; Gehr et al. 1995; Holysz et al. 2007; Madsen 1995; Higashitani et al. 1993). A review by Baker and Judd investigates numerous claims on this matter (1996). Their view is that contamination effects are the main contributor and therefor the results obtained in some experiments will not generalize to more general situations as has been claimed. In particular, they note several experiments where magnetically-enhanced corrosion likely created Fe^{2+} ions which are known to retard the growth rate of calcite scale deposition. They also note that more successful results are obtained with magnetic fields oriented orthogonally to flowing water within recirculated systems. This implies that Lorentz forces acts on particulates in the water rather than the water itself. That is, forces are exerted on charged particles passing through the magnetic field.

Differences in infrared and Raman spectra in magnetically treated water have been interpreted as implying the development of quasi-stable water clusters in a magnetic field which somehow persist after the magnetic field is switched off (a variant of the "water memory" idea). This is very hard to believe given that hydrogen bond lifetimes are around 1 ps in room temperature water and the Debye relaxation time is ~8–9 ps (Elton 2017). Interestingly, Ozeki et al. found that the effect of magnetic treatment on IR absorption increases with increased dissolved oxygen and water which was fully degassed does not show any changes after treatment (Ozeki and Otsuka 2006). They theorize that magnetic treatment leads to the formation of oxygen clathrate-like hydrates which influence the H-bond network of water (Ozeki and Otsuka 2006). Additionally, both Lee et al. (2013) and Szcześ et al. (2011) have also reported that the concentration of dissolved gases significantly affected their results. One of us, (Peter D. Spencer) performed a simple experiment which found no effect of a magnetic field on UV-vis absorption (Fig. 8.3) (Spencer 2018). The field was weak (0.63 T) but typical of the field strength employed in many studies of magnetic water treatment.

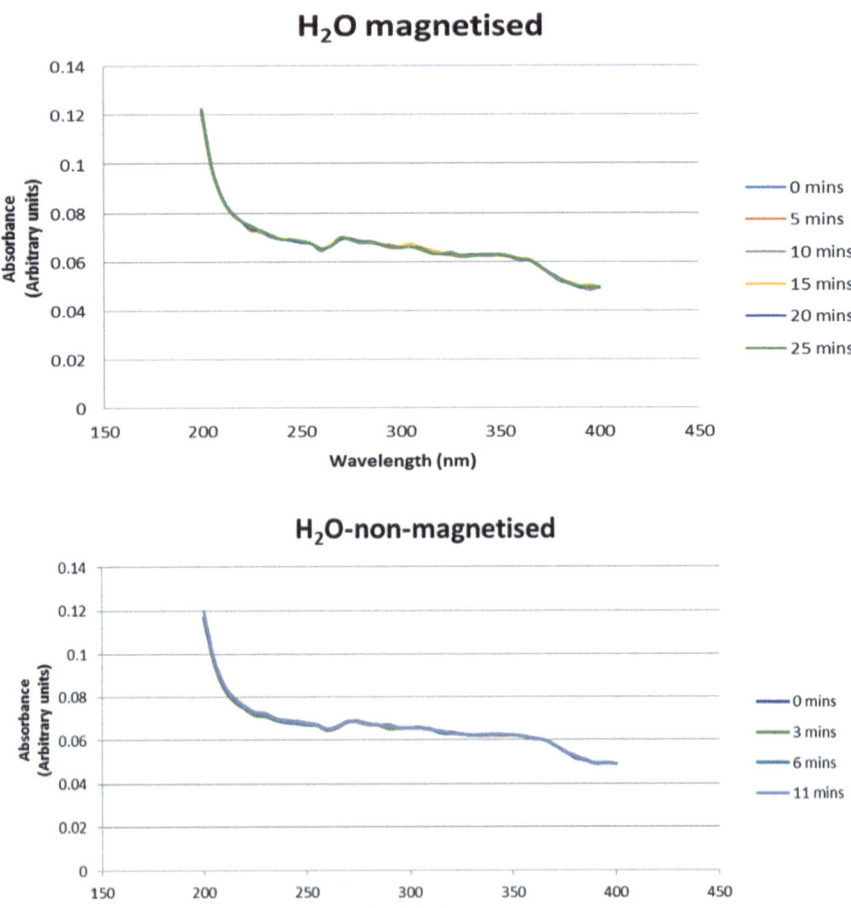

Fig. 8.3 Measurements from a simple experiment one of the authors (Peter D. Spencer) performed which looked at the UV-vis absorbance of water under a magnetic field and no magnetic field (Spencer 2018). The magnetic field was weak (0.63 T) but typical of the field strength found in studies of magnetic water treatment systems

6 Commonalities and Concluding Thoughts

In this work we reviewed four areas of pathological science. Due to time and space limitations we did not discuss another major area – water memory. The interested reader can consult the chapter in this very book by Yuvan and Bier (this volume) which discusses it in some detail. Additionally, much water memory research, in our view, often crosses out of the domain of pathological science into pseudoscience. Water memory research has its origin in *Nature* paper from 1988 and has been thoroughly debunked (Maddox et al. 1988). Still, much work continues on water memory partially because it is a mechanism those working in the lucrative homeopathy industry have latched onto as a means of scientifically justifying their work. Homeopathy has been thoroughly debunked many times and many places (for instance a metareview of metareviews found no effect (Ernst 2002)).

In the course of this work we have noticed a few different common features of pathological water science. The main one is improper removal of confounding factors. It is very difficult to remove dissolved solutes from water, and work indicates they were responsible for most of the experimental results in the four areas explored here. More research is needed on nanobubbles which are a possible confound and are very hard to remove from water (Jadhav and Barigou 2020; Ball 2012; Michailidi et al. 2020). Referring to microsphere suspensions, Horinek et al. note "these systems are notoriously plagued by secondary effects, such as bubble adsorption and cavitation effects or compositional rearrangements" (Horinek et al. 2008). Dissolved gases (not bubbles) can also be a confound as was seen in the spectral analysis of EZ water. To give another example, in 2010 Jansson et al. measured the dielectric function of water at very low frequencies and reported an "ultra-slow" Debye relaxation at 5 MHz (Jansson et al. 2010). Later work has indicated that this peak was due to microscopic bubbles in the liquid (Richert et al. 2011). Alternatively, it has been suggested that the low frequency peak is due to volatile non-polar contaminants (Casalini and Roland 2011). It is possible both mechanisms were at play in Casalini & Roland's experiment since they observed two ultra-low frequency Debye peaks.

Another thing we noticed is lack of precision in defining the phenomena being measured. In postmodernist literature and other fields a misleading and faulty type of argument called the "motte and baily" fallacy has been identified (Boudry and Braeckman 2011). In the "motte and bailey" style of argument, a proponent argues for a strong claim but then retreats to a much weaker claim under pressure from counterarguments. The weaker claim is then later conflated with the stronger one, sowing confusion and putting critics in a difficult position. For instance, researchers may proclaim their work shows "structure change in cellular water" or that "hot water freezes faster than cold" but under pressure will retreat to a weaker and much less interesting claim such as "proteins can reorient waters near their surface affecting 1–3 layers of water" or "hot water sometimes is observed to supercool more than cold water". We suggest that researchers focus on developing precision in their statements about experimental measurements and what they show, with a particular focus on *effect size*, which is often omitted by popular press coverage of research and thus misleads the public in unhealthy ways.

Researchers should also make sure their work is reproducible by explaining how the experiment was carried out, utilizing supplementary information if necessary to list all relevant details. Ideally, all raw data generated should be made publicly available so that the data analysis methods employed can also be reproduced. Making a full description of experimental methods and the raw data available also helps other researchers identify errors and questionable research practices (Gadomski et al. 2017).

Finally, we suggest that all researchers should evaluate their "evidence threshold" to immunize themselves from falling into believing in pathological science. One's evidence threshold is the threshold needed to believe that a proposed phenomenon is real. Dr. Steven Novella, author of the blog *Science Based Medicine*, has suggested four criteria for a good evidence threshold, the statement of which we believe is a fitting way to conclude this chapter (Novella 2013):

1 – *"Methodologically rigorous, properly blinded, and sufficiently powered studies that adequately define and control for all relevant variables (confirmed by surviving peer-review and post-publication analysis)."*
2 – *"Positive results that are statistically significant."*
3 – *"A reasonable signal to noise ratio."*
4 – *"Independently reproducible (and reproduced). No matter who repeats the experiment, the effect is reliably detected."*

Acknowledgements Daniel C. Elton contributed this article in his personal capacity. The opinions expressed in this article are the author's own and do not reflect the view of the National Institutes of Health, the Department of Health and Human Services, or the United States government. Peter D. Spencer was partially supported by the Research Training Program (Stipend) funded by Department of Education and Training (Australia). Daniel C. Elton acknowledges his PhD. thesis advisor, Prof. Marivi Fernández-Serra, and Prof. Martin Bier for inspiring this line of inquiry.

References

Aleksandrov, V. D., Barannikov, A. A., & Dobritsa, N. V. (2000). Effect of magnetic field on the supercooling of water drops. *Inorganic Materials, 36*(9), 895–898.
Ambashta, R. D., & Sillanpää, M. (2010). Water purification using magnetic assistance: A review. *Journal of Hazardous Materials, 180*(1–3), 38–49.
Auerbach, D. (1995). Supercooling and the Mpemba effect: When hot water freezes quicker than cold. *American Journal of Physics, 63*(10), 882–885.
Baker, J. S., & Judd, S. J. (1996). Magnetic amelioration of scale formation. *Water Research, 30*(2), 247–260.
Ball, P. (1999). *H2O: A biography of water*. London: Weidenfeld & Nicolson.
Ball, P. (2012). Nanobubbles are not a Superficial Matter. *ChemPhysChem, 13*(8), 2173–2177.
Barrett, R. A., & Parsons, S. A. (1998). The influence of magnetic fields on calcium carbonate precipitation. *Water Research, 32*(3), 609–612.
Bennion, B. C., & Neuton, L. A. (1976). The epidemiology of research on "Anomalous Water". *Journal of the American Society for Information Science, 27*(1), 53–56.

Biryukov, A. S., Gavrikov, V. F., Nikiforova, L. O., et al. (2005). New physical methods of disinfection of water. *Journal of Russian Laser Research, 26*(1), 13–25.

Boudry, M., & Braeckman, J. (2011). Immunizing strategies and epistemic defense mechanisms. *Philosophia, 39*(1), 145–161.

Brownridge, J. D. (2010). When does hot water freeze faster then cold water? A search for the Mpemba effect. *American Journal of Physics, 79*(1), 78–84.

Burridge, H. C., & Linden, P. F. (2016). Questioning the Mpemba effect: Hot water does not cool more quickly than cold. *Scientific Reports, 6*(1), 37665.

Cai, R., Yang, H., He, J., et al. (2009). The effects of magnetic fields on water molecular hydrogen bonds. *Journal of Molecular Structure, 938*(1–3), 15–19.

Casalini, R., & Roland, C. M. (2011). On the low frequency loss peak in the dielectric spectrum of glycerol. *The Journal of Chemical Physics, 135*(9), 094502.

Chai, B., & Pollack, G. H. (2010). Solute-free interfacial zones in polar liquids. *The Journal of Physical Chemistry. B, 114*(16), 5371–5375.

Chai, B.-H., Zheng, J.-M., Zhao, Q., et al. (2008). Spectroscopic studies of solutes in aqueous solution. *The Journal of Physical Chemistry. A, 112*(11), 2242–2247.

Chai, B., Yoo, H., & Pollack, G. H. (2009). Effect of radiant energy on near-surface water. *The Journal of Physical Chemistry B, 113*(42), 13953–13958.

Chang, K.-T., & Weng, C.-I. (2006). The effect of an external magnetic field on the structure of liquid water using molecular dynamics simulation. *Journal of Applied Physics, 100*(4), 043917.

Coey, J. M. D., & Cass, S. (2000). Magnetic water treatment. *Journal of Magnetism and Magnetic Materials, 209*(1), 71–74.

Dibble, W. E., Kaszyk, J., & Tiller, W. A. (2014). Bulk water with exclusion zone water characteristics: Experimental evidence of interaction with a non-physical agent. *WATER Journal, 6*, 35–44.

Eisenberg, D. (1981). A scientific gold rush. *Science, 213*(4512), 1104–1105.

Elton, D. C. (2017). The origin of the Debye relaxation in liquid water and fitting the high frequency excess response. *Physical Chemistry Chemical Physics, 19*(28), 18739–18749.

Elton, D. C., Spencer, P. D., Riches, J. D., et al. (2020). Exclusion zone phenomena in water—A critical review of experimental findings and theories. *International Journal of Molecular Sciences, 21*(14), 5041.

Ernst, E. (2002). A systematic review of systematic reviews of homeopathy. *British Journal of Clinical Pharmacology, 54*(6), 577–582.

Florea, D., Musa, S., Huyghe, J. M. R., et al. (2014a). Long-range repulsion of colloids driven by ion exchange and diffusiophoresis. *Proceedings of the National Academy of Sciences, 111*(18), 6554–6559.

Florea, D. D., Musa, S. S., Huyghe, J. J., et al. (2014b). Long-range repulsion of colloids driven by ion-exchange and diffusiophoresis. *Proceedings of the National Academy of Sciences of the United States of America, 111*(18), 6554–6559.

Gadomski, A., Ausloos, M., & Casey, T. (2017). Dynamical systems theory in quantitative psychology and cognitive science: A fair discrimination between deterministic and statistical counterparts is required. *Nonlinear Dynamics, Psychology, and Life Sciences, 21*(2), 129–141.

Gehr, R., Zhai, Z. A., Finch, J. A., et al. (1995). Reduction of soluble mineral concentrations in CaSO 4 saturated water using a magnetic field. *Water Research, 29*(3), 933–940.

Ghauri, S. A., & Ansari, M. S. (2006). Increase of water viscosity under the influence of magnetic field. *Journal of Applied Physics, 104*(6), 066101–066102.

Hasted, J. B. (1971). Water and 'polywater'. *Contemporary Physics, 12*(2), 133–152.

Higashitani, K., Kage, A., Katamura, S., et al. (1993). Effects of a magnetic field on the formation of $CaCO_3$ particles. *Journal of Colloid and Interface Science, 156*(1), 90–95.

Holysz, L., Szcześ, A., & Chibowski, E. (2007). Effects of a static magnetic field on water and electrolyte solutions. *Journal of Colloid and Interface Science, 316*(2), 996–1002.

Horinek, D., Serr, A., Geisler, M., et al. (2008). Peptide adsorption on a hydrophobic surface results from an interplay of solvation, surface, and intrapeptide forces. *Proceedings of the National Academy of Sciences, 105*(8), 2842–2847.

Hosoda, H., Mori, H., Sogoshi, N., et al. (2004). Refractive indices of water and aqueous electrolyte solutions under high magnetic fields. *The Journal of Physical Chemistry. A, 108* (9), 1461–1464.

Inaba, H., Saitou, T., Tozaki, K.-I., et al. (2004). Effect of the magnetic field on the melting transition of H2O and D2O measured by a high resolution and supersensitive differential scanning calorimeter. *Journal of Applied Physics, 96*(11), 6127–6132.

Jadhav, A. J., & Barigou, M. (2020). Bulk Nanobubbles or not Nanobubbles: That is the question. *Langmuir, 36*(7), 1699–1708.

Jansson, H., Bergman, R., & Swenson, J. (2010). Hidden slow dynamics in water. *Physical Review Letters, 104*(1), 017802.

Jeng, M. (2006). The Mpemba effect: When can hot water freeze faster than cold? *American Journal of Physics, 74*(6), 514–522.

Kahneman, D. (2011). *Thinking, fast and slow*. Farrar: Straus and Giroux.

Katz, J. I. (2008). When hot water freezes before cold. *American Journal of Physics, 77*(1), 27–29.

Knez, S., & Pohar, C. (2005). The magnetic field influence on the polymorph composition of CaCO3 precipitated from carbonized aqueous solutions. *Journal of Colloid and Interface Science, 281*(2), 377–388.

Langmuir, I., & Hall, R. N. (1989). Pathological science. *Physics Today, 42*(10), 36–48.

Lee, S. H., Jeon, S. I., Kim, Y. S., et al. (2013). Changes in the electrical conductivity, infrared absorption, and surface tension of partially-degassed and magnetically-treated water. *Journal of Molecular Liquids, 187*, 230–237.

Lippincott, E. R., Stromberg, R. R., Grant, W. H., et al. (1969). Polywater. *Science, 164*(3887), 1482–1487.

Maddox, J., Randi, J., & Stewart, W. W. (1988). "High-dilution" experiments a delusion. *Nature, 334*(6180), 287–290.

Madsen, H. E. L. (1995). Influence of magnetic field on the precipitation of some inorganic salts. *Journal of Crystal Growth, 152*(1), 94–100.

Mahmoud, B., Yosra, M., & Nadia, A. (2016). Effects of magnetic treatment on scaling power of hard waters. *Separation and Purification Technology, 171*, 88–92.

Michailidi, E. D., Bomis, G., Varoutoglou, A., et al. (2020). Bulk nanobubbles: Production and investigation of their formation/stability mechanism. *Journal of Colloid and Interface Science, 564*, 371–380.

Mpemba, E. B., & Osborne, D. G. (1969). Cool? *Physics Education, 4*(3), 172–175.

Musa, S., Florea, D., van Loon, S., Wyss, H., & Huyghe, J. (2013). Interfacial Water: Unexplained Phenomena. *Paper presented at the Poromechanics V: Proceedings of the Fifth Biot*. Conference on Poromechanics. Vienna, Austria: American Society of Civil Engineers.

Nakagawa, J., Hirota, N., Kitazawa, K., et al. (1999). Magnetic field enhancement of water vaporization. *Journal of Applied Physics, 86*(5), 2923–2925.

Novella, S. (2013). *Evidence Thresholds*. Available at: https://sciencebasedmedicine.org/evidence-thresholds/

Oehr, K., & LeMay, P. (2014). The case for tetrahedral oxy-subhydride (TOSH) structures in the exclusion zones of anchored polar solvents including water. *Entropy, 16*(11), 5712–5720.

Otero, L., Rodríguez, A. C., Pérez-Mateos, M., et al. (2016). Effects of magnetic fields on freezing: Application to biological products. *Comprehensive Reviews in Food Science and Food Safety, 15*(3), 646–667.

Ozeki, S., & Otsuka, I. (2006). Transient oxygen clathrate-like hydrate and water networks induced by magnetic fields. *The Journal of Physical Chemistry. B, 110*(41), 20067–20072.

Ozeki, S., Wakai, C., & Ono, S. (1991). Is a magnetic effect on water adsorption possible? *The Journal of Physical Chemistry, 95*(26), 10557–10559.

Pollack, G. H. (2013). The fourth phase of water: Beyond solid, liquid, and vapor. In D. Scott (Ed) *Kindle* ed. Seattle: Ebner and Sons Publishers.

Richert, R., Agapov, A., & Sokolov, A. P. (2011). Appearance of a Debye process at the conductivity relaxation frequency of a viscous liquid. *The Journal of Chemical Physics, 134*(10), 104508.

Rousseau, D. L. (1971). "Polywater" and sweat: Similarities between the infrared spectra. *Science, 171*(3967), 170–172.

Rousseau, D. L., & Porto, S. P. S. (1970). Polywater: Polymer or Artifact? *Science, 167*(3926), 1715–1719.

Schurr, J. M. (2013). Phenomena associated with gel–water interfaces. Analyses and alternatives to the long-range ordered water hypothesis. *The Journal of Physical Chemistry. B, 117* (25), 7653–7674.

Segarra-Martí, J., Roca-Sanjuán, D., & Merchán, M. (2014). Can the hexagonal ice-like model render the spectroscopic fingerprints of structured water? Feedback from quantum-chemical computations. *Entropy, 16*(7), 4101–4120.

Shin, P. (2006). *Water, everywhere, Caveat Emptor (Buyer Beware)!* Available at: http://www.csun.edu/~alchemy/Caveat_Emptor.pdf

Silva, I. B., Queiroz Neto, J. C., & Petri, D. F. S. (2015). The effect of magnetic field on ion hydration and sulfate scale formation. *Colloids and Surfaces A: Physicochemical and Engineering Aspects, 465*, 175–183.

Smothers, K. W. C., Charles, D., Gard, B. T., Strauss, R. H., & Hock, V. F. (2001). *Demonstration and evaluation of magnetic descalers.* Available at: http://oai.dtic.mil/oai/oai?verb=getRecord&metadataPrefix=html&identifier=ADA399455. Accessed 22 March.

So, E., Stahlberg, R., & Pollack, G. (2012). *Exclusion zone as intermediate between ice and water.* Southampton: WIT Press.

Spencer, P. D. (2018). *Examining claims of long-range molecular order in water molecules.* Queensland University of Technology.

Spencer, P. D., Riches, J. D., & Williams, E. D. (2018). Exclusion zone water is associated with material that exhibits proton diffusion but not birefringent properties. *Fluid Phase Equilibria, 466*, 103–109.

Szcześ, A., Chibowski, E., Hołysz, L., et al. (2011). Effects of static magnetic field on water at kinetic condition. *Chemical Engineering and Processing, 50*(1), 124–127.

Tagami, M., Hamai, M., Mogi, I., et al. (1999). Solidification of levitating water in a gradient strong magnetic field. *Journal of Crystal Growth, 203*(4), 594–598.

Toledo, E. J. L., Ramalho, T. C., & Magriotis, Z. M. (2008). Influence of magnetic field on physical–chemical properties of the liquid water: Insights from experimental and theoretical models. *Journal of Molecular Structure, 888*(1–3), 409–415.

Wowk, B. (2012). Electric and magnetic fields in cryopreservation. *Cryobiology, 64*(3), 301–303.

Yuvan, S., & Bier, M. (this volume). Sense and nonsense about water. In A. Gadomski (Ed.), *Water in biomechanical and related systems* (Biologically-Inspired System, Vol. 17).

Zheng, J.-M., & Pollack, G. H. (2003). Long-range forces extending from polymer-gel surfaces. *Physical Review E, 68*(3 Pt 1), 031408.

Zheng, J. M., Chin, W. C., Khijniak, E., et al. (2006). Surfaces and interfacial water: Evidence that hydrophilic surfaces have long-range impact. *Advances in Colloid and Interface Science, 127*(1), 19–27.

Chapter 9
Powdery Mildew Fungus *Erysiphe Alphitoides* Turns Oak Leaf Surface to the Highly Hydrophobic State

Elena V. Gorb and Stanislav N. Gorb

Abstract Young leaves of the common oak *Quercus robur* are susceptible to the powdery mildew fungus *Erysiphe alphitoides* that can induce necrosis on the leaf surface. It is known that the fungal mycelium changes the wettability of the leaf surface. The present study performed a series of measurements of contact angles of water, ethylene glycol and diiodomethane on the oak leaf surface. Based on the obtained contact angle data, we calculated surface free energy on the adaxial surface of fresh healthy leaves and compared it with that of leaves infected by the fungus. Both types of leaf surfaces were also studied using cryo scanning electron microscopy, to examine a correlation between healthy/infected leaf surface structure and measured contact angles and surface free energy values. Our microscopical and experimental data clearly demonstrate that the powdery mildew turns the oak leaf surface to the highly hydrophobic state. The fungal mycelium shows three-dimensional hierarchical structure consisting of microscopic hyphae having cylindrical cross section with conidiomata (the first hierarchical level) and nanoscopic papillae and scales on the hyphal surface (the second hierarchical level). Such hierarchy in combination with the hydrophobic nature of the surface substances of the fungus leads to the highly hydrophobic properties (and low surface free energy) of the infected surface of the leaf.

Keywords Powdery mildew · Oak leaf surface · Hydrophobicity · Water based biomimetics

E. V. Gorb (✉) · S. N. Gorb (✉)
Department of Functional Morphology and Biomechanics, Kiel University, Kiel, Germany
e-mail: egorb@zoologie.uni-kiel.de; sgorb@zoologie.uni-kiel.de

© The Author(s), under exclusive license to Springer Nature
Switzerland AG 2021
A. Gadomski (ed.), *Water in Biomechanical and Related Systems*,
Biologically-Inspired Systems 17, https://doi.org/10.1007/978-3-030-67227-0_9

1 Introduction

Oak powdery mildew is one of the most common fungal diseases in European forests. Young developing leaves of the common oak *Quercus robur* are susceptible to the fungus *Erysiphe alphitoides* (Fig. 9.1a, b) that can induce necrosis on the leaf surface. Since the disease develops in late spring, after the first leaves of oak seedlings have developed, it strongly infects the second and third flushes of leaves developing in July and August (Mougou et al. 2008; Szewczyk et al. 2015). The study of the effects of *E. alphitoides* on the *Q. robur* leaves by Hajji et al. (2009) has found a decreased stomatal conductance by 15–30%, decreased nitrogen content, reduced carbon fixation by about 40–50% in fully infected leaves and increased dark respiration. Since the infected leaf surface is covered by the fungal structures, one can also

Fig. 9.1 Adaxial side of healthy (**a**) and infected (**d**) leaves of *Q. robur* and the corresponding cryo-SEM micrographs of the leaf surface (**b**, **c** – healthy, **e**, **f** – infected). *CN* conidioma, *HP* hyphae, *WX* wax projections

assume that water storage capacity and wettability of the plant leaf surface must be also affected, because both properties are considerably dependent on the condition of the leaf surface (Jeffree 1986, 2006; Barthlott et al. 1998; Gorb and Gorb 2013; Klamerus-Iwan and Witek 2018; Fernández et al. 2014).

One can assume that the fungal structures should increase leaf wettability, in order to collect more water for their own water management. This hypothesis is especially attractive due to the fact that upper (adaxial) leaf surface of *Q. robur* is covered by three-dimensional epicuticular waxes projections well known for their highly hydrophobic and even superhydrophobic (120° < water contact angle < 150° and water contact angle > 150°, respectively, according to Zisman (1964)) properties (Jeffree et al. 1975; Jetter and Riederer 1994, 1995; Barthlott et al. 1998; Gorb and Gorb 2006; Koch and Ensikat 2008; Gorb et al. 2013). Previous studies that analysed the influence of the oak powdery mildew infection on the hydrological properties of oak leaves and on related seasonal changes demonstrated that the oak powdery mildew has great influence on the wettability of oak leaves and on the canopy water storage capacity (Klamerus-Iwan and Witek 2018), which is indeed not surprising. The authors showed that the increase in leaf area covered with the oak powdery mildew, building a hydrophilic mycelium on the leaf, increases wettability of leaves. However, the question about the relationship between the leaf surface structure and the corresponding contact angle remained unresolved.

This study performed a series of measurements of contact angles of water, ethylene glycol and diiodomethane on the leaf surface of the common oak *Q. robur*. Based on the obtained contact angle data, we calculated surface free energy on the adaxial surface of fresh intact leaves and compared it with that determined for leaves infected by *E. alphitoides*. Both types of surfaces were also studied using cryo scanning electron microscopy (cryo-SEM), to examine a correlation between intact and infected leaf surface structure and measured contact angles and surface free energy values.

2 Materials and Methods

2.1 Samples

2.1.1 Plant Species

The common oak *Q. robur* L. (Fagaceae) is a large, long-lived deciduous tree with lobed and nearly sessile, 7–14 cm long leaves. It is native to most of Europe west of the Caucasus and is also widely cultivated in temperate regions (description after https://en.wikipedia.org/wiki/Quercus_robur).

2.1.2 Fungus Species

The powdery mildew fungus *E. alphitoides* (Griffon & Maubl.) U. Braun & S. Takam. (Erysiphaceae) causes powdery mildew on oak trees. Today, it is one of the most common diseases in European forests (description after https://en.wikipedia.org/wiki/Erysiphe_alphitoides). The fungus forms a white mycelium (vegetative part) on the oak leaf surface, ultimately covering the whole leaf area, from which root-like haustoria penetrate leaf cells. The mycelium also produces white, powder-looking asexual, non-motile spores called conidia. Infected leaves finally curl, become brown and fall down untimely (after https://de.wikipedia.org/wiki/Eichenmehltau).

2.1.3 Sampling

Both healthy and infected leaves were collected in August from young trees growing along the hiking trail near Flemhude (Rendsburg-Eckernförde district, Schleswig-Holstein, Germany, 54° 19′ 0″ N, 9° 59′ 0″ E).

2.2 Contact Angle Measurements and Estimation of the Surface Free Energy

Physicochemical properties of healthy and infected leaves were characterized by applying a high-speed optical contact angle measuring device OCAH 200 (DataPhysics Instruments GmbH, Filderstadt, Germany). Freshly collected, untreated leaves were attached with their lower (abaxial) side to a glass slide using a double-sided adhesive tape. Measurements were carried out on the adaxial side of the leaves. On infected leaves, we selected modified (i.e. infected) surface regions for measurements.

2.2.1 Contact Angle Measurements

A detailed description of the method is given in Gorb et al. (2004). We measured apparent contact angles of three well-characterized liquids (double-distilled water: density = 0.998 kg m^{-3}, surface tension = 72.1 mN m^{-1}, dispersion component = 19.9 mN m^{-1}, polar component = 52.2 mN m^{-1}; diiodomethane: density = 3.325 kg m^{-3}, surface tension = 50.0 mN m^{-1}, dispersion component = 47.4 mN m^{-1}, polar component = 2.6 mN m^{-1} (Busscher et al. 1984); ethylene glycol: density = 1.113 kg m^{-3}, surface tension = 48.0 mN m^{-1}, dispersion component = 29.0 mN m^{-1}, polar component = 19.0 mN m^{-1} (Erbil 1997)) using 1 μl droplets according to the sessile drop method with circle fitting.

For each leaf type (healthy leaf and infected leaf), we examined three leaves/ samples. On each sample, from 5 to 10 contact angle measurements were taken with each liquid. In all, 111 measurements were executed.

2.2.2   Surface Free Energy Estimation

Surface free energy with its dispersion and polar components was calculated for the two above leaf types with three samples each, one per sample according to the Owens, Wendt and Kaelble method (Owens and Wendt 1969). This method suggested initially for polymers is also applicable for wax-covered plant surfaces (e.g. Gorb and Gorb 2006). Altogether, six calculations of the surface free energy were performed.

2.3   Statistics

Data are presented in the text as mean ± s.d. Contact angles of different liquids within each leaf type were analyzed with one-way ANOVA followed by the post-hoc pairwise Tukey test for multiple comparisons of means (software SigmaStat 3.5, Systat Software, Point Richmond, CA, USA). For examining differences in contact angles between leaf types, we applied t-test for both water and diiodomethane and Mann-Whitney rank sum test for ethylene glycol. To determine whether there is a significant difference in surface free energy, its dispersion and polar components between healthy and infected leaves, data were tested using t-test.

2.4   Cryo Scanning Electron Microscopy

Small pieces (1 cm × 1 cm) were cut out from the healthy and infected leaves using a razor blade, attached mechanically with the adaxial side facing up to a small vice on a metal holder and frozen in a cryo stage preparation chamber at -140 °C (Gatan ALTO 2500 cryo preparation system, Gatan Inc., Abingdon, UK). Frozen samples were sputter coated with gold–palladium (thickness 6 nm) and examined in the frozen condition in a cryo-SEM Hitachi S-4800 (Hitachi High-Technologies Corporation, Tokyo, Japan) at 3 kV accelerating voltage and -120 °C temperature.

Types of wax projections were identified according to Barthlott et al. (1998). Morphometrical variables of wax projections as well as of hyphae and conidiomata of *E. alphitoides* fungus were measured from digital images using SigmaScan Pro 5 software (SPSS Inc., Chicago, USA). These data are presented in the text as mean ± s.d. for n = 10.

3 Results

3.1 Physicochemical Properties of Healthy and Infected Leaves

3.1.1 Contact Angles

Both healthy and infected leaves showed significantly different contact angles of different fluids (Fig. 9.2) (healthy: $F_{2,50} = 13.941$; infected: $F_{2,53} = 90.635$; both $P < 0.001$, one-way ANOVA). Healthy leaves being wetted well by all three fluids (contact angles $< 90°$ for all, Fig. 9.3a) were significantly stronger wetted by polar water than by both other fluids, whereas contact angles of polar ethylene glycol and non-polar diiodomethane were rather similar (Table 9.1). On the infected leaves showing unwettable properties in the cases of all fluids (contact angles $> 90°$ for all, Fig. 9.3b), contact angles differed significantly in the following order of fluids: water − ethylene glycol − diiodomethane (Table 9.1). Each tested fluid formed significantly higher contact angle on infected leaves compared to healthy ones (Fig. 9.3c; Table 9.2).

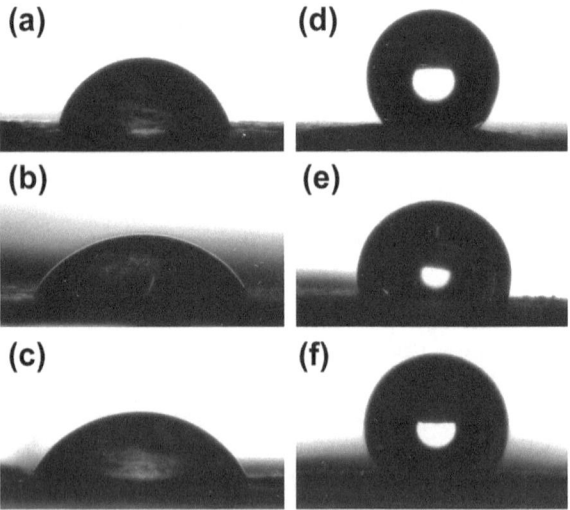

Fig. 9.2 Droplets of double-distilled water (**a, d**), diiodomethane (**b, e**), and ethylene glycol (**c, f**) on healthy (**a − c**) and infected leaves (**d − f**) of *Q. robur*

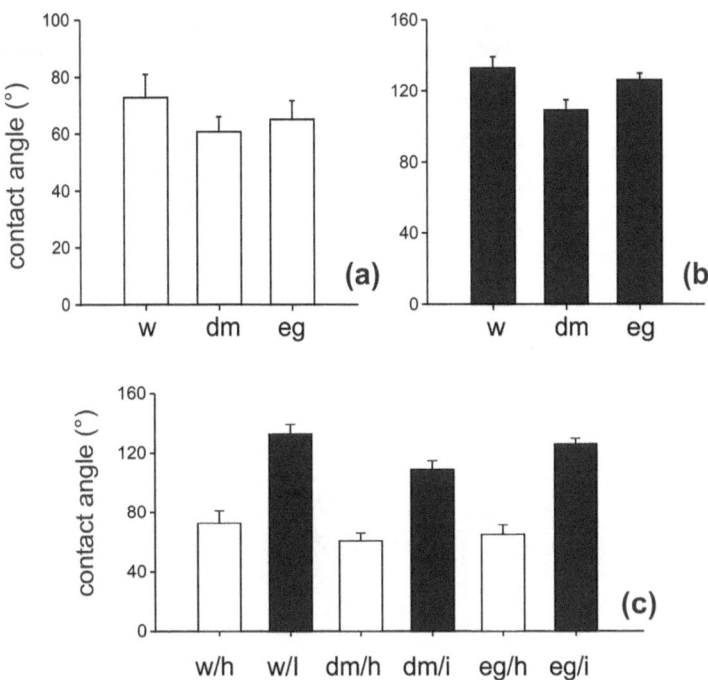

Fig. 9.3 Contact angles of double-distilled water, diiodomethane, and ethylene glycol on healthy (**a**) and infected leaves (**b**) of *Q. robur* and comparison between angles on two leaf types (**c**). *dm* diiodomethane, *eg* ethylene glycol, *h* healthy leaves, *i* infected leaves, *w* double-distilled water

Table 9.1 Results of the post-hoc pairwise multiple comparisons (Tukey test) between contact angles of different fluids on healthy and infected leaves of *Q. robur*. *df* difference of means, *P* probability value, *q* Tukey test statistics

Comparison	df	q	P
Healthy			
Water vs diiodomethane	11.987	7.221	< 0.001
Water vs ethylene glycol	7.633	4.598	0.006
Diiodomethane vs ethylene glycol	4.354	2.384	0.221
Infected			
Water vs diiodomethane	23.928	18.919	< 0.001
Water vs ethylene glyco	6.779	5.254	0.002
Diiodomethane vs ethylene glycol	17.150	12.074	< 0.001

3.1.2 Surface Free Energy

Both leaves types showed rather low surface free energy values (< 30 mN m^{-1}) with prevailing dispersion components, whereas polar parts constituted less than one fifth of the total surface free energy values (Fig. 9.4). In infected leaves, the surface free energy was extremely low (< 30 mN m^{-1}). Both the total surface free energy

Table 9.2 Results of the comparisons for the contact angles of double-distilled water, diiodomethane (both t-test), and ethylene glycol (Mann-Whitney rank sum test) between healthy and infected leaves of *Q. robur*. *P* probability value, *t* t-test statistics, *U* Mann-Whitney rank sum test statistics

Comparison	Test statistics	P
Water	t = −28.470	< 0.001
Diiodomethane	t = −24.272	< 0.001
Ethylene glycol	U = 225.000	< 0.001

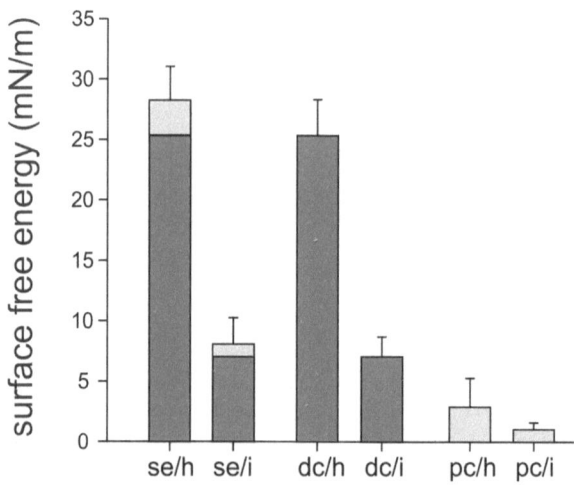

Fig. 9.4 Surface free energy, its dispersion and polar components of healthy and infected leaves of *Q. robur. dc* dispersion component, *h* healthy leaves, *i* infected leaves, *pc* polar component, *se* surface free energy

and its dispersion component were significantly higher in healthy leaves compared to infected ones, while polar component values were similar in both leaf types tested (Fig. 9.4; Table 9.3).

3.2 Microstructure of Healthy and Infected Leaf Surfaces

3.2.1 Healthy Leaves

Healthy adaxial side of the *Q. robur* leaf (Fig. 9.1a) showed slightly uneven surface topography caused by the convex shape of epidermal cells (Fig. 9.1b). At higher magnification, the surface appeared to be densely covered by microscopic (length: $0.53 \pm 0.07\mu m$; width: $0.34 \pm 0.10\mu m$) three-dimensional, flat and thin epicuticular wax projections called platelets (Fig. 9.1c). These thin platelets had entire margins,

Table 9.3 Results of the comparisons (t-test) for the surface free energy, its dispersion and polar components between healthy and infected leaves of *Q. robur*. *P* probability value, *t* t-test statistics

Comparison	t	P
Surface free energy	9.751	< 0.001
Dispersion component	9.341	< 0.001
Polar component	1.309	0.261

protruded almost perpendicular from the surface and were arranged in groups (rosettes). Also large areas with destroyed, probably melted (according to Gülz and Boor (1992)) projections, forming lumps of wax material, could be often seen.

3.2.2   Fungal Structures

In infected leaves, the surface beared *E. alphitoides* colonies (Fig. 9.1d). Mycelium (vegetative part of a fungus) was mostly superficial and highly branched, being composed of a dense network of microscopic (diameter: $0.53 \pm 0.07\mu m$) fungal filaments called hyphae (Fig. 9.1e, f). The hyphal surface was rarely smooth, but usually micro/nanostructured (Fig. 9.5). Numerous pycnidial (flask-shaped) conidiomata (specialized fruiting structures containing masses of conidia, i.e. asexual non-motile spores) were densely scattered (Fig. 9.1e, f). These separate, elongated, with 1.6–2.6 length/diameter ratio, picnidia (length: $26.08 \pm 2.49\mu m$; max. diameter: $12.79 \pm 1.47\mu m$) were sitting on $26.95 \pm 0.88\mu m$ long conidiophores (stalks).

4   Discussion

Our microscopical and experimental data clearly demonstrate that the powdery mildew turns the oak leaf surface to the highly hydrophobic state (Fig. 9.6c). Interestingly, the contact angles of polar and non-polar fluids are very high on the leaves covered with the fungal mycelium, which in turn lead to rather high dispersion component of the surface free energy. Since the original wax coverage of intact leaves has strong dispersion component, one may assume that the chemical background of the surface free energy is similar in both cases. In other words, it means that the fungal surface contains hydrophobic substances that might be similar to the lipophilic coverage of intact leaves. Furthermore, then main differences in contact angles and the surface free energy between intact leaves and those covered by mycelium are presumably based on differences in the surface structure. Indeed, the leaves collected in the midseason bear surfaces covered by three-dimensional wax projections that partially degraded, which is well known for different plant species (Tranquada and Erb 2014) including oak (Gülz and Boor 1992). Such degraded wax

Fig. 9.5 Cryo-SEM
micrographs of the hyphae
of *E. alphitoides* on the
infected leaves of *Q. robur*.
Note nanostructures on the
hyphal surface in
(**b**) and (**c**)

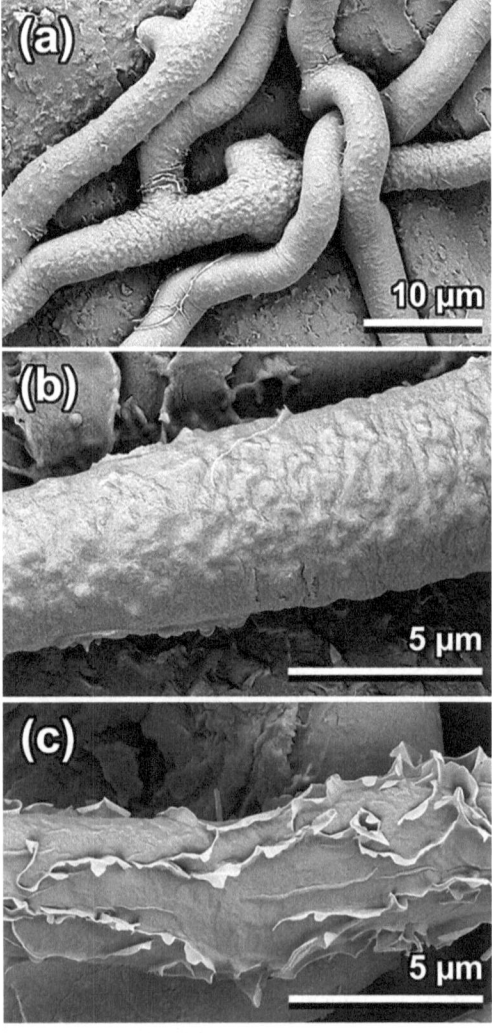

coverage on the non-infected leaves leads to a decrease of contact angles (Klamerus-Iwan and Witek 2018) and to an increase of the surface free energy.

Careful observation of the non-infected leaf surfaces in the native condition at the high resolution of the cryo-SEM detected some degree of degradation of three-dimensional wax projections (Fig. 9.6b), as previously shown (Tranquada and Erb 2014). The sharp edges of wax platelets seem to melt away on the leaf surface. It has been previously assumed that the content of the glandular trichomes empties over the wax layer making wax projections practically dissolved (Gülz and Boor 1992). That is why in the second half of the season, wax structures are seen with several intermediate forms until the state that no more projections can be seen. In contrast, the fungal mycelium demonstrated three-dimensional hierarchical structure,

Fig. 9.6 Schematic explanation of obtained results. At the beginning of the season, oak leaves that are covered by intact three-dimensional wax projections are superhydrophobic (**a**). In the middle and at the end of the season, non-infected leaves become hydrophilic (**b**). The infection by the mildew turns the degraded wax-covered leaf surface to the highly hydrophobic state (**c**)

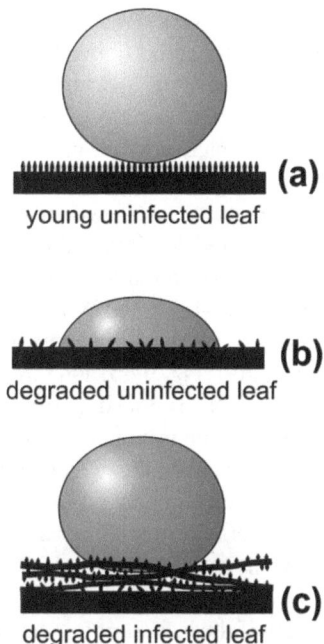

(a)
young uninfected leaf

(b)
degraded uninfected leaf

(c)
degraded infected leaf

consisting of hyphae having cylindrical cross section with some conidiomata bearing conidia on the top (the first hierarchical level) and little nanoscale papillae and scales on the hyphal surface (the second hierarchical level). Such hierarchy in combination with the hydrophobic nature of the surface substances of the fungus leads to the highly hydrophobic state (and low surface free energy) of the infected surface of the leaf (Fig. 9.6c).

Knowing that fungi rely in their growth and development on the moisture (Liyanage et al. 2018), one might see some contradiction revealed by our data showing highly hydrophobic nature of the fungal mycelium. Possible explanation of this fact is that the fluid water presumably does not present optimal condition for the fungal development, but rather high humidity of environmental air. In the case of the mildew, water can be additionally obtained from the host leaves and that is why water is not a limiting factor that should be harvested from environmental fluid water (rain, dew). Furthermore, considering huge overall surface area of the fungal mycelium, there should be some mechanisms preventing water lost due to evaporation from the surface, which might be an additional reason for having lipophilic/hydrophobic surface coating of the hyphae. Fungal mycelium uses water from the host and from environmental humidity for its growth and chemically protects itself from evaporation.

It has been previously found that fungal proteins hydrophobins self-assemble at hydrophobic/hydrophilic interfaces (Wösten 2001; Linder et al. 2005). Due to hydrophobic and hydrophilic amino-acid patches, creating an amphiphilic protein surface, they are able to invert the polarity of surfaces, on which they self-assemble very effectively (Bonazza et al. 2015). Their surface activity-altering properties are relevant for aerial fungal hyphae that produce spores, which are covered with a non-wettable layer of hydrophobins. This also facilitates the dispersal of fungal conidia. The common feature of all these layers is that they consist of amphiphilic mole-cules, which are uniformly oriented at hydrophilic/hydrophobic interfaces and therefore, their growth generates monolayers (Bonazza et al. 2015).

Previous data showed that at the beginning of the season (May), oak leaves that are covered by intact three-dimensional wax projections are superhydrophobic (Fig. 9.6a) and infection by the mildew slightly reduces such hydrophobicity (Klamerus-Iwan and Witek 2018). However, at the end of the season (September), non-infected leaves become hydrophilic (the fact that is also supported by the pres-ent study) (Fig. 9.6b) and infection by the mildew still holds the surface hydrophilic (the fact that is not supported by the present study). The possible explanation of this contradiction might be in the definition of the "infected" surface, which in our case was any area on the leaf covered by fungal mycelium, whereas in the previous study it was a certain percentage of the leaf area covered by the mycelium.

Superhydrophobic surfaces are self-cleaning, due to the so-called "Lotus effect" (Barthlott and Neinhuis 1997), and small particles including conidia would have a low chance to adhere to intact superhydrophobic surfaces of young leaves, espe-cially if those are not in a horizontal position and remain under continuous oscilla-tions due to surrounding air movements. When water drops contact such a surface, they will roll along it and clean it up from the contaminating particles. As a mixture of very long-chain (> C20) fatty acids and their derivatives, epicuticular waxes play an important role as a plant physical barrier limiting the entry of pathogens. Therefore, these waxes have gained increasing attention in the study of plant disease resistance (Aragón et al. 2017; Ziv et al. 2018; Wang et al. 2020). However, the exact role and mechanisms of epicuticular waxes in regulating plant-pathogen inter-actions remain to be explored in future research.

Taking into account above mentioned one may ask the following question: How do conidia generate an initial adherence to such a challenging surface of young intact leaves? In reality, there is no ideal superhydrophobic surface on the leaf. The surface usually has some minute defects, which will be sites of the dew condensa-tion and sites, where rain water sticks to the leaf. This means that rain drops collect-ing conidia elsewhere will deposit them exactly at the sites having defects in the three-dimensional wax coverage and at these sites conidia will have better chances to adhere mechanically and will have higher chances to have more humidity/mois-ture than on other micro-sites of the leaf.

In strongly infected leaves that produce numerous conidia of the fungus, water drops will take conidia from the highly hydrophobic fungal surface, transport and then deposit them on more hydrophic sites (defects, which number is continuously increasing on a non-infected leaf surface with time). That is why another function of high hydrophobicity of the mycelium might be a potential optimization of conidia dispersal.

Our previous study on the insect attachment to micro- and nano-structured sam-ples revealed extremely poor attachment of ladybird beetles to these substrates due to possible absorption of the secretion fluid from insect adhesive pads by the porous media (Gorb et al. 2010, 2017). In the case of the mildew mycelium, due to possible high affinity between the fungal surface and oily insect pad secretion, the fluid might be quickly removed from the contact. That is why all kinds of forces contributing to the wet adhesion (capillary interactions and viscous adhesion) will be reduced or even completely eliminated. This will cause a great reduction in the attachment of insect to such substrates. From the biological point of view, it would be interesting to test whether the mildew mycelium is in general anti-adhesive for insects, which rely on wet adhesion. This will provide new insights into interactions between different organisms associated with a single plant species.

Some aspects of the above discussion may appear rather speculative, but we hope that they might stimulate further studies on the mechanical interactions between the powdery mildew and leaf surface. Following research questions might be of a high importance. (1) How do the defects in the superhydrophobic coating of young intact leaves influence distribution (and further development) of fungal conidia? (2) What is the added effect of the hierarchical structure of mycelium on the highly hydrophobic properties of the fungal surface? (3) What is the role of water in transport and distribution of conidia on the leaf surface? (4) How does the hierarchical structure of mycelium affect insect adhesion?

Since the powdery mildew generates the highly hydrophobic protective layer on the surface, it might represent potential biotechnological route for producing highly hydrophobic and superhydrophobic industrial surfaces and coatings.

References

Aragón, W., Reina-Pinto, J. J., & Serrano, M. (2017). The intimate talk between plants and microorganisms at the leaf surface. *Journal of Experimental Botany, 68*, 5339–5350.

Barthlott, W., & Neinhuis, C. (1997). Purity of the sacred lotus, or escape from contamination in biological surfaces. *Planta, 202*, 1–8. https://doi.org/10.1007/s004250050096.

Barthlott, W., Neinhuis, C., Cutler, D., Ditsch, F., Meusel, I., Theisen, I., & Wilhelmi, H. (1998). Classification and terminology of plant epicuticular waxes. *Botanical Journal of the Linnean Society, 126*, 237–260.

Bonazza, K., Gaderer, R., Neudl, S., Przylucka, A., Allmaier, G., Druzhinina, I. S., Grothe, H., Friedbachera, G., & Seidl-Seiboth, V. (2015). The fungal cerato-platanin protein EPL1 forms highly ordered layers at hydrophobic/hydrophilic interfaces. *Soft Matter, 11*, 1723–1732. https://doi.org/10.1039/c4sm02389g.

Busscher, H. J., van Pelt, A. W. J., de Boer, P., de Jong, H. P., & Arends, J. (1984). The effect of surface roughening of polymers on measured contact angles of liquids. *Colloids and Surfaces, 9*(4), 319–331. https://doi.org/10.1016/0166-6622(84)80175-4.

Erbil, H. J. (1997). Surface tension of polymers. In K. S. Birdi (Ed.), *Handbook of surface and colloid chemistry* (pp. 265–312). Boca Raton: CRC Press.

Fernández, V., Sancho-Knapik, D., Guzmán, P., Peguero-Pina, J. J., Gil, L., Karabourniotis, G., Khayet, M., Fasseas, C., Heredia-Guerrero, J. A., Heredia, A., & Gil-Pelegrín, E. (2014). Wettability, polarity, and water absorption of holm oak leaves: Effect of leaf side and age. *Plant Physiology, 166*, 168–180. https://doi.org/10.1104/pp.114.242040.

Gorb, E. V., & Gorb, S. N. (2006). Physicochemical properties of functional surfaces in pitchers of the carnivorous plant *Nepenthes alata* Blanco (Nepenthaceae). *Plant Biology, 8*, 841–848.

Gorb, E. V., & Gorb, S. N. (2013). Anti-adhesive surfaces in plants and their biomimetic potential. In P. Fratzl, J. W. C. Dunlop, & R. Weinkamer (Eds.), *Materials design inspired by nature: Function through inner architecture* (pp. 282–309). Cambridge: RSC Publishing.

Gorb, E. V., Kastner, V., Peressadko, A., Arzt, E., Gaume, L., Rowe, N., & Gorb, S. N. (2004). Structure and properties of the glandular surface in the digestive zone of the pitcher in the carnivorous plant *Nepenthes ventrata* and its role in insect trapping and retention. *The Journal of Experimental Biology, 207*, 2947–2963. https://doi.org/10.1242/jeb.01128.

Gorb, E. V., Hosoda, N., Miksch, C., & Gorb, S. N. (2010). Slippery pores: Anti-adhesive effect of nanoporous substrates on the beetle attachment system. *Journal of the Royal Society Interface, 7*, 1571–1579.

Gorb, E. V., Baum, M. J., & Gorb, S. N. (2013). Development and regeneration ability of the wax coverage in *Nepenthes alata* pitchers: A cryo-SEM approach. *Scientific Reports, 3*, 3078.

Gorb, E. V., Hofmann, P., Filippov, A. E., & Gorb, S. N. (2017). Oil absorbing ability of three-dimensional epicuticular wax coverages in plants. *Scientific Reports, 7*(45483), 1–11.

Gülz, P.-G., & Boor, G. (1992). Seasonal variations in epicuticular wax ultrastructures of *Quercus robur* leaves. *Zeitschrift für Naturforschung C, 47c*, 807–814.

Hajji, M., Dreyer, E., & Marçais, B. (2009). Impact of *Erysiphe alphitoides* on transpiration and photosynthesis in *Quercus robur* leaves. *European Journal of Plant Pathology, 125*, 63–72. https://doi.org/10.1007/s10658-009-9458-7.

https://de.wikipedia.org/wiki/Eichenmehltau. Accessed 1 Oct 2020.

https://en.wikipedia.org/wiki/Erysiphe_alphitoides. Accessed 1 Oct 2020.

https://en.wikipedia.org/wiki/Quercus_robur. Accessed 1 Oct 2020.

Jeffree, C. F. (1986). The cuticle, epicuticular waxes and trichomes of plants, with reference to their structure, function and evolution. In B. Juniper & R. Southwood (Eds.), *Insects and the plant surface* (pp. 23–64). London: Edward Arnold Publishers.

Jeffree, C. E. (2006). The fine structure of the plant cuticle. In M. Riederer & C. Müller (Eds.), *Biology of the plant cuticle* (pp. 11–125). Oxford: Blackwell.

Jeffree, C. E., Baker, E. A., & Holloway, P. J. (1975). Ultrastructure and recrystallisation of plant epicuticular waxes. *The New Phytologist, 75*, 539–449.

Jetter, R., & Riederer, M. (1994). Epicuticular crystals of nanocosan-10-ol: *in vitro* reconstitution and factors influencing crystal habits. *Planta, 195*, 257–270.

Jetter, R., & Riederer, M. (1995). *In vitro* reconstitution of epicuticular wax crystals: Formation of tubular aggregates by long chain secondary alkanediols. *Botanica Acta: Journal of the German Botanical Society, 108*, 111–120.

Klamerus-Iwan, A., & Witek, W. (2018). Variability in the wettability and water storage capacity of common oak leaves (*Quercus robur* L.). *Water, 10*, 695. https://doi.org/10.3390/w10060695.

Koch, K., & Ensikat, H. J. (2008). The hydrophobic coatings of plant surfaces: Epicuticular wax crystals and their morphologies, crystallinity and molecular self-assembly. *Micron, 39*, 759–772.

Linder, M., Szilvay, G., Nakari-Setälä, T., & Penttilä, M. (2005). Hydrophobins: The protein amphiphiles of fungi. *FEMS Microbiology Reviews, 29*, 877–896.

Liyanage, K. K., Khan, S., Brooks, S., Mortimer, P. E., Karunarathna, S. C., Xu, J., & Hyde, K. D. (2018). Morpho-molecular characterization of two *Ampelomyces* spp. (Pleosporales) strains mycoparasites of powdery mildew of *Hevea brasiliensis*. *Frontiers in Microbiology, 9*, 12. https://doi.org/10.3389/fmicb.2018.00012.

Mougou, A., Dutech, C., & Desprez-Loustau, M.-L. (2008). New insights into the identity and origin of the causal agent of oak powdery mildew in Europe. *Forest Pathology, 38*(4), 275–287. https://doi.org/10.1111/j.1439-0329.2008.00544.x.

Owens, D. K., & Wendt, R. C. (1969). Estimation of the surface free energy of polymers. *Journal of Applied Polymer Science, 13*, 1741–1747. https://doi.org/10.1002/app.1969.070130815.

Szewczyk, W., Kuźmiński, R., Mańka, M., Kwaśna, H., Łakomy, P., Baranowska-Wasilewska, M., & Behnke-Borowczyk, J. (2015). Occurrence of *Erysiphe alphitoides* in oak stands affected by flood disaster. *Forest Research Papers, 76*(1), 73–77. https://doi.org/10.1515/frp-2015-0008.

Tranquada, G. C., & Erb, U. (2014). Morphological development and environmental degradation of superhydrophobic aspen and black locust leaf surfaces. *Ecohydrology, 7*, 1421–1436.

Wang, X., Kong, L., Zhi, P., & Chang, C. (2020). Update on cuticular wax biosynthesis and its roles in plant disease resistance. *International Journal of Molecular Sciences, 21*, 5514. https://doi.org/10.3390/ijms21155514.

Wösten, H. A. B. (2001). Hydrophobins: Multipurpose proteins. *Annual Review of Microbiology, 55*, 625–646.

Zisman, W. A. (1964). Relation of the equilibrium contact angle to liquid and solid constitution. In F. Fowkes & American. Chemical Society (Eds.), *Contact angle, wettability, and adhesion, advances in chemistry* (Vol. 10, pp. 1–51). Washington, DC: American Chemical Society.

Ziv, C., Zhao, Z., Gao, Y. G., & Xia, Y. (2018). Multifunctional roles of plant cuticle during plant-pathogen interactions. *Frontiers in Plant Science, 9*, 1088.

Chapter 10
Physics of Suction Cups in Air and in Water

A. Tiwari and B. N. J. Persson

Abstract We present experimental results for the dependency of the pull-off time (failure time) on the pull-off force for suction cups in the air and in water. The results are analysed using a theory we have developed for the contact between suction cups and randomly rough surfaces. The theory predicts the dependency of the pull-off time (failure time) on the pull-off force, and is tested with measurements performed on suction cups made from a soft polyvinyl chloride (PVC). As substrates we used sandblasted poly(methyl methacrylate) (PMMA). The theory is in good agreement with the experiments in air, except for surfaces with the root-mean-square (rms) roughness below ≈ 1 μm, where we observed lifetimes much longer than predicted by the theory. We show that this is due to out-diffusion of plasticizer from the soft PVC, which block the critical constrictions along the air flow channels. In water, some deviation between theory and experiments is observed which may be due to capillary forces. We discuss the role of cavitation for the failure time of suction cups in water.

Keywords Contact mechanics · Suction cups · Ballistic and diffusive leakage · Viscoelasticity

1 Introduction

All solids have surface roughness which has a huge influence on a large number of physical phenomena such as adhesion, friction, contact mechanics and the leakage of seals (Barber 2018; Creton and Ciccotti 2016; Gnecco and Meyer 2015; Israelachvili 2011; Müser et al. 2017; Pastewka and Robbins 2014; Persson 2001, 2006, 2013; Spolenak et al. 2005; Tiwari 2018; Vakis et al. 2018). Thus, when two solids with nominally flat surfaces are squeezed into contact, unless the applied

A. Tiwari (✉) · B. N. J. Persson (✉)
PGI-1, Forschungszentrum Juelich, Juelich, Germany
e-mail: b.persson@fz-juelich.de; https://www.MultiscaleConsulting.com

© The Author(s), under exclusive license to Springer Nature
Switzerland AG 2021
A. Gadomski (ed.), *Water in Biomechanical and Related Systems*,
Biologically-Inspired Systems 17, https://doi.org/10.1007/978-3-030-67227-0_10

squeezing pressure is high enough, or the elastic modulus of at least one of the solids low enough, a non-contact region will occur at the interface. If the non-contact region percolate there will be open channels extending from one side of the nominal contact region to the other side. This will allow fluid to flow at the interface from a high fluid pressure region to a low-pressure region.

For elastic solids with randomly rough surfaces the contact area percolate when the relative contact area $A/A_0 \approx 0.42$ (see (Dapp et al. 2012)), where A_0 is the nominal contact area and A the area of real contact (projected on the xy-plane). When the contact area percolate there is no open (non-contact) channel at the interface extending across the nominal contact region, and no fluid can flow between the two sides of the nominal contact.

The discussion above is fundamental for the leakage of static seals (Lorenz and Persson 2010a, b; Persson and Yang 2008; Tiwari et al. 2017). Here we are interested in rubber suction cups. In this application, the contact between the suction cup and the counter surface (which form an annulus) must be so tight that negligible fluid can flow from outside the suction cup to inside it.

Suction cups find ubiquitous usage in our everyday activities such as hanging of items to smooth surfaces in our houses and cars, and for technologically demanding applications such as lifting fragile and heavy objects safely in a controlled manner using suction cups employing vacuum pumps. Suction cups are increasingly used in robotic applications, such as robots which can climb walls and clean windows. The biomimetic design of suction cups based on octopus vulgaris, remora (sucker fish), limpets and Northern Clingfish is an area of current scientific investigations whose main objectives is to manufacture suction cups exhibiting adhesion under water and on surfaces with varying degree of surface roughness (Ditsche and Summers 2019; Sandoval et al. 2019). Recently Iwasaki et al. presented the concept of magnet embedded suction cups for in-vivo medical applications (Iwasaki et al. 2020).

2   Theory: Gas Leakage

The suction cups we study below can be approximated as a truncated cone with the diameter $2r_1$. The angle α and the upper plate radius r_0 are defined in Fig. 10.1. When a suction cup is pressed in contact with a flat surface the rubber cone will make apparent contact with the substrate in an annular region, but the contact pressure will be largest in a smaller annular region of width $l(t)$ formed close to the inner edge of the nominal contact area (see Fig. 10.1). We will assume that the rubber-substrate contact pressure in this region of space is constant, $p = p(t)$, and zero elsewhere.

If we define $h_0 = r_0 tan\alpha$ the volume of gas inside the suction cup is

$$V = \pi r^2 \frac{1}{3}(h + h_0) - \pi r_0^2 \frac{1}{3} h_0$$

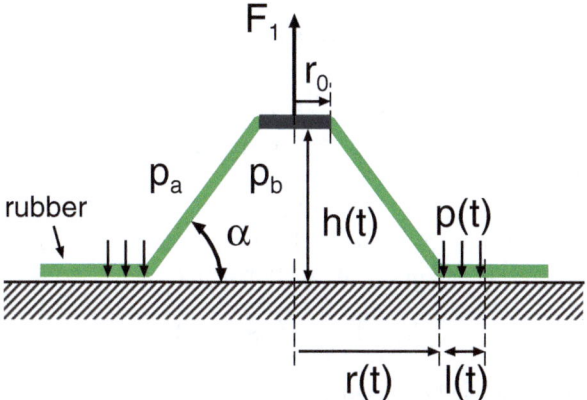

Fig. 10.1 Schematic picture of the suction cup pull-off experiment used in the present study. The container with the suction cup is either empty or filled with distilled water

Since

$$\frac{r}{r_0} = 1 + \frac{h}{h_0} \tag{10.1}$$

we get

$$V = V_0 \left[\left(\frac{r}{r_0} \right)^3 - 1 \right] \tag{10.2}$$

where $V_0 = \pi r_0^2 h_0/3$. The elastic deformation of the rubber film (cone) needed to make contact with the substrate require a normal force $F_0(h)$, which we will refer to as the cup (non-linear) spring-force. This force result from the bending of the film and to the (in-plane) stretching and compression of the film needed to deform (part of) the conical surface into a flat circular disc or annulus. The function $F_0(h)$, can be easily measured experimentally (see below).

We assume that the rubber cup is in repulsive contact with the substrate over a region of width $l(t)$. Since the thickness of the suction cup material decreases as r increases, we expect that l decreases as r increases. From optical pictures of the contact we have found that to a good approximation

$$l = l_0 + l_a \left(1 - \frac{r}{r_0} \right) = l_0 - l_a \frac{h}{h_0} \tag{10.3}$$

where $l_a = (l_1 - l_0)/(1 - r_1/r_0)$ where l_1 is the width of the contact region when $r = r_1$, and l_0 the width of the contact region when $r = r_0$. The contact pressure $p = p(t)$ in the circular contact strip is assumed to be constant

$$p \approx \frac{F_0(h)}{2\pi r l} + \beta(p_a - p_b)$$

$$(10.4)$$

where β is a number between 0 and 1, which takes into account that the gas pressure (in the non-contact region) in the strip $l(t)$ changes from p_a for $r > r(t) + l(t)$ to p_b for $r < r(t)$, while the outside pressure is p_a.

Assume that the pull-force F_1 act on the suction cup (see Fig. 10.1). The sum of F_1 and the cup spring-force F_0 must equal the force resulting from the pressure difference between outside and inside the suction cup, i.e.

$$F_0 + F_1 = \pi r^2(p_a - p_b)$$

$$(10.5)$$

We assume that the air can be treated as an ideal gas so that

$$p_b V_b = N_b k_B T$$

$$(10.6)$$

The number of molecules per unit time entering the suction cup $dN_b(t)/dt$, is given by

$$\frac{dN_b}{dt} = f(p, p_a, p_b)\frac{L_y}{L_x}$$

$$(10.7)$$

Here L_x and L_y are the lengths of the sealing region along and orthogonal to the gas leakage direction, respectively. In the present case

$$\frac{L_y}{L_x} = \frac{2\pi r}{l}$$

The (square-unit) leak-rate function $f(p, p_a, p_b)$ will be discussed below.

The Eqs. (10.1, 10.2, 10.3, 10.4, 10.5, 10.6, and 10.7) constitute 7 equations from which the following 7 quantities can be obtained: $h(t)$, $r(t)$, $l(t)$, $V(t)$, $p_b(t)$, $p(t)$ and $N_b(t)$. The Eqs. (10.1, 10.2, 10.3, 10.4, 10.5, 10.6, and 10.7) can be easily solved by numerical integration.

The suction cup stiffness force $F_0(h)$ depends on the speed with which the suction cup is compressed (or decompressed). The reason for this is the viscoelastic nature of the suction cup material. To take this effect into account we define the contact time state variable $\phi(t)$ as (Persson 2013; Rice and Ruina 1983; Ruina 1983)

$$\frac{d\phi}{dt} = 1 - \frac{\phi}{l}\frac{dr}{dt}$$

$$(10.8)$$

with $\phi(0) = 0$. For stationary contact, $dr/dt = 0$, this equation gives just the time t of stationary contact, $\phi(t) = t$. When the ratio $(dr/dt)/l$ is non-zero but constant (10.8) gives

$$\phi(t) = \left(1 - e^{-t/\tau}\right)\tau,$$

where $\tau = l/(dr/dt)$. Thus, for $t >> \tau$ we get $\phi(t) = \phi_0 = \tau$, which is the time a particular point on the suction cup surface stay in the rubber-substrate contact region of width $l = l(t)$. It is only in this part of the rubber-substrate nominal contact region where a strong (repulsive) interaction occur between the rubber film and the substrate, and it is region of space which is most important for the gas sealing process.

From dimensional arguments, we expect that $F_0(h)$ is proportional to the effective elastic modulus of the cup material. We have measured $F_0(h)$ at a constant indentation speed dh/dt, corresponding to a constant radial velocity $dr/dt = [(dh/dt)(r_0/h_0)]$ (see (10.1)). In this case the effective elastic modulus is determined by the relaxation modulus $E_{eff}(t)$ calculated for the contact time $\phi_0 = l/(dr/dt)$. However, in general dr/dt may be strongly time-dependent. We can take that into account by replacing the measured $F_0(h)$ by the function $F_0(h)E_{eff}(\phi(t))/E_{eff}(\phi_0)$.

2.1 Diffusive and Ballistic Gas Leakage

The gas leakage result from the open (non-contact) channels at the interface between the rubber film and the substrate. Most of the leakage occur in the biggest open flow channels. The narrowest constriction in the biggest open channels are denoted as the critical constrictions. Most of the gas pressure drop occur over the critical constrictions, which therefore determine the leak-rate to a good approximation. The surface separation in the critical constrictions is denoted by u_c. Theory shows that the lateral size of the critical constrictions is much larger than the surface separation u_c (typically by a factor of ~ 100) (Lorenz and Persson 2010a, b; Persson and Yang 2008).

In the theory for suction cups enters the leakrate function $f(p, p_a, p_b)$ (see (10.7)). This function can be easily calculated when the gas flow through the critical constrictions occur in the diffusive and ballistic limits (see Fig. 10.2).

Here we present an interpolation formula which is (approximately) valid independent of the ratio between the gas mean free path and the surface separation at the critical constrictions:

$$\frac{dN_b}{dt} = \frac{1}{24}\frac{L_y}{L_x}\frac{\left(p_a^2 - p_b^2\right)}{k_B T}\frac{u_c^3}{\eta}\left(1 + 12\frac{\eta \bar{v}}{\left(p_a + p_b\right)u_c}\right) \qquad (10.9)$$

Here η is the gas viscosity and $k_B T$ the thermal energy. The gas leakage Eq. (10.9) is in good agreement with treatment using the Boltzmann equation, and with experiments (Nacer 2012; Tiwari and Persson 2019). To calculate u_c we need the relation between the interfacial separation u and the contact pressure p. For this we have used the Persson contact mechanics theory (Afferrante et al. 2018; Almqvist et al. 2011; Persson 2001).

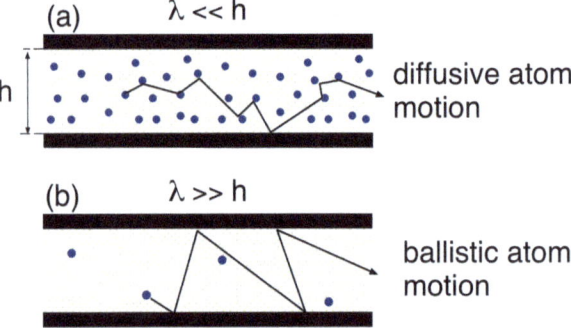

Fig. 10.2 Diffusive (**a**) and ballistic (**b**) motion of the gas atoms in the critical junction. In case (**a**) the gas mean free path λ is much smaller than the gap width u_c and the gas molecules makes many collisions with other gas molecules before a collision with the solid walls. In the opposite limit, when $\lambda > > u_c$ the gas molecules make many collisions with the solid wall before colliding with another gas molecule. In the first case (**a**) the gas can be treated as a (compressible) fluid, but this is not the case in (**b**)

3 Experimental

We carried out the leakage experiments in air and water using two suction cups, denoted A and B, made from soft PVC. These suction cups have different geometrical designs, which has an influence on the suction cup stiffness and failure time, as discussed in Sects. 3.2 and 4.1 respectively.

3.1 Viscoelastic Modulus of PVC

Viscoelastic modulus measurements of the suction cup A and B where carried out in oscillatory tension mode using a Q800 DMA instrument produced by TA Instruments. Figure 10.3 shows (a) the temperature dependency of the low strain ($\varepsilon = 0.0004$) modulus E, and (b) $\tan\delta = \mathrm{Im}E/\mathrm{Re}E$, for the frequency f = 1 Hz. Results are shown for the soft PVC of suction cup A (red lines) and B (green lines). Note that both materials exhibit very similar viscoelastic modulus. If we define the glass transition temperature as the temperature where $\tan\delta$ is maximal (for the frequency f = 1 Hz) then Tg $\approx$ 0°C.

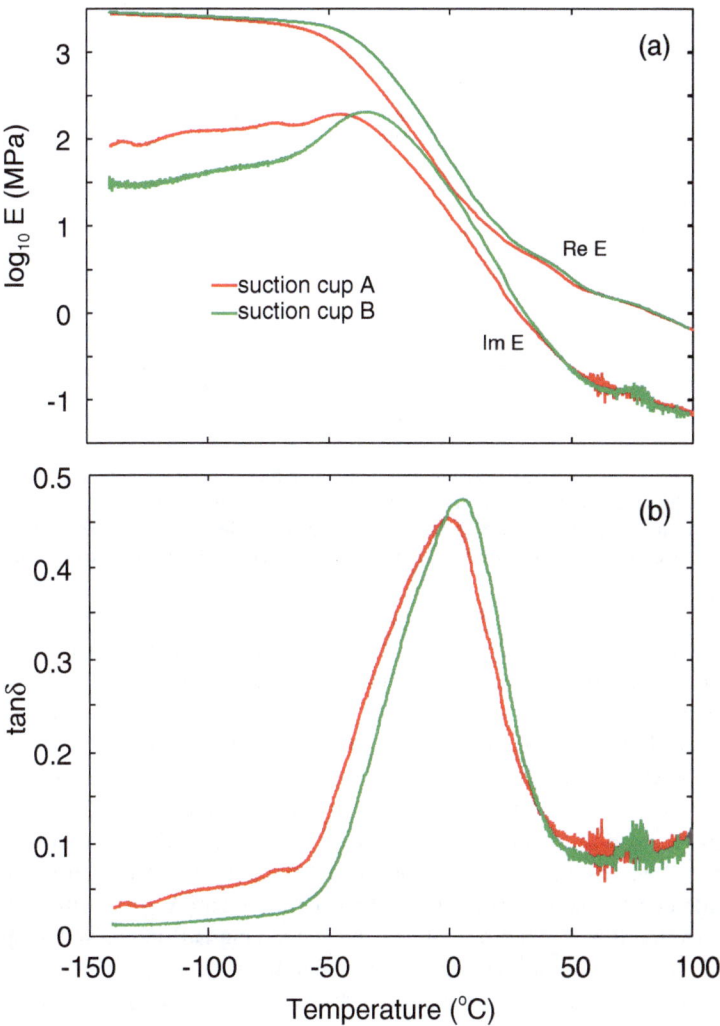

Fig. 10.3 The dependency of the low strain ($\varepsilon = 0.0004$) viscoelastic modulus (**a**), and tanδ = ImE/ReE (**b**), on the temperature for the frequency f = 1 Hz. For the soft PVC of suction cup A (red) and B (green)

3.2   Suction Cup Stiffness Force F_0

We have measured the relation between the normal force F_0 and the normal displacement of the top of a suction cup. In the experiments, we increase the displacement of the top plate (see Fig. 10.1) at a constant speed and measure the resulting force. We show results for the two-different suction cups A and B.

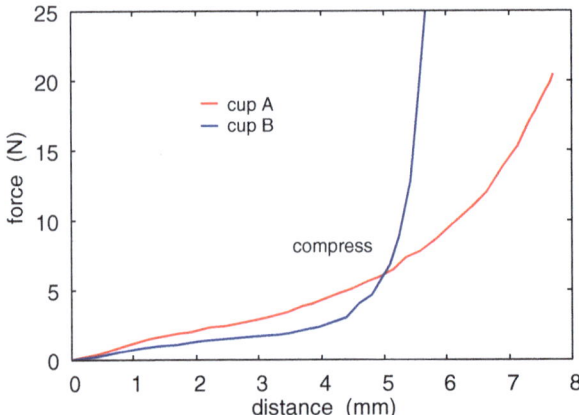

Fig. 10.4 The stiffness force $F_0(h)$ (in N) as a function of the squeezing (or compression) distance (in mm) for the suction cups A (red) and B (blue). The suction cups are squeezed against a smooth glass plate with a hole in the centre through which the air can leave so the air pressure inside the rubber suction cup is the same as outside (atmospheric pressure). The glass plate is lubricated with soap-water

We have measured the force F_0 for the suction cups squeezed against a smooth glass plate lubricated by soap water. The glass plate has a hole below the top of the suction cup; this allowed the air to leave the suction cup without any change in the pressure inside the suction cup (i.e. $p_a = p_b$ is equal to the atmospheric pressure). Figure 10.4 shows the stiffness force $F_0(h)$ (in N) as a function of the squeezing (or compression) distance (in mm) for the suction cups A (red) and B (blue). The suction cups A and B are both made from similar type of soft PVC and both have the diameter ≈ 4 cm. However, for suction cup B the angle $\alpha = 21^o$ in contrast to $\alpha = 33^o$ for suction cup A, and the PVC film is thicker for the cup A. This difference in the angle α and the film thickness influence the suction cup stiffness force as shown in Fig. 10.4. Note that before the strong increase in the $F_0(h)$ curve which result when the suction cup is squeezed into complete contact with the counter surface the suction cup A has a stiffness nearly twice as high as that of the suction cup B.

4 Experiment: Failure of Suction Cup in Air

We have studied how the failure time of a suction cup depends on the pull-off force (vertical load) and the substrate surface roughness (see Fig. 10.5). The suction cup was always attached to a horizontal surface and a mass-load was attached to the suction cup. We varied the mass-load from 0.25 kg to 8 kg. If full vacuum would prevail inside the suction cup, the maximum possible pull-off force would be $\pi r_1^2 p_a$. Using $r_1 = 19$ mm and $p_a = 100$ kPa we get $F_{max} = 113$ N or about 11 kg mass load. However, the maximum load possible in our experiments for a smooth substrate surface is about 9 kg, indicating that no complete vacuum was obtained. This may, in least in part, be due to problems to fully remove the air inside the suction cup in

Fig. 10.5 Schematic
picture of the experimental
set-up for measuring the
pull-off force in the air and
in water

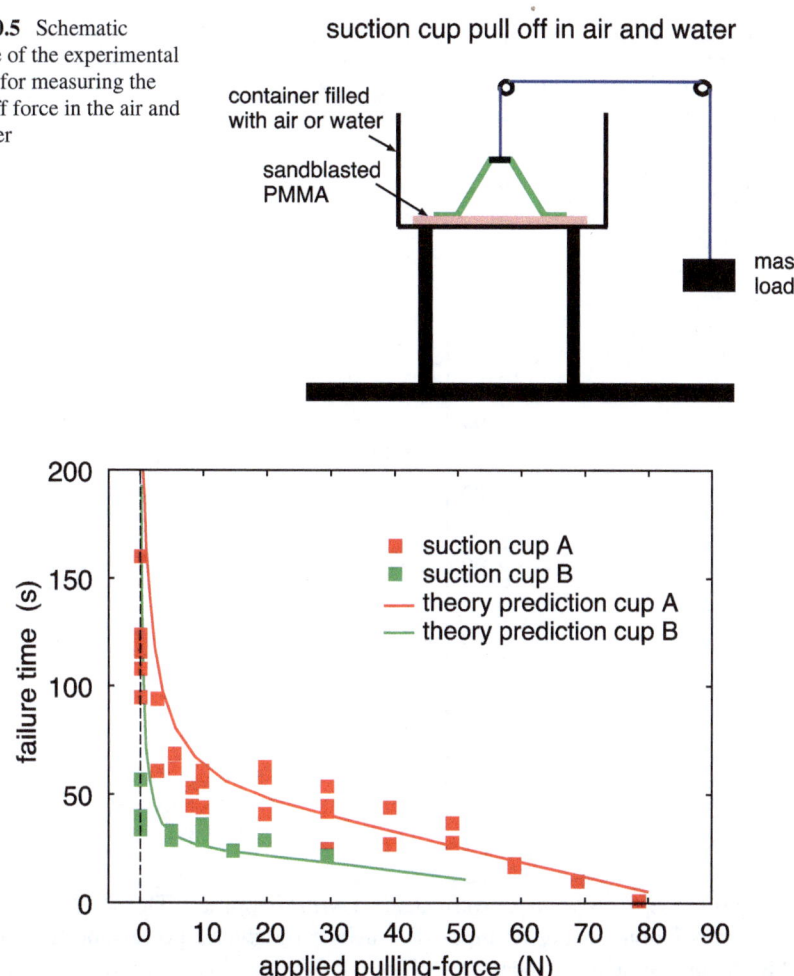

Fig. 10.6 The dependency of the pull-off time (failure time) on the applied (pulling) force. The
soft PVC suction cups A and B are in contact with a sandblasted PMMA surface with the rms
roughness 1.89 μm. All surfaces were cleaned with soap water before the experiments

the initial state. In addition, we have found that for mass loads above 8 kg the pull-
off is very sensitive to instabilities in the macroscopic deformations of the suction
cup, probably resulting from small deviations away from the vertical direction of
the applied loading force.

4.1   The Dependency of the Failure Time on the Pull off Force

Figure 10.6 shows the dependency of the pull-off time (failure time) on the applied
pulling force. The results are for the soft PVC suction cups A (red) and B (green) in
contact with a sandblasted PMMA surface with the rms-roughness 1.89 μm. Before

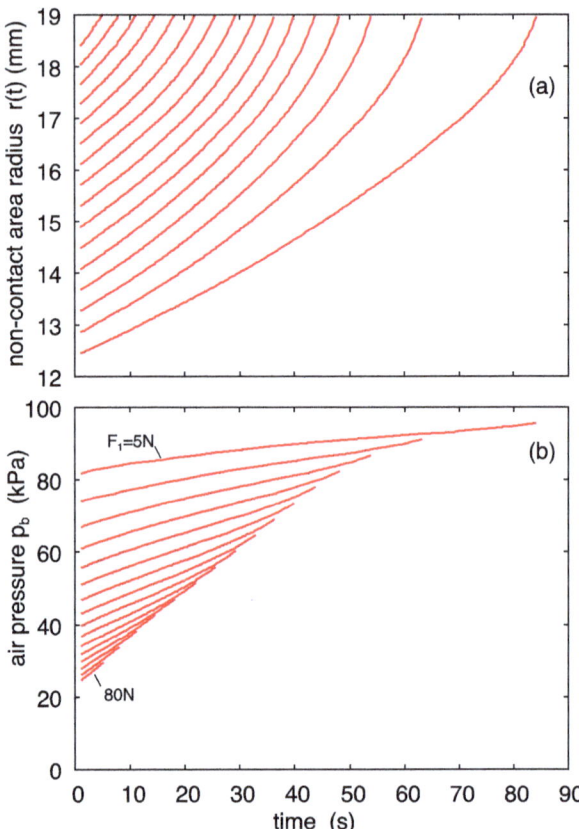

Fig. 10.7 The calculated dependency time dependency of the (**a**) radius of the non-contact region and (**b**) the pressure in the suction cup, for several pull-off forces (from $F_1 = 5$ N in steps of 5 N to 80 N). The soft PVC suction cup is in contact with a sandblasted PMMA surface with the rms roughness 1.89 μm

the measurement, all surfaces were cleaned with soap water. The solid lines are the theory predictions, using as input the surface roughness power spectrum of the PMMA surface, and the measured stiffness of the suction cup, the latter corrected for viscoelastic time-relaxation as described above. Note the good agreement between the theory and the experiments in spite of the simple nature of the theory.

Figure 10.7a shows the calculated time dependency of the radius of the non-contact region, and (b) the gas pressure in the suction cup. We show results for several pull-off forces from $F_1 = 5$ N in steps of 5 N to 80 N. The smaller angle α for suction cup B than for cup A implies that if the same amount of gas would leak into the suction cups the gas pressure p_b inside the suction cup will be highest for the suction cup B. This will tend to reduce the lifetime of cup B. Similarly, the smaller stiffness of the cup B result in smaller contact pressure p, which will increase the leakage rate and reduce the lifetime. Hence both effects will make the lifetime of the suction cup B smaller than that of the cup A.

4.2   The Dependency of the Failure Time on Surface Roughness

Figure 10.8 shows the dependency of the pull-off time (failure time) on the substrate surface roughness. The results are for the type A PVC suction cup in contact with sandblasted PMMA surfaces with different rms roughness, and a table surface. Note that for "large" roughness the predicted failure time is in good agreement with the measured data, but for rms roughness below $\approx 1 \ \mu m$ the measured failure times are much larger than the theory prediction. In addition, the dependency of the radius $r(t)$ of the non-contact region on time is very different in the two cases: For roughness, larger than $\approx 1 \ \mu m$ the radius increases continuously with time as also expected from theory (see Fig. 10.7a). For roughness below ≈ 1 μm the boundary line $r(t)$ stopped to move a short time after applying the pull-off force, and remained fixed until the detachment occurred by a rapid increase in $r(t)$ (catastrophic event).

We attribute this discrepancy between theory and experiments to transfer of plasticizer from the soft PVC to the PVC-PMMA interface; this (high viscosity) fluid will fill-up the critical constrictions and hence stop, or strongly reduce, the flow of air into the suction cup. This is consistent with many studies (Kim et al. 2003) of the transfer of plasticizer from soft PVC to various contacting materials. These studies show typical transfer rates (at room temperature) corresponding to a ~ 1–$10 \ \mu m$ thick film of plasticizer after one week waiting time. Optical pictures of the rough

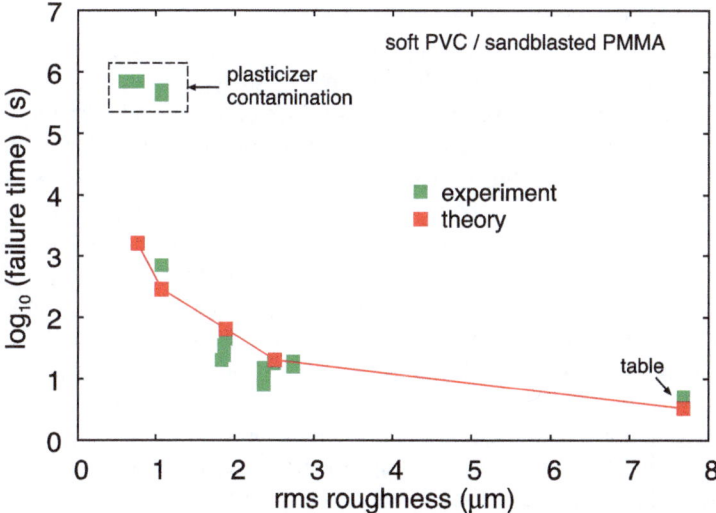

Fig. 10.8 The dependency of the pull-off time (failure time) on the substrate surface roughness. For soft PVC suction cups in contact with sandblasted PMMA surfaces with different rms-roughness, and a table surface. The pull-off force $F_1 = 10$ N. All surfaces were cleaned with soap water before the experiments, and a new suction cup was used for each experiment

PMMA surface after long contact with the suction cup A also showed darkened (and sticky) annular regions indicating transfer of material from the PVC to the PMMA surface (Tiwari and Persson 2019).

5 Theory: Liquid (Water) Leakage

Assuming that the water can be treated as an incompressible Newtonian liquid, and assuming laminar flow, the volume of liquid flowing per unit time into the suction cup is given by

$$\frac{dV_b}{dt} = \frac{1}{12\eta}\frac{L_y}{L_x}u_c^3\left(p_a - p_b\right) \tag{10.10}$$

This equation replaces (10.9), which is valid for gas flow, but all the other equations are unchanged. In the experiments reported on below, the suction cup is squeezed vertically into contact with the counter surface (here a PMMA sheet). Even if no gas (air) can be detected inside the suction cup before squeezing it in contact with the PMMA surface, after removing the squeezing force we always observe a gas filled region at the top of the suction cup (see Fig. 10.9).

We believe this result from the spring force generating a reduced pressure inside the suction cup, which result in cavitation. When loaded with a pull-off force this gas region expand and result in a pressure inside the suction cup which is somewhere between atmospheric pressure and perfect vacuum (zero pressure). If all the air would have been removed, the pressure in the water could, at least initially, be negative (below vacuum), where the liquid is under mechanical tension. In fact, pressures as low as $p_b \approx -20$ MPa has been observed for water at short times (Herbert et al. 2006). However, this state is only metastable and after long enough time one would expect the nucleation of a gas bubble in the liquid, and the liquid pressure would increase above zero. In fact, water in thermal (or kinetic) equilibrium with the normal atmosphere will have dissolved air molecules (the volume ratio of dissolved gas (at atmospheric pressure) to water is about 0.04 at room temperature), and is unstable against cavitation whenever the pressure falls below the atmospheric pressure. Thus, the problem of determining the pressure in the fluid occurring inside the suction cup is nontrivial, and depend on trapped air bubbles and on how long time the reduced (compared to the atmospheric pressure) pressure prevail.

When a suction cup is used in water, when the pull-off force $F_1 \to 0$, at pull-off the suction cup contains water of atmospheric pressure. When a suction cup is used in the normal atmosphere the suction cup is instead filled with air of atmospheric pressure. Assuming an ideal gas the volume of air of atmospheric pressure flowing per unit time into the suction cup is determined by $p_a(dV/dt) = (dN_b/dt)k_BT$. Using (10.9) this gives

(a) Distilled but not degased water

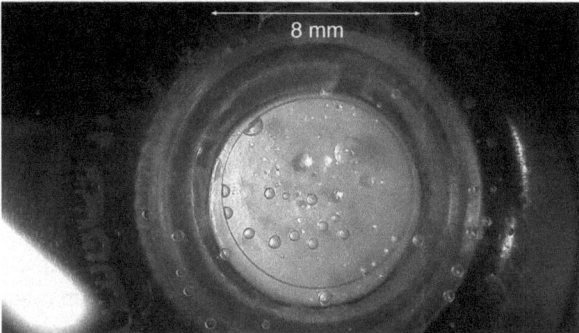

(b) Distilled and degased water

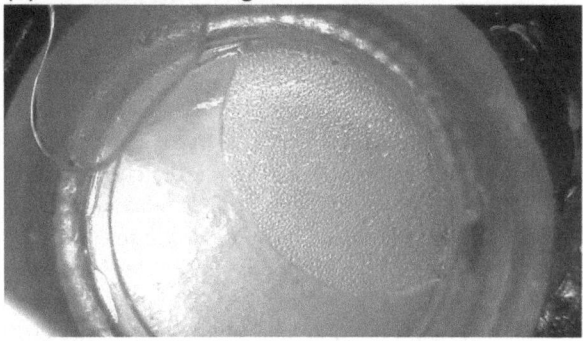

Fig. 10.9 Optical pictures of the center region of a suction cup after squeezing it against a smooth glass plate in (**a**) distilled but not degassed water, and (**b**) distilled and degassed (at 0.2 atmosphere pressure) water. In (**a**) we squeezed the cup with about 15 N against the glass plate, and then pulled it with a similar force for 10 s and then removed the pulling force and waited ∼ 5 min before taking the picture shown. In (**b**) we did the same but using a bigger squeezing and pulling force, about 60 N. In both cases cavitation has occurred, and we observed a slow evolution in time of the gas covered region. The central (or top) part of the suction cup is a flat circular metal disc (diameter 8 mm) covered by a PVC film. The magnification in (**b**) is higher than in (**a**). Note the condensation of small water droplets on the glass surface in the cavity region in (**b**). These droplets were growing with increasing time while the size of the gas bubble decreased slowly

$$\frac{dV}{dt} = \frac{1}{24}\frac{L_y}{L_x}\frac{\left(p_a^2 - p_b^2\right)}{p_a}\frac{u_c^3}{\eta}\left(1 + \frac{12\eta\bar{v}}{\left(p_a + p_b\right)u_c}\right) = \frac{L_y}{12\eta L_x}\left(p_a - p_b\right)u_c^3 Q \quad (10.11)$$

Where

$$Q = \frac{p_a + p_b}{2p_a}\left(1 + 12\frac{\eta\bar{v}}{\left(p_a + p_b\right)u_c}\right) \quad (10.12)$$

Thus, the gas leakage rate differs from the liquid leakrate by a factor Q, which is a product of a factor $(p_a + p_b)/(2p_a)$, derived from the fact that the gas is a compressible fluid, and another factor arising from ballistic air flow. The latter factor does not exist in the liquid case because of the short molecule mean free path in the liquid.

6 Experiment: Suction Cup in Water

We have studied the failure time for suction cup A immersed in distilled water. The experimental set up is shown in Fig. 10.5. Here, the suction cup is squeezed against the rough PMMA countersurface under water. The water level is at least 2 cm above the contacting interface. Figure 10.10 shows the surface roughness power spectrum of the sandblasted PMMA surface used in water. The surface has the rms roughness amplitude 3.8 μm and the rms slope 0.3.

Using the power spectrum in Fig. 10.10 and the Persson contact mechanics theory, in Fig. 10.11 we show the calculated average surface separation $\bar{u}$, and the separation u_c at the critical constriction, as a function of the nominal contact pressure in units of the modulus E (with the Poisson ratio $\nu = 0.5$). The Young's modulus of the suction cup (soft PVC) is of order (depending on the relaxation time) 2–4 MPa, and since the contact pressure is of order ≈ 0.1 MPa, we are interested in p/E below 0.1 and the separation at the critical constriction in most cases is of order a few μm.

At the start of a pull-off experiment the suction cup was squeezed against the PMMA surface with maximum possible hand-force (about 100 N). The squares in Fig. 10.12 shows the dependency of the failure time (time to pull-off) in water on the applied (pulling) force. It is remarkable that, on the average, the failure time increases with increasing pull-off force for F_1 between ≈ 30 N and ≈ 60 N (see also dashed line in Fig. 10.13).

The solid lines in Fig. 10.12 are the theory predictions assuming that in the initial state, before applying the pull-off force, there is only trapped water (but no air) (red line), or trapped air (water with an air bubble) (green line) inside the suction cup.

Fig. 10.10 The surface roughness power spectrum of sandblasted PMMA. Sandblasting was carried out for 10 min at a pressure of 8 bars. The root mean square roughness is 3.8 μm

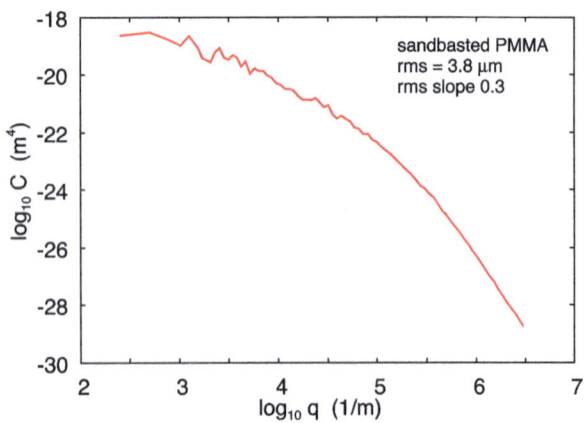

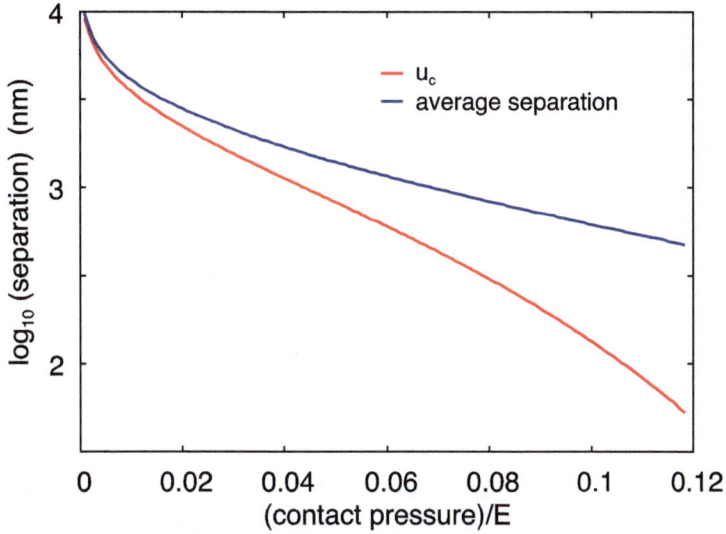

Fig. 10.11 The average surface separation $\bar{u}$, and the separation u_c at the critical constriction, as a function of the nominal contact pressure in units of the modulus E. For the surface with the power spectrum given in Fig. 10.10

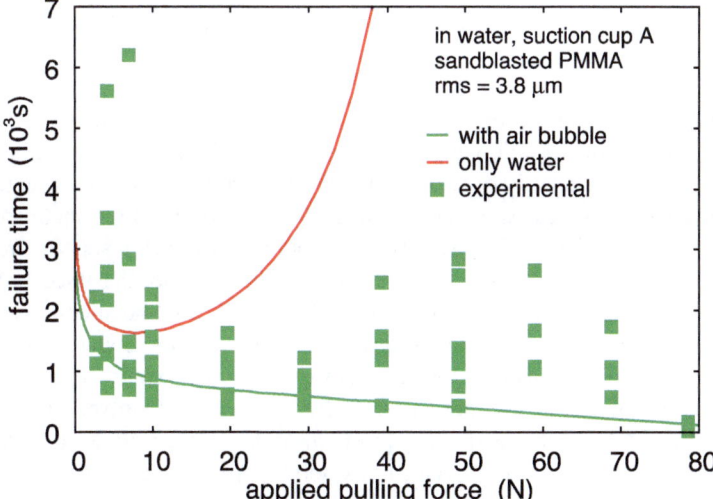

Fig. 10.12 Squares: The dependency of the failure time (time to pull-off) on the applied (pulling) force for the (soft PVC) suction cup in contact with a sandblasted PMMA surface (with the rms roughness 3.8 μm) in water. The red and green solid lines are the theory predictions assuming that in the initial state, before applying the pull-off force, there is some trapped water (but no air), or trapped air (but no water) inside the suction cup, so the initial radius of the detached region for no external load is about 10 mm. When only water occur, if the applied pull-force is big enough, a negative pressure (mechanical tension) develop in the water which will pull the surfaces further together in the contact strip $l(t)$. This reduces the water leakage and increases the failure time. When air occur in the suction cup the pressure is always positive, but below the atmospheric pressure

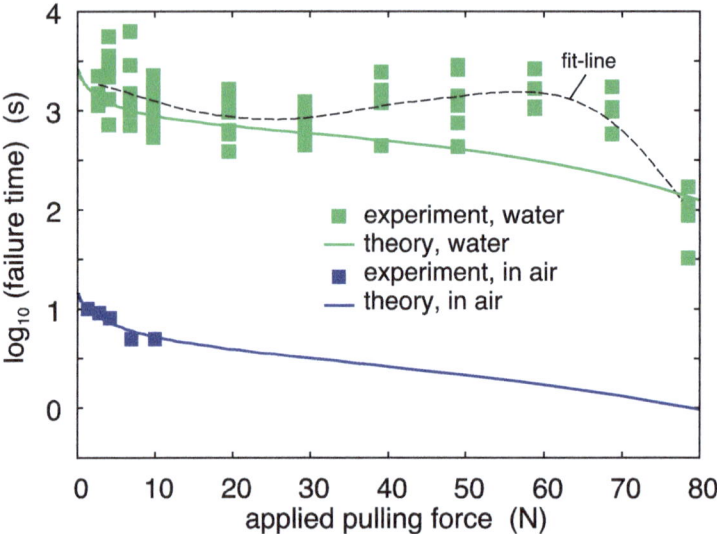

Fig. 10.13 Squares: The dependency of the logarithm of the failure time (time to pull-off) on the applied (pulling) force for the (soft PVC) suction cup in contact with a sandblasted PMMA surface (with the rms roughness 3.8 μm) in water (green) and in the air (blue). The green and blue solid lines are the theory predictions assuming that in the initial state, before applying the pull-off force, there is trapped air inside the suction cup, giving an initial radius of the detached region for no external load of $\approx$ 10 mm as observed experimentally. The black dashed line is a fit-line to the experimental data in water

When only water occur, if the applied pull-force is big enough, a negative pressure (mechanical tension) develop in the water which will pull the surfaces further together in the contact strip of width $l(t)$ (see last term in (10.4)). This reduces the water leakage and increases the failure time. When air occur in the suction cup the pressure in the suction cup is always positive, but below the atmospheric pressure.

In the present experiments, we have used distilled but not degassed water. We observed that even if no air bubbles can be detected inside the suction cup when immersed in the water, after squeezing it in contact with the counter surface and removing the applied squeezing force, we always observed a cavity region (gas bubble) at the top of the suction cup. That is, for water with dissolved air cavitation always occurred inside the suction cup due to the reduced pressure resulting from the spring force. We observed this even for a smooth glass substrate where no trapped micro or nano bubbles of gas is expected before application of the suction cup. (For the sandblasted PMMA, gas (air) may be trapped in roughness cavities.) Consequently, some air is always trapped inside the suction cup before application of the pull-off force F_1. Thus, the experimental results should be compared to the green line which assumes trapped air. The deviation between theory and experiments, observed mainly for applied forces above 30 N, may be due to capillary forces.

Figure 10.13 shows the dependency of the logarithm of the failure time (time to pull-off) on the applied (pulling) force in water (green squares, from Fig. 10.12) and in the air (blue squares). The green and blue solid lines are the theory predictions assuming that in the initial state, before applying the pull-off force, there is some air inside the suction cup. The black dashed line is a fit-line to the experimental data in water.

In the air, the failure time decreased from ≈ 10 s for the pull-off $F_1 = 1.4$ N to $\approx$ 5 s for $F_1 = 10$ N. In water, the failure time was typically ≈ 100 times longer (see Fig. 10.13). This is also predicted by the theory (solid lines in 13) and is mainly due to the change in the viscosity which amount to a factor of ≈ 56 ($\eta = 1 \times 10^{-3}$Pas for water and $\approx 1.8 \times 10^{-5}$ Pas for air). The factor Q, which is due to the finite compressibility of air and to ballistic air flow, is close to unity in the present case. Since the rubber-substrate contact time is longer in water, and since the relaxation modulus $E(t)$ decreases with increasing time (see Ref. (Tiwari and Persson 2019)), in water the surfaces approach each other more closely in the contact strip $l(t)$ then in air, which also tend to increase the failure time in water as compared to in air.

To get a deeper understanding of the failure process, in Fig. 10.14 we show the pressure p_b in the water inside the suction cup as a function of the applied pulling force. The vertical lines give the water pressure as the time changes from start of pull-off ($t = 0$) to detachment ($t = t_{pull-off}$). The red lines assume only water inside

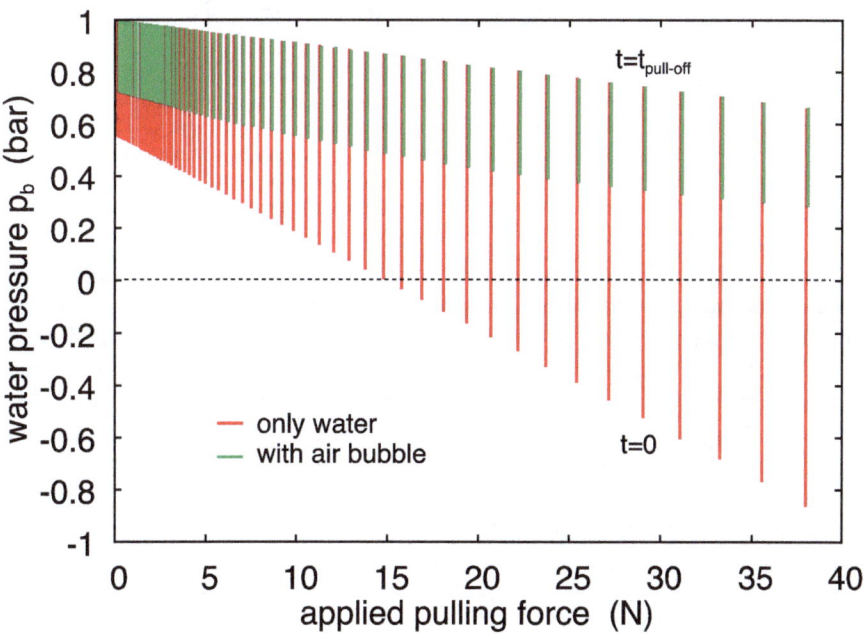

Fig. 10.14 The pressure p_b in the water inside the suction cup as a function of the applied pulling force. The vertical lines give the water pressure as the time changes from start of pull-off ($t = 0$) to detachment ($t = t_{pull-off}$). The red lines assume only water inside the suction cup while the green lines assume a small air bubble trapped in the initial state. In the latter case the water (and air) pressure inside the suction cup is always above the vacuum pressure $p = 0$, while in the first case, when the pull-off force is bigger than ≈ 15 N, the water pressure is negative for some initial time period

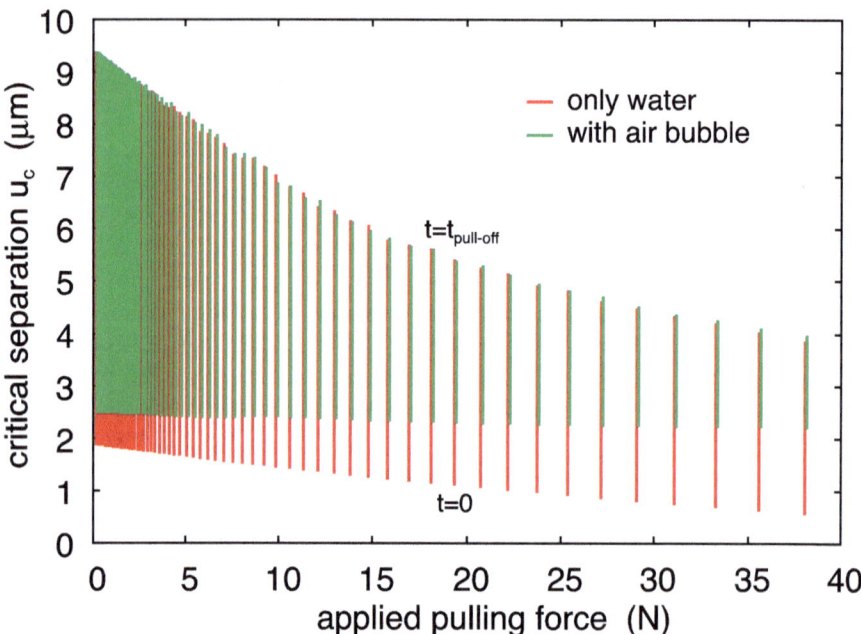

Fig. 10.15 The separation at the critical junction, u_c, as a function of the applied pulling force. The vertical lines give the critical separation as the time changes from start of pull-off ($t = 0$). to detachment ($t = t_{pull - off}$). The red lines assume only water inside the suction cup, while the green lines assume a small air bubble trapped in the initial state. The critical separation for short contact times is smaller in the former case owing to the lower fluid pressure in the suction cup, which pull the surfaces in the nominal contact strip $l(t)$ closer together (see last term in (10.4))

the suction cup, while the green lines assume a small air bubble trapped in the initial state. In the latter case the water (and air) pressure inside the suction cup is always above the vacuum pressure $p = 0$, while in the first case, when the pull-off force is bigger than ≈ 15 N, the water pressure becomes negative for some initial time period.

Figure 10.15 shows the separation at the critical junction, uc, as a function of the applied pulling force. The vertical lines give the critical separation as the time changes from start of pull-off ($t = 0$) to detachment ($t = t_{pull - off}$). Again, the red lines assume only water inside the suction cup, while the green lines assume a small air bubble trapped in the initial state. The critical separation for short contact times is smaller in the former case owing to the lower fluid pressure in the suction cup, which pull the surfaces in the nominal contact strip $l(t)$ closer together (see last term in (10.4)).

Figure 10.16 shows a schematic picture illustrating the influence of cavitation on the failure time. Here we assume that for small applied pulling force, the pressure in the water in the suction cup is not low enough to induce cavitation, but for large enough pulling-force cavitation occur. In this case, since cavitation is a stochastic process involving thermally activated nucleation of an air bubble, when the experiment is repeated many times large fluctuations in the failure time may occur (dotted region).

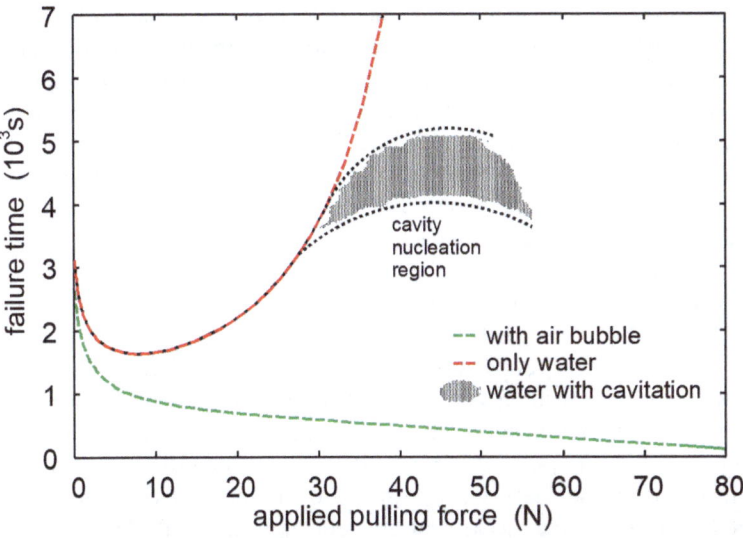

Fig. 10.16 A schematic picture illustrating the influence of cavitation on the failure time. For small applied pulling force, the pressure in the water in the suction cup is not low enough to induce cavitation, but for large enough pulling-force cavitation occur. In this case, since cavitation is a stochastic process involving thermally activated nucleation of an air bubble, when the experiment is repeated many times large fluctuations in the failure time may occur (dotted region)

7   Discussion

Many animals have developed suction cups to adhere to different surfaces in water, e.g., the octopus (Karlsson Green et al. 2013; Smith 1991; Tinnemann et al. 2019; Tramacere et al. 2014) or northern clingfish (Wainwright et al. 2013). Studies have shown that these animals can adhere to much rougher surfaces then man-made suction cups. This is due to the very low elastic modulus of the material covering the suction cup surfaces. Thus, while most man-made suction cups are made from rubber-like materials with a Young's modulus of order a few MPa, the suction cups of the octopus and the northern clingfish are covered by very soft materials with an effective modulus of order 10 kPa.

When a block of a very soft material is squeezed against a counter surface in water it tends to trap islands of water which reduces the contact area and the friction (Hays 2013; Persson and Mugele 2004; Roberts and Tabor 1971). For this reason, the surfaces of the soft adhesive discs in the octopus and the northern clingfish have channels which allow the water to get removed faster during the approach of the suction cup to the counter surface. However, due to the low elastic modulus of the suction cup material, the channels are "flattened-out" when the suction cup is in adhesive contact with the counter surface, and negligible fluid leakage is expected to result from these surface structures.

There are two ways to attach a suction cup to a counter surface. Either a squeezing force is applied, or a pump must be used to lower the fluid (gas or liquid) pressure inside the suction cup. The latter is used in some engineering applications. However, it is not always easy for the octopus to apply a large normal force when attaching a suction cup to a counter surface, in particular before any arm is attached, and when the animal cannot wind the arm around the counter surface as may be the case in some accounts with sperm whale. Similarly, the northern clingfish cannot apply a large normal force to squeeze the adhesive disk in contact with a counter surface. So how can they attach the suction cups? We believe it may be due to changes in the suction cup volume involving muscle contraction as discussed in Ref. (Tiwari and Persson 2019).

For adhesion to very rough surfaces, the part of the suction cup in contact with the substrate must be made from an elastically very soft material. However, using a very soft material everywhere result in a very small suction cup stiffness. We have shown (see in Ref. (Tiwari and Persson 2019)) that a long lifetime requires a large enough suction cup stiffness. Only in this case will the contact pressure p be large enough to reduce the water leakrate to small enough values. Based on this, we have proposed a biomimetic design of an artificial suction cup having a elastically stiff membrane covered with a soft layer with a potential to stick to very rough surfaces under water, see ref. (Tiwari and Persson 2019) for more details. Recently two groups (Ditsche and Summers 2019; Sandoval et al. 2019), working on manufacturing of suction cups inspired from Northern clingfish, were successful in attaining high pull-off forces for rough surfaces in water. These devices utilize a relatively stiff material for the suction cup chamber, and a soft layer at the disc rim (with and without hierarchical structures), which increases the contact area with rough surfaces, and reduce the leakage of the fluid into the suction cups. Sandoval et al. (2019) also varied the design of suction cups, which included radial slits to remove water from the contact. It was suggested (Ditsche and Summers 2019) that the use soft layer increases the friction on a rough substrate, and that this helps in reducing leakage. However, we believe that friction in itself is not very important for the water leakrate, but the elastically soft layer reduces the surface separation and the leakage across the interface.

For suction cups used in the air the lowest possible pressure inside the suction cup is $p_b = 0$, corresponding to perfect vacuum. In reality, it is usually much larger, $p_b \approx 30–90$ kPa. For suction cups in water the pressure inside the suction cup could be negative where the liquid is under mechanical tension. This state is only metastable, and negative pressures have been observed for short times (Herbert et al. 2006). For water in equilibrium with the normal atmosphere one expects cavitation to occur for any pressure below the atmospheric pressure, but the nucleation of cavities may take long time, and depends strongly on impurities and imperfections. For example, crack-like surface defects in hydrophobic materials may trap small (micrometer or nanometer) air bubbles which could expand to macroscopic size when the pressure is reduced below the atmospheric pressure.

Negative pressures have been observed inside the suction cups of octopus. Thus, in one study (Kier and Smith 2002) it was found that most suction cups have pressures above zero, but some suction cups showed pressures as low as ≈ -650 kPa. We find this observation remarkable because for the suction cups we studied cavitation is always observed and the water pressure is therefore always positive.

Trapped air bubbles could influence the water flow into the suction cup by blocking flow channels. For not degassed water, whenever the fluid pressure falls below the atmospheric pressure cavitation can occur, and gas bubbles could form in the flow channels and block the fluid flow due to the Laplace pressure effect. For the suction cups studied above, in the initial state the Laplace pressure is likely to be smaller than the fluid pressure difference between inside and outside the suction cup, in which case the gas bubbles would get removed, but at a later stage in the detachment process this may no longer be true. This is similar to observations in earlier model studies of the water leakage of rubber seals, where strongly reduced leakage rates was observed for hydrophobic surfaces when the water pressure difference between inside and outside the seal become small enough (Tiwari et al. 2017).

8   Summary and Conclusion

We have studied the leakage of suction cups both in air and water. The experimental results were analysed using a newly developed theory of fluid leakage valid in diffusive and ballistic limits combined with Persson contact mechanics theory. In these experiments, the suction cups (made of soft PVC) were pressed against sandblasted PMMA sheets. We found that the measured failure times of suction cups in air to be in good agreement with the theory, except for surfaces with rms-roughness below $\approx$ 1 μm, where diffusion of plasticizer occurred, from the PVC to the PMMA counterface resulting in blocking of critical constrictions. For experiments in water, we found that the failure times of suction cup were $\approx$100 times longer than in air, and this could be attributed mainly to the different viscosity of air and water.

References

Afferrante, L., Bottiglione, F., Putignano, C., Persson, B. N. J., & Carbone, G. (2018). Elastic contact mechanics of randomly rough surfaces: An assessment of advanced asperity models and Persson's theory. *Tribology Letters, 66*(2), 75.

Almqvist, A., Campana, C., Prodanov, N., & Persson, B. N. J. (2011). Interfacial separation between elastic solids with randomly rough surfaces: Comparison between theory and numerical techniques. *Journal of the Mechanics and Physics of Solids, 59*(11), 2355–2369.

Barber, J. R. (2018). *Contact mechanics* (Vol. 250). Cham: Springer.

Creton, C., & Ciccotti, M. (2016). Fracture and adhesion of soft materials: A review. *Reports on Progress in Physics, 79*(4), 46601.

Dapp, W. B., Lücke, A., Persson, B. N. J., & Müser, M. H. (2012). Self-affine elastic contacts: Percolation and leakage. *Physical Review Letters, 108*(24), 244301.

Ditsche, P., & Summers, A. (2019). Learning from northern clingfish (Gobiesox maeandricus): Bioinspired suction cups attach to rough surfaces. *Philosophical Transactions of the Royal Society B, 374*(1784), 20190204.

Gnecco, E., & Meyer, E. (2015). *Elements of friction theory and nanotribology.* Cambridge: Cambridge University Press.

Hays, D. (2013). *The physics of tire traction: Theory and experiment.* New York: Springer.

Herbert, E., Balibar, S., & Caupin, F. (2006). Cavitation pressure in water. *Physical Review E, 74*(4), 41603.

Israelachvili, J. N. (2011). *Intermolecular and surface forces.* London: Academic.

Iwasaki, H., Lefevre, F., Damian, D. D., Iwase, E., & Miyashita, S. (2020). Autonomous and reversible adhesion using elastomeric suction cups for in-vivo medical treatments. *IEEE Robotics and Automation Letters, 5*(2), 2015–2022.

Karlsson Green, K., Kovalev, A., Svensson, E. I., & Gorb, S. N. (2013). Male clasping ability, female polymorphism and sexual conflict: Fine-scale elytral morphology as a sexually antagonistic adaptation in female diving beetles. *Journal of the Royal Society Interface, 10*(86), 20130409.

Kier, W. M., & Smith, A. M. (2002). The structure and adhesive mechanism of octopus suckers. *Integrative and Comparative Biology, 42*(6), 1146–1153.

Kim, J.-H., Kim, S.-H., Lee, C.-H., Nah, J.-W., & Hahn, A. (2003). DEHP migration behavior from excessively plasticized PVC sheets. *Bulletin of the Korean Chemical Society, 24*(3), 345–349.

Lorenz, B., & Persson, B. N. J. (2010a). Leak rate of seals: Effective medium theory and comparison with experiment. *The European Physical Journal E, 31*(2), 159–167.

Lorenz, B., & Persson, B. N. J. (2010b). On the dependence of the leak rate of seals on the skewness of the surface height probability distribution. *EPL (Europhysics Letters), 90*(3), 38002.

Müser, M. H., Dapp, W. B., Bugnicourt, R., Sainsot, P., Lesaffre, N., Lubrecht, T. A., Persson, B. N. J., Harris, K., Bennett, A., Schulze, K., & Others. (2017). Meeting the contact-mechanics challenge. *Tribology Letters, 65*(4), 118.

Nacer, M. H. (2012). *Tangential momentum accommodation coefficient in microchannels with different surface materials.* PhD thesis, 2012. University of Aix Marseille.

Pastewka, L., & Robbins, M. O. (2014). Contact between rough surfaces and a criterion for macroscopic adhesion. *Proceedings of the National Academy of Sciences, 111*(9), 3298–3303.

Persson, B. N. J. (2001). Theory of rubber friction and contact mechanics. *The Journal of Chemical Physics, 115*(8), 3840–3861.

Persson, B. N. J. (2006). Contact mechanics for randomly rough surfaces. *Surface Science Reports, 61*(4), 201–227.

Persson, B. N. J. (2013). *Sliding friction: Physical principles and applications.* Berlin: Springer.

Persson, B. N. J., & Mugele, F. (2004). Squeeze-out and wear: Fundamental principles and applications. *Journal of Physics: Condensed Matter, 16*(10), R295.

Persson, B. N. J., & Yang, C. (2008). Theory of the leak-rate of seals. *Journal of Physics: Condensed Matter, 20*(31), 315011.

Rice, J. R., & Ruina, A. L. (1983). Stability of steady frictional slipping. *Journal of Applied Mechanics, 50,* 343–349.

Roberts, A. D., & Tabor, D. (1971). The extrusion of liquids between highly elastic solids. *Proceedings of the Royal Society of London. A. Mathematical and Physical Sciences, 325*(1562), 323–345.

Ruina, A. (1983). Slip instability and state variable friction laws. *Journal of Geophysical Research: Solid Earth, 88*(B12), 10359–10370.

Sandoval, J. A., Jadhav, S., Quan, H., Deheyn, D. D., & Tolley, M. T. (2019). Reversible adhesion to rough surfaces both in and out of water, inspired by the clingfish suction disc. *Bioinspiration & Biomimetics, 14*(6), 66016.

Smith, A. M. (1991). Negative pressure generated by octopus suckers: A study of the tensile strength of water in nature. *Journal of Experimental Biology, 157*(1), 257–271.

Spolenak, R., Gorb, S., Gao, H., & Arzt, E. (2005). Effects of contact shape on the scaling of biological attachments. *Proceedings of the Royal Society A: Mathematical, Physical and Engineering Sciences, 461*(2054), 305–319.

Tinnemann, V., Hernández, L., Fischer, S. C. L., Arzt, E., Bennewitz, R., & Hensel, R. (2019). Adhesion: In situ observation reveals local detachment mechanisms and suction effects in micropatterned adhesives (Adv. Funct. Mater. 14/2019). *Advanced Functional Materials, 29*(14), 1970091.

Tiwari, A. (2018). *Adhesion, friction and leakage in contacts with elastomers.* NTNU.

Tiwari, A., & Persson, B. N. J. (2019). Physics of suction cups. *Soft Matter, 15*(46), 9482–9499.

Tiwari, A., Dorogin, L., Tahir, M., Stöckelhuber, K. W., Heinrich, G., Espallargas, N., & Persson, B. N. J. (2017). Rubber contact mechanics: Adhesion, friction and leakage of seals. *Soft Matter, 13*(48), 9103–9121.

Tramacere, F., Appel, E., Mazzolai, B., & Gorb, S. N. (2014). Hairy suckers: The surface microstructure and its possible functional significance in the Octopus vulgaris sucker. *Beilstein Journal of Nanotechnology, 5*(1), 561–565.

Vakis, A. I., Yastrebov, V. A., Scheibert, J., Nicola, L., Dini, D., Minfray, C., Almqvist, A., Paggi, M., Lee, S., Limbert, G., Molinari, J. F., Anciaux, G., Aghababaei, R., Echeverri Restrepo, S., Papangelo, A., Cammarata, A., Nicolini, P., Putignano, C., Carbone, G., et al. (2018). Modeling and simulation in tribology across scales: An overview. *Tribology International, 125*, 169–199.

Wainwright, D. K., Kleinteich, T., Kleinteich, A., Gorb, S. N., & Summers, A. P. (2013). Stick tight: Suction adhesion on irregular surfaces in the northern clingfish. *Biology Letters, 9*(3), 20130234.

Chapter 11
Water Transport Through Synthetic Membranes as Inspired by Transport Through Biological Membranes

Anna Strzelewicz, Gabriela Dudek, and Monika Krasowska

Abstract Water is regarded as the "solvent of life" because all vertebrates, invertebrates, microbes, and plants are also primarily water. Each living cell of those organisms requires a constant supply of both water and various substances from the external environment. Water is the medium in which living cells and tissue work. It can be said that the plumbing systems for cells are aquaporins (AQPs) that are transmembrane proteins forming channels in the cell membranes. The biological roles of these proteins have been thoroughly investigated in the past several years. Scientists have gained substantial knowledge on the structure, cellular localization, biological function, and potential pathophysiological significance of mammalian aquaporins. The extremely efficient AQPs with the knowledge of water structure remain an important exploratory challenge and are inspiration to study the new artificial membrane materials. The development of bioinspired synthetic systems is based on a good understanding of the complexity of natural systems. Researchers, following the example of biological membranes, study the new artificial membrane materials and their application in water removal processes or preferential water transport. Biological membranes, which are membranes of eukaryotic cells, thanks to their structure, allow some components to penetrate and retain others. Such membrane's property is used in the separation of liquid mixtures in membrane technology especially in pervaporation. In this process one substance preferentially permeates through the membrane and the other remains before membrane due to the pressure difference on both sides of the membrane. Our main thrust of research efforts has been concentrated on the polymeric membranes: investigation of new polymeric materials, modeling of component transport through dense polymers and description of their application. In this chapter the general overview of the comparison biological and synthetic membranes, preparation and application of different kinds of polymer membranes in the separation of water/alcohol mixture is presented.

A. Strzelewicz (✉) · G. Dudek (✉) · M. Krasowska (✉)
Faculty of Chemistry, Silesian University of Technology, Gliwice, Poland
e-mail: Anna.Strzelewicz@polsl.pl; Gabriela.Maria.Dudek@polsl.pl;
Monika.Krasowska@polsl.pl

© The Author(s), under exclusive license to Springer Nature
Switzerland AG 2021
A. Gadomski (ed.), *Water in Biomechanical and Related Systems*,
Biologically-Inspired Systems 17, https://doi.org/10.1007/978-3-030-67227-0_11

The separation of water/alcohol mixture is of extreme importance in a variety of industries ranging from medical, biology and pharmaceutical to food, chemical and alternative fuel source. The progress in transport modelling and structure analysis is also discussed.

Keywords Water transport · Artificial membranes · Biomimetic issues · Aquaporines · Water transport through membranes · Separation of water alcohol mixtures

1 Introduction

Water is the most important component for sustaining life of all organisms. It participates in the most metabolic reactions and is a transport medium for metabolism products, nutrients, hormones and enzymes (Chaplin 2001). It regulates temperature and participates in hydrolysis reactions. It constitutes a liquid environment necessary for the removal of final metabolic products. Water makes up 60–75% of human body weight and a person wouldn't survive three days without water. Futhermore, water which people use for different purposes should be very often free of germs and chemicals and be clear. Nowadays, water scarcity and problems related to lack of clean water are well known and scientists continue to conduct research focusing on developing new or improved optimal fresh water production technologies (Kocsis et al. 2018; Eliasson 2015). Several years ago it was projected that fresh water demand globally would increase by 55% between 2000 and 2050 and much of the demand would be driven by agriculture, which accounts for 70% of global freshwater use (GlobalWaterForum; Amy et al. 2017; Qasim et al. 2019). Efficiency of fresh water production may increase in the coming years by changing the process conditions, or by changing the membranes used in the cleaning water processes. Need for membrane innovations have lead to the emerging research fields of thin-film nanocomposites and biomimetic membranes and thus led to a deeper understanding of material structure. It seems to be highly important for the upscale of the materials from the laboratory toward membranes, modules and further process designs (Kocsis et al. 2018).

Water in biological systems has specific functions, it is the medium in which living cells and tissue work (Agre 2004). The transport of water across the cell membrane occurs along aquaporins (AQPs), that are transmembrane proteins that form channels in the cell membranes, known for their high osmotic water permeability and perfect rejection of ions. It can be said that aquaporins are the plumbing systems for cells. These properties have inspired the incorporation of AQPs into membrane materials for the design of bio-assisted membranes. In recent years, interest in the biomimetic approach to the development of membranes for separation has reappeared (Shen et al. 2014; Fuwad et al. 2019; Li et al. 2019). Biometric membranes contain biological elements or take their inspiration from biological systems. For example, contemporary water purification technologies can profit from the development of membranes with specialized pores that mimic highly efficient and water selective biological proteins such as Aquaporin-1 (AQP1). The key structural feature that allows efficient transport in AQP1 is its hydrophobic and narrow channel that forces water to translocate in a single-file arrangement. In article (Tunuguntla et al. 2017) we can read that fast water transport through carbon nanotube pores has raised the possibility to use them in the next generation of water treatment technologies. Computer simulations and experimental studies of water transport through carbon nanotubes reports enhanced water flux in these channels. Researchers keep working on the nanotube transport efficiency and ionic selectivity, which can disappear at higher salinities.

Understanding of extremely efficient AQPs with the knowledge of water structure remain an important exploratory challenge and are inspiration to study the new artificial membrane materials not only for desalination of water. In general, membrane techniques are used to separate gases and liquid components. Depending on the size of the separated particles, different membranes can be used. For separation of large particles (50-500 nm), i.e. suspensions, following processes are used: microfiltration, ultrafiltration and dialysis. Using the solid (nonporous) membranes, it is possible to separate the components of a very similar size, such as ions, vapours or gases in the processes called reverse osmosis, nanofiltration, pervaporation or permeation of gases/vapours (Nath 2017; Noble and Stern 1995; Ismail et al. 2018; Uragami 2017; Baker 2012). From this group of aforementioned processes pervaporation (PV) seems to be a promising technique employed for liquid separation. It is used to remove water from organic solvents, purify aqueous streams and separate organic-organic mixtures (Nath 2017; Noble and Stern 1995; Ismail et al. 2018).

In the PV technique, two different types of membranes can be used: hydrophilic or hydrophobic, depending on the process to which the membrane is to be applied. For the separation of water-alcohol mixture, hydrophilic membranes are used because of their preferential transportation of water. In recent years, the most popular membrane materials were those obtained from polyvinyl alcohol (PVA), sodium alginate and chitosan (Crespo et al. 2015; Wijmans and Baker 1995; Dudek et al. 2017). Despite of their many advantages, they are not effective in an unmodified form in separation processes. One of their main disadvantages is their low stability, especially in the case of solutions containing significant amount of water. In order to increase the stability of hydrophilic membranes, scientists use different modifications including membrane crosslinking (Dudek et al. 2014, 2018; Qu et al. 2010), blends of polymers with different properties (Zhu et al. 2011; Nawawi et al. 2013), organic-inorganic composites (hybrid membranes) (Dudek et al. 2017, 2018), polymer grafting and surface modification (Sajjan et al. 2015; Wang et al. 2014) and the production of asymmetric membranes with a thin active layer deposited on a porous substrate (Zhu et al. 2010; Huang et al. 2000).

The field of enormous importance, where pervaporation and membranes can be applied, is the water purification. Wastewater generated from various chemical products including commonly used refrigerants, plastics, adhesives, paints, petroleum products, and pharmaceutics, as well chemical plants is a serious environmental problem (Fernández et al. 2016; Li et al. 2015). The wastewater mainly contains of volatile organic compounds (VOCs), which are particularly the halogenated hydrocarbons known for their negative effects on the environment and human health. The most common ones are the depletion of stratospheric ozone, formation of the ground level smog, origination of the annoying odours and induction of the chronic toxicity (Ozturk et al. 2015; Ramaiah et al. 2013). Traditional VOCs control technologies such as conventional distillation process (Kiss 2013; Saidur et al. 2011), adsorption (Maddah 2019; Liu et al. 2015) and biological treatment (Padhi and Gokhale 2014; Muñoz et al. 2013) do not always provide a complete and economic reasonable solution, because usually the concentration of VOCs in the wastewater is very low. The pervaporation process is appropriate for the removing of VOCs' traces from the wastewater (Shah et al. 2004; Bruggen and Luis 2015). Membranes with high organic permeability and low water permeability are required in these processes. A typical organic permselective material for pervaporative membranes is poly(dimethylsiloxane) (PDMS) also known as silicone rubber (Ramaiah et al. 2013; Kim et al. 2002) or zeolite membranes (Aguado et al. 2005; Drobek et al. 2012).

2 Water Transport Through Biological Membranes

Nature inspired modern technology is transforming our world. Researchers have been increasingly keen to study nature in search of new innovations. Focusing on the microscopic layers, scientists became concerned with uncovering how biological machineries work, in hopes to regulate and replicate their functions. One representative class of these biological machineries are the proteins belonging to the aquaporin family (Agre 2004; Tunuguntla et al. 2017). In the past several years, scientists have gained substantial knowledge on the structure, cellular localization, biological function, and potential pathophysiological significance of mammalian AQPs. Aquaporins explain how brains secrete and absorb spinal fluid, how tears are secreted, saliva, sweat, and bile, and how kidneys can concentrate urine so effectively. The common efforts of many labs have led to the molecular identification of 12 mammalian aquaporin homologues. Additionally several hundred related proteins have been recognized in other vertebrates as well as invertebrates, plants, and unicellular microorganisms (Agre 2004).

The main role of the aquaporin proteins is to transport water through the membrane of cells, only a handful of AQPs are specific water transporters, other members being able to transport small solutes such as ammonia, carbon dioxide, urea or glycerol (Kocsis et al. 2018; Murata et al. 2000).

The membranes of a subset of cells with aquaporin proteins have a very high capacity for permeation by water and this permeability is selective, i.e. water (H_2O) passes through the membranes with almost no resistance, while the hydronium ion (H_3O^+) does not permeate the proteins. This distinction is crucial to life. Aquaporins form a simple pore that allows water to rapidly pass through membranes by osmosis (Kocsis et al. 2018). They are not pumps or exchangers and the water movement is directed by osmotic gradients.

Peter Agre in his article (Agre 2004) states that the AQP1 molecule has the hourglass structure i.e. it has an extracellular vestibule and an internal vestibule where water is in bulk solution. These vestibules are separated by a distance of approximately 20 Å, and linked through a channel which is so narrow that water moves through in single file while protons are reflected. Near the top of this bridge the channel reaches its narrowest constriction of 2.8 Å. This narrow diameter of the pore is large enough for the water molecules to pass, but theoretically not restrictive enough to block the passage of dehydrated Na^+ or Mg^{2+} cations, with ionic radii of about 2 and 1.6 Å. There are two mechanisms that prohibit the passage of cations. One is related to the water-solute pair interaction. Both Na^+ and Mg^{2+} have >3 Å hydration shells in water solutions but the structure of the pore itself doesn't favor the dehydration of the cations, leaving them too large to pass through the pore (de Groot et al. 2003; Tajkhorshid et al. 2002).

Near the narrowest diameter of the pore, the side chain of a perfectly conserved arginine residue following the NPA motif (Asparagine-Proline-Alanine) forms a fixed positive charge, and a conserved histidine residue on the other wall forms a partial positive charge. Together they serve to repel protons.

In a further part, there is another barrier to protons where a single water molecule will transiently undergo a transient dipole reorientation as it forms at the same time hydrogen bonds with the side chains of the two asparagine residues in the juxtaposed NPA motifs. Furthermore, the nonbilayer-spanning α-helices contribute partial positive charges that further serve to block proton conduction.

The sagittal cross-section of AQP1 reveals bulk water in the extracellular and intracellular vestibules of the hourglass structure (Agre 2004). These vestibules are separated by a 20 Å channel through which water passes in a single file with transient interactions with pore-lining residues that prevent the formation of hydrogen bonds between the water molecules. It is considered that two structural elements prevent permeation by protons (H_3O^+): (1) electrostatic repulsion is created by a fixed positive charge from the pore-lining arginine residue at a 2.8 Å narrowing in the channel; (2) reorientation of the water dipole occurs from the simultaneous hydrogen bonding of water molecule with the side chains of two asparagine residues in NPA motifs. Two partial positive charges at the center of the channel result from the orientation of two nonmembrane-spanning α-helices distal to the NPA motifs.

3 Water Transport Through Synthetic Membranes

3.1 Transport Through Biomimetic Membranes

The high permeability and selectivity of aquaporins motivated researchers to incorporate them into membranes for desalination and water purification. The membranes are called aquaporin biomimetic membranes and these membranes consisted of three main components: (1) aquaporin proteins (2) liposomes in which the aquaporins are embedded for protection, and (3) a polymeric support. The liposomes were chosen by the fact that the aquaporins are transmembrane proteins and their native environment is the hydrophobic region of the cellular membranes. The supported lipid bilayer (SLBs) membranes were tested in order to check that they are suitable impermeable platforms for incorporation of aquaporins for water filtration (Kaufman et al. 2010). However no steady results have been obtained with such AQP/ SLBs. Another group of researchers (Ding et al. 2015) showed the fabrication of covalently bounded SLBs on a polymeric supports that can make membranes with higher stability. This was achieved through the covalent attachment of SLBs to a polydopamine coated porous polyethersulfone support. The covalent bonding between the bilayer and the support prevents the easy desorption of the lipid matrix, offering stability to the active layer of the membrane. However, the SLB is susceptible to degradation being in contact with solutions containing detergents, which can readily disrupt the lipid bilayer. The performances of these SLB membranes are remarkable in water filtration, but their reliability remains an issue for long term applications and scalability. The production of AQP proteins is a challenge that is not trivial because biological protocols and reagents are used, as well as the purification setup still represents an expensive and time-consuming obstacle. A possible way to further improve biomimetic membranes is to replace AQPs with synthetic channels.

3.2 Water Transport Through Artificial Water Channels

Developing of synthetic artificial water channels is a promising strategy for innovative technologies based on selective water transport across a membrane barrier (Kocsis et al. 2018). Artificial systems are attractive, because of their increased stability, scalability and easier fabrication process than their protein counterparts. However, in order to achieve similar efficiency using artificial systems, it is important that they are able to imitate the protein's key structural features and functions using a simple as possible synthetic production approach. A very promising and quite characteristic type of synthetic porous systems, which can be considered highly permeable water channels, are carbon nanotubes (Siwy and Fornasiero 2017; Tunuguntla et al. 2017). They are widely applied in nanotechnology and they have been recently applied in nanofluidic devices for separation and translocation. Since pore sizes started to approaching the nanoscale a new discovery was made regarding the mass-transfer behaviour across carbon nanotubes . Simple diffusion doesn't apply when water cluster are broken down on entering the nanotubes. Under 0.9 nm, water adopts a single file structure very similar to the one found in theaquaporins. The change in structuration in the confinement of the pore makes water molecules to pass through with increased speed due to a boundary slip mechanism. Carbon nanotubes can be easily insert into both biological and synthetic lipid bilayer membranes.

4   Transport Through Synthetic Membranes – Characterization and Selection of Proper Membrane

In the membrane processes, membrane plays a key role, so the characterization of its physico-chemical and separation properties seems to be the most important. The idea of the gases/vapours membrane separation relies on the continuous supply of a vapours/gases mixture (feed) to the membrane module, separation of the mixture with the use of the membrane and the reception of permeating components through the membrane. Mass flow in the membrane module is distributed into a retentate and a permeate. The permeate is enriched with the substance preferentially penetrating the membrane, whereas the retentate is enriched with the substance retained on the membrane (Fang et al. 1975; Crespo et al. 2015; Wijmans and Baker 1995; Crank and Park 1968; Dudek et al. 2017).Depending on the application, the membranes can be divided into porous and non-porous membranes. Porous membranes are used in the processes where separated molecules are different in their size. In case of similar size of molecules, non-porous membranes are applicate. Despite of the fact that the mechanism of the separation of molecules mixture is different for porous (sieve mechanism) and non-porous (solution – diffusion mechanism) membranes, the idea of choosing the appropriate membrane with suitable structure and physico-chemical properties is similar. In the paper, as a representative model for describing the influences of membrane properties on the membrane separation process, the membrane uses for pervaporation process are describing.

To study the separation of liquid mixture by pervaporation process, the apparatus shown in Fig. 11.1 is used. The feed (separated mixture solution) is poured to the feed tank (1). Circulating pump (2) is used to supply the solution into the high pressure top side of membrane placed inside the permeation cell assembled from two half-cells made of stainless steel and fastened with bolted clamps. The membrane is placed on a finely porous stainless steel plate support. After liquid mixture separation into membrane module, retentate is recirculated to the feed tank and permeate is received and froze in the cold trap condenser unit (5). The reduced pressure on the permeate side is produced by an vacuum pump (6) and controls with a vacuum gauge (4). The efficiency of the pervaporation process is evaluated on the basis of parameters characterizing the transport of substances permeating through the membrane, i.e. diffusion, permeability and solubility coefficients and a series of parameters characterizing the efficiency of the pervaporation process, i.e. selectivity, separation coefficients and pervaporative separation index (*PSI*). The total permeate flux is calculated from the mass of liquid collected in the cold traps during certain time intervals at steady-state condition. The feed, permeate and retentate composition is analysed by a gas chromatography.

Based on the first Fick's law, it is possible to obtain the equation for **the permeation coefficient**. This coefficient determines the speed at which particles pass through the membrane under the influence of the difference in vapour pressure on both sides of the membrane (Dudek et al. 2017; Toledo et al. 2008; Kanti et al. 2004):

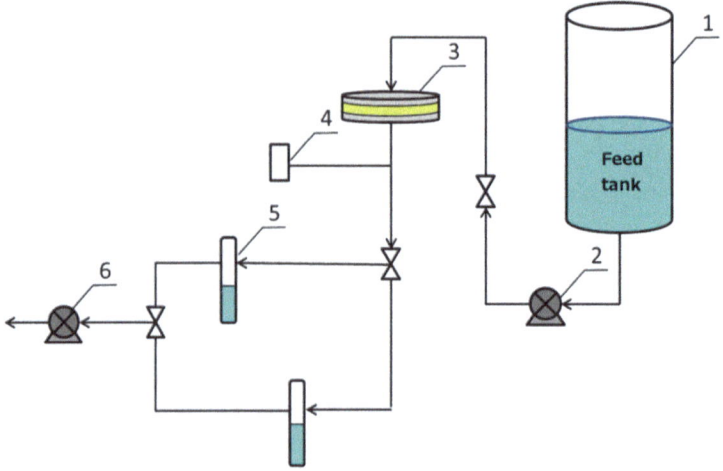

Fig. 11.1 Scheme of pervaporation setup: 1—feed tank, 2—circulation pump, 3—separation chamber, 4—vacuum gauge, 5—cooled collection traps, 6—vacuum pump

$$P_i = \frac{J_i \cdot l}{\Delta p} \tag{11.1}$$

where:

J_i – diffusive mass flux, $cm^3_{STP}\cdot cm^{-2}\cdot s^{-1}$
P_i – permeation coefficient, Barrer = $cm^3_{STP}\cdot cm\cdot cm^{-2}\cdot s^{-1}\cdot cmHg^{-1}$
Δp – difference in vapour pressure on both sides of the membrane, cmHg
l – membrane thickness, cm

   The solubility coefficient defines the equilibrium ratio of the penetrant present within the membrane and in the external phase. This parameter is calculated from the following formula:

$$S = \frac{P_i}{D_i} \tag{11.2}$$

where:

D_i – diffusion coefficient, $cm^2\cdot s^{-1}$

   The transport of molecules permeating through the membrane can be described quantitatively by means of a **diffusion coefficient**. Its value is proportional to the speed at which the diffusing molecule can move in the surrounding environment. The diffusion coefficient has a significant influence on the permeability of components through the membrane, and depends on the size of the permeating molecule and the material of the membrane.

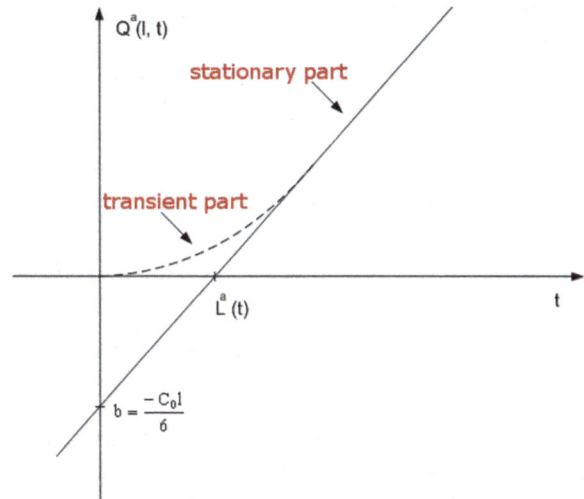

Fig. 11.2 Schematic view of a downstream absorption permeation curve. Dashed line presents mass of the penetrant transported through the outgoing side of membrane i.e. for x = 1. Solid line is an asymptote to the curve $Q^a(l,t)$ at stationary state, and it determines time lag $L^a(t)$

Commonly, the diffusion coefficient is determined on the basis of the permeation speed using the "time lag" method, i.e. time delay (Wijmans and Baker 1995; Crank and Park 1968), in which the time lag (L^a) is determined from the penetrant's permeation plot through a membrane with l thickness at time t. (Fig. 11.2).

In the case of experimental data, this curve can be determined by measuring the loss of mass or the value of flow rates over time, using the following relationship:

$$D_L = \frac{l^2}{6 \cdot L^a} \tag{11.3}$$

where:

D_L – diffusion coefficient, $cm^2 \cdot s^{-1}$
l – membrane thickness, cm
L^a – time lag, s

Because hydrophilic membranes are stored in water or feed solution before experiment, it is necessary to consider that before the measurement, the internal circulation has been activated causing the sorption of feed molecules into the membrane. It implies that when turning on the vacuum pump the initial concentration C_0 in the membrane is not zero. In such case, the Eq. (11.3) cannot be applied. In the paper (Dudek et al. 2019) a new, simple method of determining the diffusion coefficient is described. It is based on the solution of a general permeation problem, in which the analyzed membranes do not have to be empty in the initial phase of measurement. The calculations also take into account the effect of the length of the tube connecting the membrane module with the cold trap, causing a delay in the mass flow in the first moments of measurement.

Solutions to various transport problems in the field of diffusion are presented in J. Crank's book (Crank and Park 1968). In the fourth chapter, Crank considered different cases of one-dimensional diffusion in a medium limited by two parallel surfaces, e.g. the planes at x = 0, x = 1. The total amount of substance Q_t penetrating through the membrane at time t can be determined from the relation:

$$Q_t = D(C_1 - C_2)\frac{t}{l} + \frac{2l}{\pi^2}\sum_{\infty}^{n=1}\frac{C_1\cos n\pi - C_2}{n^2}\left\{1 - \exp\left(-Dn^2\pi^2 t / l^2\right)\right\}$$
$$+ \frac{4C_0 l}{\pi^2}\sum_{\infty}^{m=0}\frac{1}{(2m+1)^2}\left\{1 - \exp\left(-D(2m+1)^2\pi^2 t / l^2\right)\right\}$$

(11.4)

where: Q_t – total amount of diffusing substance per unit surface of the membrane; C_0 – the initial concentration in the membrane; C_1 – the concentration on the feed side of the membrane; C_2 – the concentration on the permeate side of the membrane; t – the time of pervaporation experiment; l – the thickness of membrane and D – the diffusion coefficient.

When the membrane is filled with feed solution, the initial concentration in the membrane is the same as the concentration on the feed side of the membrane $(C_0 = C_1)$. The concentration on the permeate side is equal to zero. In this case, the Eq. (11.4) takes the form:

$$Q_t = C_0[\frac{Dt}{l} + \frac{2l}{\pi^2}\sum_{\infty}^{n=1}\frac{\cos n\pi}{n^2}\left\{1 - \exp\left(-Dn^2\pi^2 t / l^2\right)\right\}$$
$$+ \frac{4l}{\pi^2}\sum_{\infty}^{m=0}\frac{1}{(2m+1)^2}\left\{1 - \exp\left(-D(2m+1)^2\pi^2 t / l^2\right)\right\}]$$

(11.5)

When the time tends to infinity, the exponentials in Eq. (11.5) vanish, allowing to write the equation as:

$$Q_t = C_0\left[\frac{Dt}{l} + \frac{2l}{\pi^2}\sum_{\infty}^{n=1}\frac{\cos n\pi}{n^2} + \frac{4l}{\pi^2}\sum_{\infty}^{m=0}\frac{1}{(2m+1)^2}\right]$$

(11.6)

Hence, the asymptotic total amount of diffusing substance takes the form:

$$Q_{tas} = C_0\left[\frac{Dt}{l} + \frac{2l}{\pi^2}\cdot 1.644929\right] = C_0\left[\frac{Dt}{l} + \frac{l}{3}\right]$$

(11.7)

and the diffusion coefficient can be calculated using the following equation:

$$D = \frac{-l^2}{3L^a}$$

(11.8)

where: L^a is the point at which the asymptote intersects with the time axis, in this case time lag is negative.

This equation is only valid if the additional time lag introduced by the tubing, that connects the membrane to the cold trap, is negligible. From numerical simulations performed for different pipe lengths and diffusion coefficients, it turns out that the asymptote shift is related to time delay L^a in a following manner:

$$\Delta t_{asymptote} = 6,5 \cdot L_1^a \tag{11.9}$$

Therefore, the relation allowing to determine the diffusion coefficient takes the form:

$$D = \frac{-l^2}{3\left(L_2^a - 6,5 \cdot L_1^a\right)} \tag{11.10}$$

For example, for a chitosan membrane crosslinked with glutaraldehyde, the method of determining L_1^a and L_2^a parameters is shown in Fig. 11.3.

For describing the separation properties of the membrane, following parameters are used: flux, separation factor (α_{AB}), selectivity coefficient (Sc_{AB}) and pervaporation separation index (*PSI*) (Dudek et al. 2017; Crank and Park 1968).

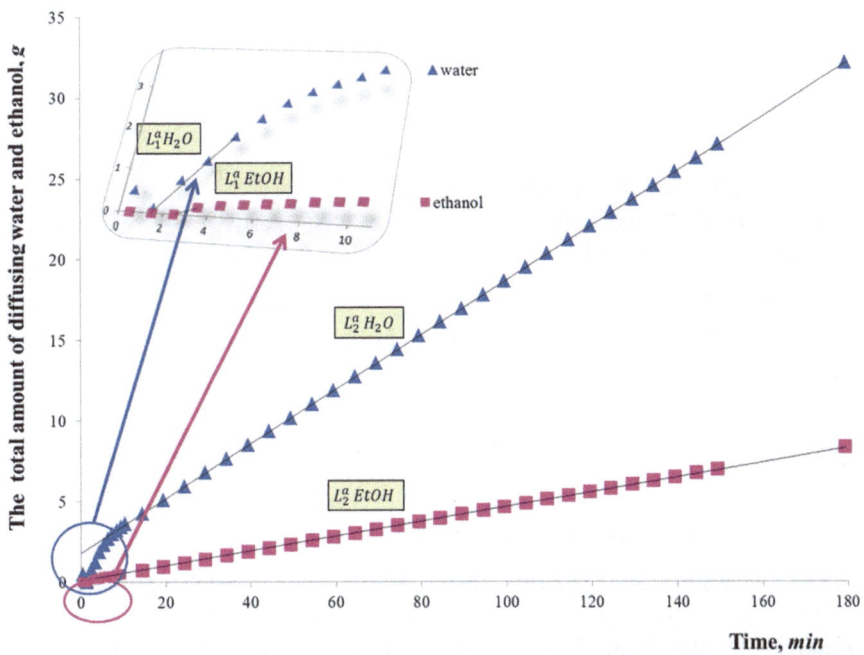

Fig. 11.3 Relationship of mass of water (denoted by triangles) and ethanol (denoted by squares) penetrating through the membrane as a function of time for chitosan membrane crosslinked with glutaraldehyde. Magnification shows a non-stationary state

Flux, J_i characterizes the amount of permeate that penetrates through the unit area of the membrane in one unit of time. This parameter can be determined from the following equation:

$$J_i = \frac{m_i}{A \cdot t}$$

(11.11)

where:

m_i – mass of the component i, kg
A – area of the membrane, m^2
t – permeation time, h

For homogeneous membranes, the flux is normalised to a membrane thickness l of 1 μm. This parameter allows a comparison of the transport properties of membranes with different thicknesses (Crank 1975). The normalized flux can be calculated using the formula:

$$J_i = \frac{J_i}{l}$$

(11.12)

The separation factor is a measure of the possibility of separating two components A and B. This parameter can be calculated from the following equation:

$$\alpha_{AB} = \frac{y_A / y_B}{x_A / x_B}$$

(11.13)

where:

y_A, y_B – weight fraction of components A, B in permeate [wt%].
x_A, x_B – weight fraction of component A, B in the feed [wt%]
$y_A + y_B = 1$ and $x_A + x_B = 1$

Selectivity coefficient is used to describe the separation properties of the membrane, and is determined according to the formula:

$$S_{AB} = \frac{P_A}{P_B}$$

(11.14)

where:

P_A, P_B – permeation coefficients of components A and B

The pervaporation separation index (*PSI*) is a measure of the ability of the membrane to separate components of the mixture, and is expressed as a product of selectivity and flux. It allows to compare the separation efficiency of a selected liquid mixture with the use of different membranes. In the case of two membranes which differ in permeate flux and selectivity, the more efficient and effective will be the membrane for which *PSI* has a higher value.

$$PSI = J_i \cdot (\alpha - 1)$$

(11.15)

4.1 Membrane Characterization

The structure and properties of the membrane impacts on the efficiently of the mixture separation, and therefore, the full characterization of the membranes should be performed. The membrane characterization includes spectroscopic techniques (NMR, ESR, Raman) to determine chemical structure and properties, DMA to study mechanical and rheological properties, and DSC and TGA to explore thermodynamic properties (Hilal et al. 2017), SEM, TEM, optical microscopy and goniometry to examine surface properties and morphology of membranes. Microscope techniques are a powerful tool for visualizing the morphology of membranes.. Atomic force microscope generates a topographic image by directly probing the membrane surface and recording surface waviness. The analysis of atomic force microscopy allows the quantification of surface roughness and membrane thickness (Kaupp 2006). Collecting cross-section and surface images of membranes from microscopy is enable to perform basic and fractal analysis of materials (Strzelewicz et al. 2020). Membrane morphology affects work output of membrane and the efficiently of the mixture separation. Examination of morphology may also greatly assist developing optimal membranes and manufacturing procedures for a particular application.

4.1.1 Membrane Morphology – Basic Analysis

The qualitative analysis of the microscopic image of a membrane sometimes shows features that are impossible to notice despite very careful visual observation (Russ 2008). The analysis of membrane requires the description of the morphology of individual elements of the system and their mutual connections as well as the characterization of the structure as a whole. The basic analysis of membrane morphology consists in determining a number, size, shape and distribution of elements (pores or particles) in the image. It enables the comparison of a particular image to other available images. The size of elements in the image is closely related to their shape. Although every person can intuitively define the shapes of surrounding objects, in the image analysis this property is difficult to characterize due to the lack of an unambiguous definition. The basic difficulty in determining the shape is the size of a given object (Russ 2008; Wojnar 1999; Krasowska et al. 2019).

Individual and separate objects or profiles can be characterized by selected geometrical parameters. These include:

- object specific area
- total object specific area
- profile specific perimeter
- object boundary specific perimeter
- convex perimeter
- length of the skeletonised object
- maximal diameter
- diameter of the circumscribing circle
- Feret diameters

- length of object
- thickness (width) of object
- diameter of area equivalent circle)
- Martin's radii

These measures are not always enough because they depend on the magnification of the analyzed image. Moreover structural elements of membrane morphology are not figures with strictly defined shapes like squares, triangles or circles. They are usually irregular and asymmetrical. To determine the shape of such objects, it is necessary to provide quantitative measures allowing for their comparison with each other. Contemporary image analysis proposes a description by means of so-called shape coefficients, which are dimensionless (the value remains unchanged with the change of the object size). Moreover, the above-mentioned parameters are easy to interpret using a reference point (e.g. a square, a circle, a rectangle) (Wojnar 1999; Krasowska et al. 2019).

The shape factors for any object are determined based on the determination of rectangles with the smallest area around the profile of the tested object (pores, particles). The construction of the rectangles with the smallest area around the particle profile enables the determination of mathematical descriptors of particle shapes. Therefore, if A is the projected area of an object and L is the actual perimeter of the profile and a and b are the lengths of the sides of the minimum area of the embracing rectangle, then the following particle shape descriptors are obtained. More popular parameter is elongation factor (or elliptic shape factor) of the following formula:

$$f_1 = \frac{a}{b} \tag{11.16}$$

Elongation factor determines the degree of elongation of the examined element in relation to a circle.

The next parameter is surface factor (or circularity)

$$f_2 = \frac{L^2}{4\pi A} \tag{11.17}$$

Circularity for the circle is equal to 1, for other shapes its value is greater than 1. This parameter is very sensitive to the occurrence of large irregularities, while the elongation of the object does not significantly affect its value. The values of this coefficient for a circle and an ellipse with the same surface area are similar. Another parameter is irregularity parameter f_3, which is a coefficient both sensitive to profile irregularities and surface elongation. It is defined as follows:

$$f_3 = \frac{d_1}{d_2} \tag{11.18}$$

where d_1 and d_2 are the diameters of the maximum inscribed and minimum circumscribed circles, respectively.

Moreover, it is also possible to give a quantitative description of the whole analyzed image. The cumulative characteristics are called total quantities for all or a specific class of objects visible in a defined measurement field (ROI). They include: count, numerical density, field area, class area, field perimeter, class perimeter, etc. Regardless of the basic cumulative characteristics listed above, average values are calculated for the previously mentioned individual object parameters (Jähne 2004).

In order to describe quantitatively the morphology of real hybrid membrane, image is characterized using the following parameters (based on the previously described measures):

- observed amount of polymer matrix, which is defined as the ratio of the polymer matrix area visible in the picture to the total image area,
- average size of polymer matrix domains,
- average number of obstacles in the proximity of each polymer matrix pixel n (Krasowska et al. 2019; Strzelewicz et al. 2020).

The exemplary morphology of real hybrid membrane is shown in Fig. 11.4. The figure shows the cross-section of alginate membrane crosslinked by orthophosphoric containing 20 wt% magnetite particles (Fe_3O_4). The image of the membrane texture was acquired using optical microscopy magnification 5700×. The obtained image was transferred to a black-white format for quantitative analysis of morphology of whole membrane or individual particles (Fig. 11.4).

4.1.2 Fractal Analysis

The basic morphological analysis presented in Sect. 4.1.1. is often insufficient for a precise description of the studied object. Such a situation takes place particularly when the analyzed structure is self-similar. Self-similarity is defined as a property

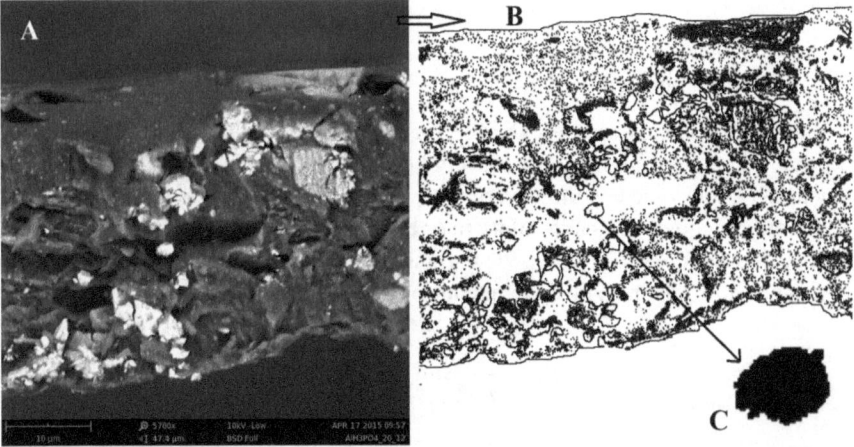

Fig. 11.4 Membrane morphology. (**a**) – Scanning electron microscopy of hybrid alginate membranes (crosslinking with orthophosphoric acid with the addition of 20% magnetite Fe_3O_4), (**b**) – Image in a black-white format, (**c**) – Shape of selected particle

of a set of points consisting in the fact that any small fragment of this set is similar to a larger fragment of itself in a certain, defined scale. Membrane structures exhibit this feature very often. In order to describe this unique property, the description based on fractal dimension is applied. The measure characterizing a self-similar set (for instance, a set of numerical data, a structure of geometrical object or a microscopic image), specifies its complexity as the ratio of the size change of its elements to the entirety of the object, depending on the scale in which the object is observed. The fractal dimension allows a quantitative characteristic of the object and is presented in the form of a fractional number. The total part signifies the geometrical dimension of a point, curve, figure or solid, whereas the fractional part describes the complexity of the fractal structure and traces to what extent it expands to the next geometrical dimension (Mandelbrot 1982; Bassingthwaite et al. 1994). There is no one universal method of the calculation of the fractal dimension. The most popular and frequent practice adopted in the description of membrane morphology is the box counting method. To calculate fractal dimension, the analyzed structure must be split into foursquare boxes with a side length of ε. Then, the fractal structure must be covered with the above-mentioned boxes, which will be counted as N. Fractal dimension (Mandelbrot 1982; Bassingthwaite et al. 1994; Grzywna et al. 2001; Krasowska et al. 2009) results from the relation of scaling of a number of covers of a given object to the size of this cover

$$N(\varepsilon) \sim \varepsilon^{-d_f} \tag{11.19}$$

and can be defined in the following way:

$$d_f = \lim_{\varepsilon \to 0} \frac{\ln N(\varepsilon)}{\ln\left(\dfrac{1}{\varepsilon}\right)} \tag{11.20}$$

where:

ε – size of an element of the cover
$N(\varepsilon)$ – number of elements of the cover

Fractal dimension describes an object using only one value to it. Because of the fact that we add and sum up only the number of non-empty boxes without checking how much of our object is immersed in it, we can find different objects whose fractal dimension equals the same value. In order to specify the quantitative description of fractals more precisely with a view to their differentiation, a notion of a generalized dimension was introduced by means of the following formula:

$$D_q = \frac{1}{q-1} \lim_{\varepsilon \to 0} \frac{\ln \sum_{i=1}^{N(\varepsilon)} P_i^q}{\ln \varepsilon} \tag{11.21}$$

where:

q – real number
ε – size of an element of the cover

P_i – probability of finding a point in a given element of the cover
$N(\varepsilon)$ – number of elements of the cover

Using generalized dimension we may characterize an object by means of an infinite quantity of numbers. These numbers create a set which is called a multifractal spectrum. It includes numerical values which describe structural elements of the object, groups of such elements and their mutual relation. For instance, for $q = 0$ we will obtain fractal dimension D_0, previously called d_f, which is responsible for scaling mass. The higher it gets, the denser the object is.

Multifractal spectra are most often presented in a graphic form. In order to facilitate the analysis and comparison of the spectra, function D_q is transformed by means of the Legendre's transformation in the form of function $f(\alpha)$ from α.

$$f(\alpha) = \alpha q - \lim_{\varepsilon \to 0} \frac{\ln \sum_{i=1}^{N(\varepsilon)} P_i^q}{\ln \varepsilon} \tag{11.22}$$

where α is a new index of dimension (after the Legendre's transformation performed for D_q) (Avnir 1989; Grzywna et al. 2001; Krasowska et al. 2009, 2012).Description of morphology based on fractal analysis (d_f, D_q, $f(\alpha)$) in relation to both natural and synthetic membranes is more and more often found in the literature (Krasowska et al. 2009, 2012, 2019; Bitler et al. 2012; Agboola et al. 2016; Dudek et al. 2019; Strzelewicz et al. 2020).

One of the most important parameters in the examination of morphology is degree of multifractality (Krasowska et al. 2019). The degree of multifractality ΔD is related to the deviation from simple self similarity and is the difference of the maximum and minimum dimension associated with the least dense and most dense points in the sets, as you can see in this formula:

$$\Delta D = D_{-\infty} - D_{\infty} \tag{11.23}$$

The lower value of ΔD is related to a more homogeneous and self-similar membrane structure. In order to predict the membrane's performance and to design membranes for specific applications, a fundamental understanding of the transport phenomena is required.

4.2 Hybrid Membranes Used in the Pervaporative Dehydration of Ethanol Process

4.2.1 The Influence of Membrane Properties on Transport Parameters

The kind and the properties of investigated membranes has a significant impact on separation process. The transport properties of the membranecan be characterized by three coefficients: diffusion, solubility and permeability coefficients. Dudek et al. (2014) checked the influence of polymer matrix, crosslinking agent and the type and amount of filleron the transport properties of the chitosan membranes crosslinked with two crosslinking agents, i.e. sulphuric acid or glutaraldehyde, filled with iron oxide (II, III) nanoparticles. The results show that a pristine sulphuric acid crosslinked chitosan membrane is not effective in the separation of ethanol/water mixtures. All the transport parameters determined for both components have similar values, meaning that neither water nor ethanol penetrates preferentially. The addition of magnetite nanoparticles influences the transport properties of investigated membranes. The evaluated diffusion coefficient of ethanol increases slightly (about five times), while the diffusion coefficient of water increases more than seventy times. The solubility coefficients of water and ethanol decrease significantly. In the case of pristine chitosan membrane crosslinked with glutaraldehyde, both diffusion and solubility coefficients are higher than the corresponding values estimated for ethanol, hence the water permeability coefficient is about six times higher than for ethanol. The presence of magnetite in the membrane structure further deepens the difference in permeability of both components. Since solubility coefficients decrease slightly for both components, the observed increase in water permeability coefficient results from faster diffusion of this component in the presence of magnetite nanoparticles (Fig. 11.5).

In another paper Dudek et al. (2020a) analysed the mechanism of water and ethanol transport through PVA membranes filled with five different chitosan particles (neat (CS), phosphorylated (CS-P), crosslinked with glutaraldehyde (CS-GA), glycidol-modified (CS-G) and sulphated (CS-SO₃)). Analysing the particular graphs presenting the changes in the diffusion and solubility coefficients (Figs. 11.6 and 11.7) it may conclude that in almost all cases the diffusion mechanism is responsible for the permeation process. Moreover, the shape of the curves suggests

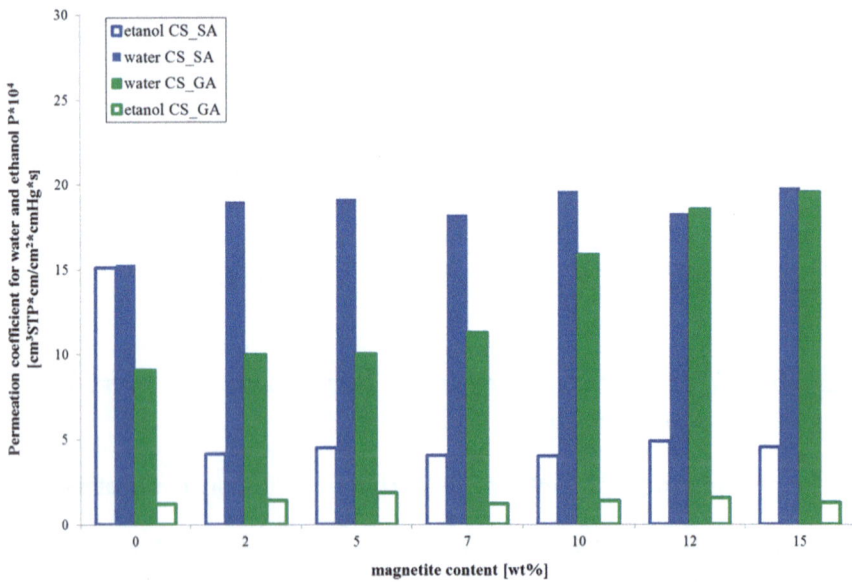

Fig. 11.5 Water and ethanol permeation coefficients as the function of the amount of iron oxide particles (0–15 wt%) added to the chitosan matrix crosslinked with sulphuric acid (CS_SA) and glutaraldehyde (CS_GA)

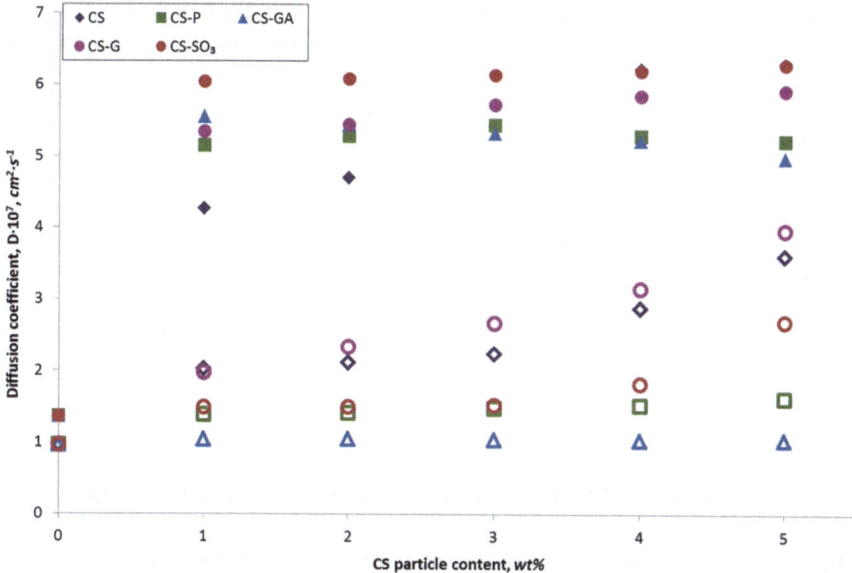

Fig. 11.6 The variation of evaluated diffusion coefficients of water (filled symbol) and ethanol (blank symbol) for hybrid poly (vinyl alcohol) membranes with increasing content of unmodified and four modified chitosan particles: phosphorylated (CS-P), crosslinked with glutaraldehyde (CS-GA), glycidol-modified (CS-G) and sulphated (CS-SO3))

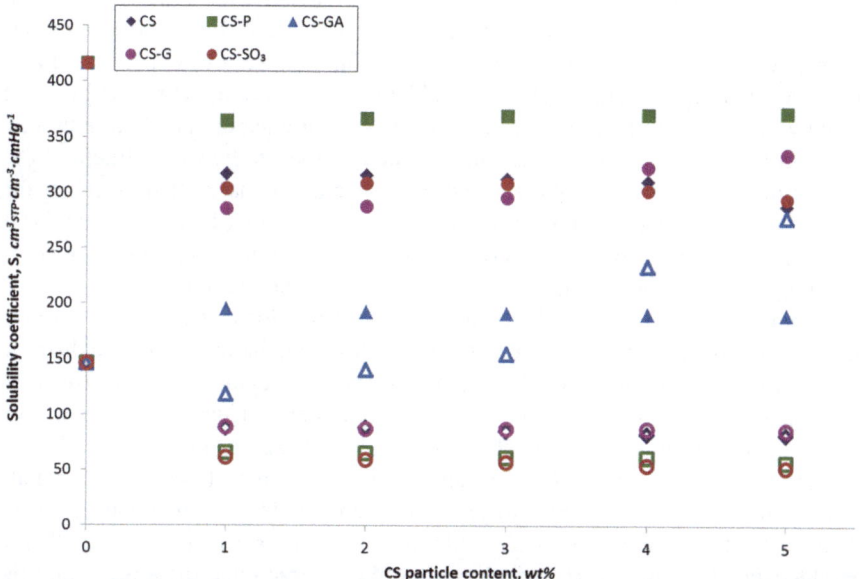

Fig. 11.7 The variation of evaluated solubility coefficients of water (filled symbol) and ethanol (blank symbol) for hybrid poly (vinyl alcohol) membranes with increasing content of unmodified and four modified chitosan particles: phosphorylated (CS-P), crosslinked with glutaraldehyde (CS-GA), glycidol-modified (CS-G) and sulphated (CS-SO3))

that the diffusion takes place mainly through the free volume generated near the filler surface and is limited by the interactions between penetrant molecules and surface of CS particles. Only for the PVA_CS-GA membrane, the solubility of the penetrant, especially ethanol, limits the permeation process. In this case, the final value of solubility coefficient of ethanol at 5 wt% exceeds the relevant water solubility coefficient. In contrary, the addition of other type of fillers into PVA matrix causes the decrease in water and ethanol solubility coefficients, which are nearly independent on the amount of filler. The values of water solubility coefficient correspond to the membranes surface hydrophilicity and are well correlated with the estimated contact angle. As a consequence, for more hydrophilic material the water solubility coefficient is bigger. The observed increase in ethanol diffusion coefficient starting from the 3 wt% filler content for several types of chitosan particles is the reason of the drop in the selectivity of the membranes.

4.2.2 The Influence of Membrane Structure on Membrane Properties

The structure of hybrid membranes has the great impact on the separation properties of investigated membranes. Hybrid membranes, as the connection of polymer matrix and filler is characterized by the properties of filler particles and their compatibility with the polymer matrix. In literature two different types of fillers are used: inorganic and organic. The most popular inorganic fillers are Fe_3O_4, Fe_2O_3, TiO_2, Ag_2O, Cr_2O_3, zeolites, silica and carbonic nanotube. The most common organic fillings are chitosan particles and metal organic frameworks. In the papers (Dudek et al. 2017, 2018), the influence of the structure of hybrid alginate membranes crosslinked with four different crosslinking agents: calcium chloride (AlgCa), phosphoric acid (AlgP), glutaraldehyde (AlgGA) and citric acid (AlgC), filled with magnetite particles on the effectiveness of ethanol dehydration process was studued. The obtained results showed the relationship between structural factors of the investigated membranes and their separation characteristics. All evaluated shape descriptors, *i.e.* elongation factor, bulkiness, surface factor and irregularity parameter are practically independent on either the type of a crosslinking agent or the magnetite content, indicating the similar value for all membranes. Moreover, it was found that alginate hybrid membranes have a fractional structure with a fractional dimension ΔD between 2.65 and 2.76. This value indicates that the distribution of iron oxide particles in alginate membranes is irregular, which causes their heterogeneity. For a homogeneous, ideally self-similar structure, the value of fractional dimension is equal to 0. The lowest value of ΔD is obtained for alginate membranes crosslinked with orthophosphoric acid containing 15 or 20 wt% of magnetite. These membranes are the most homogeneous and self-similar among others. The most heterogeneous group of membranes are membranes crosslinked with glutaraldehyde. They are characterized by the highest fractional dimension, and the distribution of filler is the least regular.

Comparing the max. *PSI* and the corresponding ΔD for all examined hybrid membranes, it can see that maximum value of *PSI* is always reached for the smallest

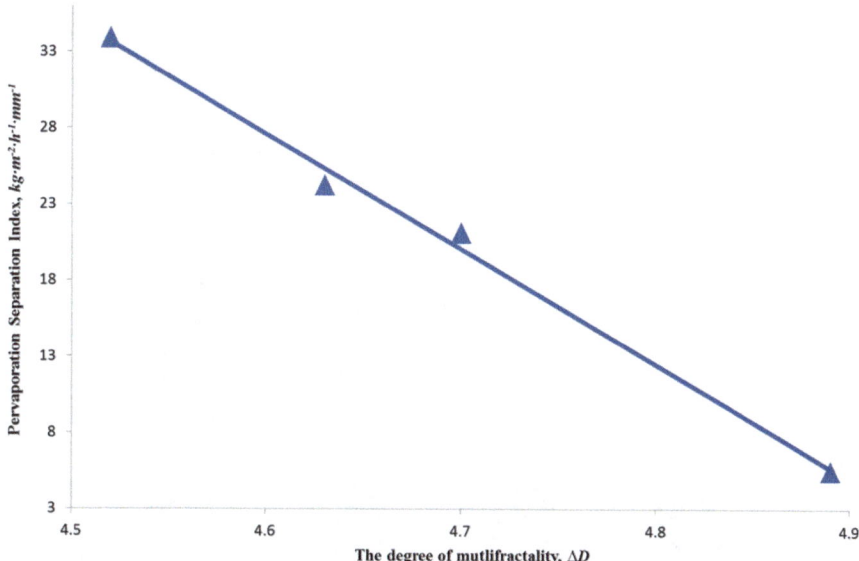

Fig. 11.8 The plot of max Pervaporation Separation Index (PSI) as a function of corresponding ΔD for hybrid alginate membranes crosslinked by the following agents: calcium chloride (AlgCa), orthophosphoric acid (AlgP), glutaraldehyde (AlgGA) and citric acid (AlgC)

values of fractal dimension. Among all investigated hybrid membranes, AlgP has the lowest fractional dimension $\Delta D = 4.52$ and at the same time the highest value of *PSI* = 33.86. Moreover, this relation has a linear character, what suggests that ΔD fractal parameter can be regarded as a measure of separation efficiency of the series of Alg membranes (Fig. 11.8).

In another work Dudek et al. (2020b) looked for the correlations between the experimental data and the results obtained from the fractal and random walk analysis for alginate membranes filled with neat chitosan (CS) and modified CS microparticles; phosphorylated chitosan (CS-P), glutaraldehyde crosslinked chitosan (CS-GA) and glycidol-modified chitosan (CS-G) The results showed (Fig. 11.9) the different value of the ratio of experimental and theoretical water/ethanol diffusion coefficients for the investigated membranes. The smallest and the same value (i.e. 14) of these parameters are obtained for Alg membranes loaded with CS and CS-P particles. It means that the diffusion of water and ethanol particles through the investigated membranes is not very selective, so taking into account only this parameter, Alg membranes with CS and CS-P particles have the worst pervaporation performance. The significantly bigger values of the experimental and theoretical diffusion ratios are observed for membranes with CS-G and CS-GA particles. The ratio in the diffusion coefficients of water and ethanol particles grows as a consequence of the structure of such membranes. Considering the channel length, obtained as a ratio of the average size of the void domain cross-section to the average pore diameter, it can be seen that Alg membranes with CS-G and CS-GA particles are less crowded than membranes with CS and CS-P particles. The longer

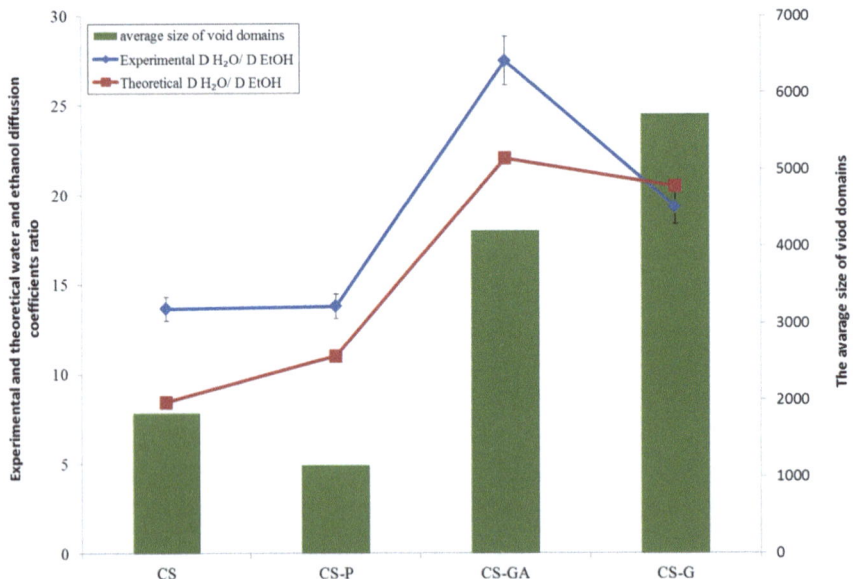

Fig. 11.9 Comparison of the intercorrelations between the experimental data and the results obtained from the fractal and random walk analysis of the alginate membranes filled with neat chitosan (CS) and modified CS microparticles; phosphorylated chitosan (CS-P), glutaraldehyde crosslinked chitosan (CS-GA) and glycidol-modified chitosan (CS-G)

chains that appear in membranes with chitosan particles modified with glycidol and glutaraldehyde cause fewer obstacles to the movement of the particles along the membrane than in the case of shorter chains. Additionally, the simulation shows that the size of channels which are created in the alginate matrix is not sufficient for easy penetration of ethanol molecules because of their greater size when compared to water molecules. Such a situation favours the separation of water from ethanol.

4.2.3 The Influence of Membrane Properties on Effectiveness of Water/ Ethanol Separation

The effectiveness of gases/liquid mixture separation is related to the properties of the polymer matrix, applied crosslinking agent and additional modifications that have been introduced to the membrane i.e. preparation of blends, hybrids or asymmetric membranes. In (Dudek et al. 2018) the authors compared pristine and hybrid alginate membranes crosslinked with four different agents: calcium chloride (AlgCa), phosphoric acid (AlgP), glutaraldehyde (AlgGA) and citric acid (AlgC). It is observed that citric acid (AlgC) and phosphoric acid (AlgP) crosslinked membranes have higher selectivity than those crosslinked with calcium chloride (AlgCa) or glutaraldehyde (AlgGA). Goniometric studies showed that AlgP and AlgC membranes are more hydrophilic, that is also associated with higher affinity towards water than in the case of the other two investigated kinds of membranes. Ester groups formed by crosslinking alginate with both citric acid and orthophosphoric acid are responsible for such character of membranes (Kalyani et al. 2008; Crossingham et al. 2014). Membranes crosslinked with calcium chloride or glutaraldehyde have a lower separation efficiency, which is particularly evident in the case of AlgGA membrane. As a result of crosslinking reaction, acetal rings and ether bonds are formed between hydroxyl groups of alginate and aldehyde group of glutaraldehyde, causing an increase in affinity of such membranes towards alcohol (Beppu et al. 2007).

In the case of alginate membranes filled with ferric oxide (II, III) nanoparticles, an increase in selectivity coefficient, separation factor and *PSI* is observed for three crosslinking agents, i.e. calcium chloride, phosphoric acid and citric acid. The pervaporation process is most effective for membranes filled with 15 wt% of magnetite. In the literature, the influence of the filler on the transport properties through the membrane is associated with the change of available free spaces in the polymer matrix. Since the diameter of water molecule is 0.28 nm and is smaller than the diameter of ethanol (0.44 nm), as a consequence, water molecules may more easily pass through small pores present in the alginate matrix containing small amounts of magnetite, while the transport of ethanol molecules is limited (Schatzberg 1967).

In another paper Dudek et al. (2019) looked for the relation between investigated five different metal oxides, *i.e.* iron(II, III), silver(I), chromium(III), titanium(IV) and zinc(II), used as a filler of, alginate membranes crosslinked with calcium chloride, on the efficiently of ethanol dehydration via pervaporation. The results show that the *PSI* value for alginate membranes filled with silver, titanium, chromium and zinc oxide particles is very similar and varies from 29.96 to 36.50 $kg \cdot m^{-2} \cdot h^{-1}$ for 15 wt%. A different nature of the changes can be observed in the case of iron oxide particles. For magnetite, the observed increase in *PSI* is rapid and the obtained values of this parameter is much higher than the values obtained for membranes containing other fillers (Fig. 11.10).

In this case, it is assumed that the magnetic properties of the filler are an additional factor that supports the separation process of the ethanol/water mixture. All substances have specific magnetic characteristics due the spin and orbital magnetic moment of their electrons and nuclei of the molecules. The diamagnetic characteristic of pure water and ethanol differ from each other and, furthermore, the magnetic moment of their mixture shows a concentration dependence. It is believed that water and ethanol form supramolecular structures, called clusters, which are in thermodynamically stable equilibrium state due to the presence of many hydrogen bonds. In particular, at higher concentrations of ethanol, the structure with small water clusters with restricted freedom of the spatial arrangement is formed, leading to the reduction of magnetic moment. The presence of magnetic nanoparticles in membrane modifies the dynamic magnetic properties of water and ethanol mixture. Under the influence of the weak magnetic field produced by Fe_3O_4, the rotation of water and ethanol molecules in the mixture generates a very small magnetic moment, proportional to the ring currents, which may cause a more effective separation of the ethanol/water mixture (Tsukada et al. 2017).

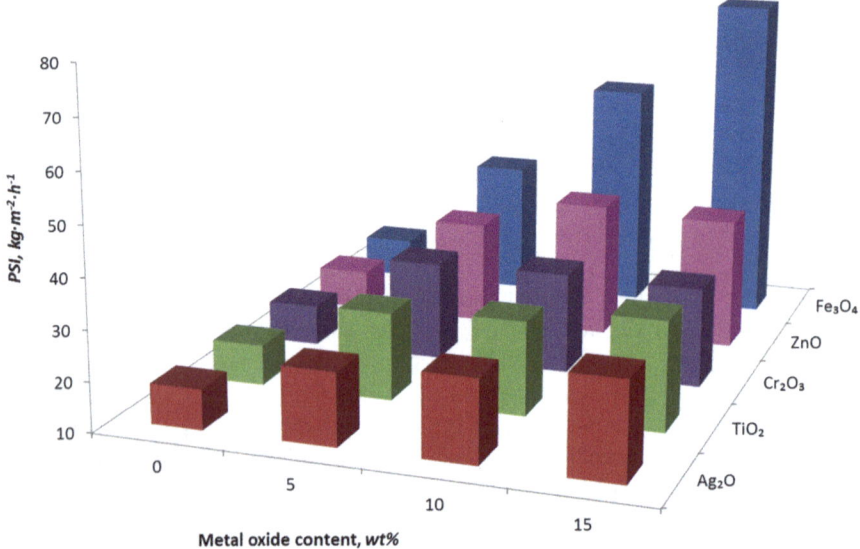

Fig. 11.10 The variation of evaluated pervaporation separation index *PSI* with increasing metal oxide (silver(I), chromium(III), titanium(IV) zinc(II) and iron(II, III) oxide) content of alginate membranes crosslinked with calcium chloride

In the papers (Dudek et al. 2018, 2020a), the results concerning the process of pervaporative dehydration of ethanol by alginate or poly(vinyl alcohol) membranes filled with unmodified and four types of modified chitosan particles (phosphory-lated (CS-P), crosslinked with glutaraldehyde (CS-GA), glycidol-modified (CS-G) and sulphated (CS-SO3)) is described. The results indicate that the new type of organic matrix/organic filler hybrid membranes show higher efficiency in the process of pervaporative ethanol dehydration than commonly used hybrid membranes filled with inorganic compounds. Furthermore, these membranes retain very high selectivity simultaneously with high fluxes. Comparing the alginate membranes with other hybrid membranes used in the process of pervaporative ethanol dehydration (Fig. 11.9), it can be seen that the newly developed membranes have signifi-cantly better separating properties than other hybrid membranes presented in the literature. The highest values of fluxes are obtained for chitosan membranes con-taining metal-organic framework (MOFs) and alginate membranes filled with phos-phorylated or crosslinked glutaraldehyde chitosan particles investigated in the work (Dudek et al. 2018). Among all membranes presented in Fig. 11.11, the highest *PSI* value have alginate membranes filled with phosphorylated chitosan particles (Alg_CS-P) which are the most effective in the process of ethanol dehydration by pervaporation.

In case of poly (vinyl alcohol) hybrid membranes filled with different chitosan particles, their separation properties are compared with other PVA-based mem-branes used in the PV dehydration process for alcohol aqueous solutions (Fig. 11.12) (Dudek et al. 2020a). Taking into account the values of membrane fluxes, the inves-tigated membranes exhibit medium values of this parameter in the range from 1.01

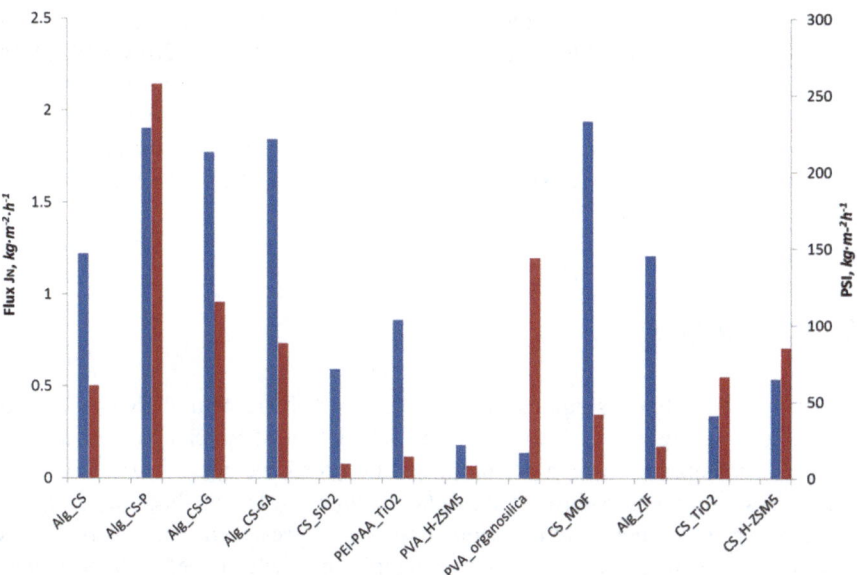

Fig. 11.11 Comparison of fluxes and the best Pervaporation Separation Index (*PSI*) of alginate hybrid membranes containing various chitosan particles: unmodified and three types of modified chitosan particles (phosphorylated (CS-P), crosslinked with glutaraldehyde (CS-GA) and glycidol-modified (CS-G) with other hybrid membranes, described in literature, used in pervaporative ethanol dehydration

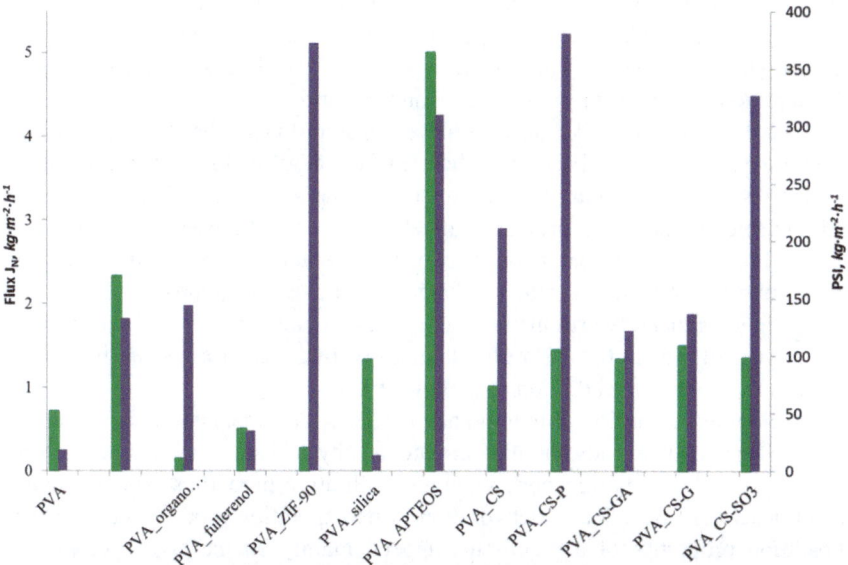

Fig. 11.12 Comparison of the highest permeation fluxes and the best *PSI* of poly(vinyl alcohol) membranes filled with unmodified and four types of modified chitosan particles (phosphorylated (CS-P), crosslinked with glutaraldehyde (CS-GA), glycidol-modified (CS-G) and sulphated (CS-SO3)) with several hybrid PVA membranes, described in literature, used for pervaporative dehydration of ethanol

to 1.50 $kg \cdot m^{-2} \cdot h^{-1}$. Concerning the separation factor, the best values are obtained in case of PVA membranes filled with organosilica (Liu and Kentish 2018) and ZIF-90 (Wei et al. 2018) (1026 and 1379, respectively). Membranes presented in the work (Dudek et al. 2020a) are characterized by the medium values of separation factor (92.1–263.3). Nevertheless, the combination of medium values of estimated fluxes and separation factors leads to the significant values of evaluated *PSI*, in contrast to the other considered membranes.

5 Conclusions

Chapter starts with a description of aquaporins, which are natural channels. Next, the authors' attention was focused on the description of water transport through biological membranes, which are the inspiration for developing new solutions in water transport processes through synthetic membranes. The methods of describing the structure and properties of these membranes were presented. The next part of the work focused on practical examples of using this knowledge to design hybrid membranes applied for ethanol dehydration.

The discovery of the aquaporins explained the selective transport of water through the plasma membranes of cells, while preventing ions from passing through the membrane. The structural models of aquaporins offer remarkable insight into the biophysical functions and give scientists a chance to understand role of aquaporins in numerous clinical disorders. Aquaporins are involved in some forms of renal vascular diseases, including nephrogenic diabetes insipidus. Aquaporins are also associated with problems of brain edema, loss of vision and with defense against thermal stress. Aquaglyceroporins are involved in the defense against starvation. These proteins are present in the whole natural world.

Natural membranes, like aquaporins are inspiration to develop new artificial membrane materials for different application from desalination to dehydration. The present research is evolving in two directions. The first group of studies includes artificial membranes with protein channels, which can be used in filtration processes. Furthermore, research on biomimetic membranes is aimed at replacing aquaporins by imitation synthetic channels. The second approach is based on designing a synthetic hybrid membrane. In these membranes, the role of the channels are the fillings in the form of nanotubes or free areas formed in the polymer matrix as a result of the addition of particles filler.

Developing new artificial membranes for particular applications is a challenge for researchers. Such study should correlate the physical and chemical properties of the membrane with its transport, separation, filtration properties. The example of membranes used for ethanol dehydration in pervaporation process shows that the separation properties of a membrane depend mainly on the choice of polymer matrix and filler, as well as the compatibility between matrix and filler, which affects the dispersion of the filler in the matrix. Depending on the process, two kind of polymer matrix hydrophilic or hydrophobic can be used. In case of fillers, the most popular are carbon nanotubes and metal or metal oxide particles. In recent times, however, organic fillings have become increasingly popular. Furthermore, the additional properties of filler such as magnetic or electric have significance influence on the properties of obtained membranes. In case of compatibility of polymer matrix with filler, structural analyses of membrane morphology including basic and fractal analysis are helpful. The study showed the correlation between effectiveness of pervaporative process and parameters that indicate the self-similar structure. Membranes that show a high degree of self-similarity are also more useful in separation processes.

An example of such a bioinspired membranes are alginate membranes filled with unmodified and modified chitosan particles where the addition of polymer powder to the alginate matrix gives the chance to make very compatible membranes with excellent separation properties. Among the four investigated membranes filled with chitosan particles, alginate membrane contains phosphorylated chitosan particles proves to be the most effective in the process of ethanol dehydration by pervaporation. It gives a chance to remove nearly whole water contained in ethanol. This chapter also presents other examples that can inspire other researchers.

Summarizing, artificial membranes inspired by biological membranes are the future of the membrane separation techniques. Understanding of structural properties and the transport mechanism of separated molecules through natural membranes give a chance to perform more efficient industrial separation processes. The design, manufacture and understanding of membrane biomimetic and bioinspired properties is one of the most important challenges of modern materials science.

References

Agboola, O., Mokrani, T., & Sadiku, R. (2016). Porous and fractal analysis on the permeability of nanofiltration membranes for the removal of metal ions. *Journal of Materials Science, 51,* 2499–2511.

Agre, P. (2004). Aquaporin water channels (Nobel lecture). *Angewandte Chemie, International Edition, 43,* 4278–4290.

Aguado, S., Coronas, J., & Santamaría, J. (2005). Use of zeolite membrane reactors for the combustion of VOCs present in air at low concentrations. *Chemical Engineering Research and Design, 83,* 295–301.

Amy, G., Ghaffour, N., Li, Z., Francis, L., Linares, R. V., Missimer, T., & Lattemann, S. (2017). Membrane-based seawater desalination: Present and future prospects. *Desalination, 401,* 16–21.

Avnir, D. (1989). *The fractal approach to heterogeneours chemistry – Surfaces, colloids, polymers.* Chichester: Wiley.

Baker, R. W. (2012). *Membrane technology and applications* (3rd ed.). New York: Wiley.

Bassingthwaite, J. B., Liebovitch, L. S., & West, B. J. (1994). *Fractal physiology.* New York: Oxford University Press.

Beppu, M. M., Vieira, R. S., Aimoli, C. G., & Santana, C. C. (2007). Crosslinking of chitosan membranes using glutaraldehyde: Effect on ion permeability and water absorption. *Journal of Membrane Science, 301,* 126–130.

Bitler, A., Dover, R., & Shai, Y. (2012). Fractal properties of macrophage membrane studied by AFM. *Micron, 43,* 1239–1245.

Bruggen, B. V., & Luis, P. (2015). Chapter four – Pervaporation. In *Progress in filtration and separation* (pp. 101–154). Amsterdam: Elsevier.

Chaplin, M. F. (2001). Water: Its importance to life. *Biochemistry and Molecular Biology Education, 29*(2), 54–59.

Costa, L. D. F., & Cesar, R. M. (2000). *Shape analysis and classification: Theory and practice* (Image Processing Series). Boca Raton: CRC Press.

Crank, J. (1975). *The mathematics of diffusion*. Oxford: Clarendon Press.

Crank, J., & Park, G. (1968). *Diffusion in polymers*. New York/London: Academic.

Crespo, J. G., & Brazinha, C. (2015). *Pervaporation, vapour permeation and membrane distillation*, Woodhead Publishing Series in Energy.

Crossingham, Y. J., Kerr, P. G., & Kennedy, R. A. (2014). Comparison of selected physico-chemical properties of calcium alginate films prepared by two different methods. *International Journal of Pharmaceutics, 473*, 259–269.

de Groot, B. L., Frigato, T., Helms, V., & Grubmuller, H. (2003). The mechanism of proton exclusion in the aquaporin-1 water channel. *Journal of Molecular Biology, 333*, 279–293.

Ding, W., Cai, J., Yu, Z., Wang, Q., Xu, Z., Wang, Z., & Gaoa, C. (2015). Fabrication of an aquaporin-based forward osmosis membrane through covalent bonding of a lipid bilayer to a microporous support. *Journal of Materials Chemistry A, 3*, 20118–20126.

Drobek, M., Figoli, A., Algieri, C., Santoro, S., Trotta, A., & Gaeta, N. (2012). Preparation of novel MFI zeolite PVDF mixed matrix membranes for potential VOCs removal. *Procedia Engineering, 44*, 1806.

Dudek, G., & Borys, P. (2019). A simple methodology to estimate the diffusion coefficient in pervaporation-based purification experiments. *Polymers, 11*, 343.

Dudek, G., & Turczyn, R. (2017). Chapter 10: Application of chitosan membranes for permeation and pervaporation. In G. L. Dotto & L. A. A. Pinto (Eds.), *Chitosan based materials and its applications* (Book Series: Frontiers in Biomaterials) (Vol. 3, pp. 281–324). Sharjah: Bentham Science Publishers Ltd.

Dudek, G., & Turczyn, R. (2018). New type of alginate/chitosan microparticle membranes for highly efficient pervaporative dehydration of ethanol. *RSC Advances, 8*, 39567–39578.

Dudek, G., Gnus, M., Turczyn, R., Strzelewicz, A., & Krasowska, M. (2014). Pervaporation with chitosan membranes containing iron oxide nanoparticles. *Separation and Purification Technology, 133*, 8–15.

Dudek, G., Krasowska, M., Turczyn, R., Gnus, M., & Strzelewicz, A. (2017). Structure, morphology and separation efficiency of hybrid Alg/Fe$_3$O$_4$ membranes in pervaporative dehydration of ethanol. *Separation and Purification Technology, 182*, 101–109.

Dudek, G., Turczyn, R., Gnus, M., & Konieczny, K. (2018). Pervaporative dehydration of ethanol/water mixture through hybrid alginate membranes with ferroferic oxide nanoparticles. *Separation and Purification Technology, 193*, 398–407.

Dudek, G., Krasowska, M., Turczyn, R., Strzelewicz, A., Djurado, D., & Pouget, S. (2019). Clustering analysis for pervaporation performance assessment of alginate hybrid membranes in dehydration of ethanol. *Chemical Engineering Research and Design, 144*, 483–493.

Dudek, G., Turczyn, R., & Konieczny, K. (2020a). Robust poly(vinyl alcohol) membranes containing chitosan/chitosan derivatives microparticles for pervaporative dehydration of ethanol. *Separation and Purification Technology, 234*, 116094.

Dudek, G., Borys, P., Strzelewicz, A., & Krasowska, M. (2020b). Characterization of the structure and transport properties of alginate/chitosan microparticle membranes utilized in the pervaporative dehydration of ethanol. *Polymers, 12*(2), 1–18.

Eliasson, J. (2015). The rising pressure of global water shortages. *Nature, 517*, 6–7.

Fang, S., Stern, H., & Frisch, L. (1975). A "free volume" model of permeation of gas and liquid mixtures through polymeric membranes. *Chemical Engineering Science, 30*, 773–780.

Feng, X., & Huang, R. Y. M. (1997). Liqiud separation by membrane perwaporation: A review. *Industrial Engineering Chemistry Research, 36*, 1048–1066.

Fernández, J. G., Almeida, C. A., Fernández-Baldo, M. A., Felici, E., Raba, J., & Sanz, M. I. (2016). Development of nitrocellulose membrane filters impregnated with different biosynthesized silver nanoparticles applied to water purification. *Talanta, 146*, 237–243.

Fuwad, A., Ryu, H., Malmstadt, N., Kim, S. M., & Jeon, T.-J. (2019). Biomimetic membranes as potential tools for water purification: Preceding and future avenues. *Desalination, 458*, 97–115.

GlobalWaterForum. https://globalwaterforum.org. Accessed 14 Jul 2020.

Grzywna, Z. J., Krasowska, M., Ostrowski, Ł., & Stolarczyk, J. (2001). Can generalized dimension (D_q) and f(α) be used in structure-morphology analysis? *Acta Physica Polonica, B, 32*(5), 1561–1577.

Hilal, N., Ismail, A. F., Matsuura, T., & Oatley-Radcliffe, D. (2017). *Membrane characterization*. Elsevier.

Huang, R. Y. M., Pal, R., & Moon, G. Y. (2000). Pervaporation dehydration of aqueous ethanol and isopropanol mixtures through alginate/chitosan two ply composite membranes supported by poly(vinylidene fluoride) porous membrane. *Journal of Membrane Science, 167*, 275–289.

Ismail, A. F., Rahman, M. A., Othman, M. H. D., & Matsuura, T. (2018). *Membrane separation principles and applications* (1st ed.). Amsterdam: Elsevier.

Jähne, B. (2004). *Practical handbook on image processing for scientific and technical applications*. Boca Raton: CRC Press.

Kalyani, S., Smitha, B., Sridhar, S., & Krishnaiah, A. (2008). Pervaporation separation of ethanol–water mixtures through sodium alginate membranes. *Desalination, 229*, 68–81.

Kanti, P., Srigowri, K., Madhuri, J., Smitha, B., & Sridhar, S. (2004). Dehydration of ethanol through blend membranes of chitosan and sodium alginate by pervaporation. *Separation and Purification Technology, 40*, 259–266.

Kaufman, Y., Berman, A., & Freger, V. (2010). Supported lipid bilayer membranes for water purification by reverse osmosis. *Langmuir, 26*, 7388–7395.

Kaupp, G. (2006). *Atomic force microscopy, scanning nearfield optical microscopy and nanoscratching. Application to rough and natural surfaces*. Berlin: Springer.

Kim, H. J., Nah, S. S., & Min, B. R. (2002). A new technique for preparation of PDMS pervaporation membrane for VOC removal. *Advances in Environmental Research, 6*, 255–264.

Kiss, A. S. (2013). *Advanced distillation technologies: Design, control and applications*. New York: Wiley.

Kocsis, I., Sun, Z., Legrand, Y. M., & Barboiu, M. (2018). Artificial water channels – Deconvolution of natural Aquaporins through synthetic design. *npj Clean Water, 1*, 13.

Krasowska, M., Grzywna, Z. J., Mycielska, M. E., & Djamgoz, M. B. A. (2009). Fractal analysis and ionic dependence of endocytic membrane activity of human breast cancer cells. *European Biophysics Journal, 38*(8), 1115–1125.

Krasowska, M., Rybak, A., Pawełek, K., Dudek, G., Strzelewicz, A., & Grzywna, Z. J. (2012). Structure morphology problems in the air separation by polymer membranes with magnetic particles. *Journal of Membrane Science, 415–416*, 864–870.

Krasowska, M., Strzelewicz, A., & Dudek, G. (2019). Stereological-fractal analysis as a tool for a precise description of the morphology of hybrid alginate membranes. *Acta Physiologica Polonica B, 50*(8), 1463–1478.

Li, P., Wang, Z., Qiao, Z., Liu, Y., Cao, X., Li, W., Wang, J., & Wang, S. (2015). Recent developments in membranes for efficient hydrogen purification. *Journal of Membrane Science, 495*, 130–168.

Li, Y., Qi, S., Tian, M., Widjajanti, W., & Wang, R. (2019). Fabrication of aquaporin-based, biomimetic membrane for seawater desalination. *Desalination, 467*, 103–112.

Liu, L., & Kentish, S. E. (2018). Pervaporation performance of crosslinked PVA membranes in the vicinity of the glass transition temperature. *Journal of Membrane Science, 553*, 63–69.

Liu, Y., Zhou, X., Wang, D., Song, C., & Liu, J. (2015). A prediction model of VOC partition coefficient in porous building materials based on adsorption potential theory. *Building and Environment, 93*, 221–233.

Maddah, H. A. (2019). Industrial membrane processes for the removal of VOCs from water and wastewater. *International Journal of Engineering and Applied Sciences, 6*, 21–26.

Mandelbrot, B. B. (1982). *The fractal geometry of nature.* New York: W. H. Freeman and Co.

Muñoz, R. T., Souza, S. O., Glittmann, L., Pérez, R., & Quijano, G. (2013). Biological anoxic treatment of O2-free VOC emissions from the petrochemical industry: A proof of concept study. *Journal of Hazardous Materials, 260*, 442–450.

Murata, K., Mitsuoka, Hirai, T., Walz, T., Agre, P., Heymann, J.B., Engel, A., Fujiyoshi, Y., (2000), Structural determinants of water permeation through aquaporin-1, Nature, 407, 599–605.

Nath, K. (2017). *Membrane separation processes* (2nd ed.). Delphi: PHI Learning.

Nawawi, M. G. M., Zamrud, Z., Idham, Z., Hassan, O., & Sakri, N. M. (2013). Blended chitosan and polyvinyl alcohol membrane for pervaporation separation methanol/methyl tert-butyl ether mixture. (II) effect of operating parameters. *Jurnal Teknologi (Sciences & Engineering), 65*, 39–43.

Noble, R. D., & Stern, S. A. (1995). *Membrane separations technology: Principles and applications.* Amsterdam: Elsevier.

Ozturk, B., Kuru, C., Aykac, H., & Kaya, S. (2015). VOC separation using immobilized liquid membranes impregnated with oils. *Separation and Purification Technology, 153*, 1–6.

Padhi, S. K., & Gokhale, S. (2014). Biological oxidation of gaseous VOCs – Rotating biological contactor a promising and eco-friendly technique. *Journal of Environmental Chemical Engineering, 2*, 2085–2102.

Qasim, M., Badrelzaman, M., Darwish, N. N., Darwish, N. A., & Hilal, N. (2019). Reverse osmosis desalination: A state-of-the-art review. *Desalination, 459*, 59–104.

Qu, H., Kong, Y., Lv, H., Zhang, Y., Yang, J., & Shi, D. (2010). Effect of crosslinking on sorption, diffusion and pervaporation of gasoline components in hydroxyethyl cellulose membranes. *Chemical Engineering Journal, 157*, 60–66.

Ramaiah, K. P., Satyasri, D., Sridhar, S., & Krishnaiah, A. (2013). Removal of hazardous chlorinated VOCs from aqueous solutions using novel ZSM-5 loaded PDMS/PVDF composite membrane consisting of three hydrophobic layers. *Journal of Hazardous Materials, 261*, 362–371.

Russ, J. C. (2008). *The image processing handbook.* Boca Raton: CRC Press.

Saidur, R., Elcevvadi, E. T., Mekhilef, S., Safari, A., & Mohammed, H. A. (2011). An overview of different distillation methods for small scale applications. *Renewable and Sustainable Energy Reviews, 15*, 4756–4764.

Sajjan, A. M., Premakshi, H. G., & Kariduraganavar, M. Y. (2015). Synthesis and characterization of GTMAC grafted chitosan membranes for the dehydration of low water content isopropanol by pervaporation. *Journal of Industrial and Engineering Chemistry, 25*, 151–161.

Schatzberg, P. (1967). Molecular diameter of water from solubility and diffusion measurements. *The Journal of Physical Chemistry, 71*, 4569–4570.

Shah, M. R., Noble, R. D., & Clough, D. E. (2004). Pervaporation–air stripping hybrid process for removal of VOCs from groundwater. *Journal of Membrane Science, 241*, 257–263.

Shen, Y. X., Saboe, P. O., Sines, I. T., Erbakan, M., & Kumar, M. (2014). Biomimetic membranes: A review. *Journal of Membrane Science, 454*, 359–381.

Siwy, Z., & Fornasiero, F. (2017). Improving on aquaporins. *Science, 357*(6353), 753.

Strzelewicz, A., Krasowska, M., Dudek, G., & Cieśla, M. (2020). Design of polymer membrane morphology with prescribed structure and diffusion properties. *Chemical Physics, 531*, 110662.

Tajkhorshid, E., Nollert, P., Jensen, M., Miercke, L. J. W., O'Connell, J., Stroud, R. M., & Schulten, K. (2002). Control of the selectivity of the aquaporin water channel family by global orientational tuning. *Science, 296*, 525–530.

Toledo, E. J. L., Ramalho, T., & Magriotis, Z. M. (2008). Influence of magnetic field on the physical–chemical properties of the liquid water: Insights from experimental to theoretical models. *Journal of Molecular Structure, 888*, 409–415.

Tsukada, K., Matsunaga, Y., Isshiki, R., Nakamura, Y., Sakai, K., & Kiwa, T. (2017). Magnetic characteristics measurements of ethanol–water mixtures using a hybrid-type high-temperature superconducting quantum-interference device magnetometer. *AIP Advances, 7*(056707), 1–5.

Tunuguntla, R. H., Henley, R. Y., Yao, Y.-C., Pham, T. A., Wanunu, M., & Noy, A. (2017). Enhanced water permeability and tunable ion selectivity in subnanometercarbon nanotube porins. *Science, 357*, 792–796.

Uragami, T. (2017). *Science and technology of separation membranes*. Somerset: Wiley.

Wang, M., Arnal-Herault, C., Rousseau, C., Palenzuela, A., Babin, J., David, L., & Jonquieres, A. (2014). Grafting of multi-block copolymers: A new strategy for improving membrane separation performance for ethyl tert-butyl (ETBE) bio-fuel purification by pervaporation. *Journal of Membrane Science, 469*, 31–42.

Wei, Z., Liua, Q., Wu, Wang, C.H. Wang, H., (2018) Viscosity-driven in situ self-assembly strategy to fabricate cross-linked ZIF90/PVA hybrid membranes for ethanol dehydration via pervaporation. Separation and Purification Technology, 201, 256–267.

Wijmans, J. G., & Baker, R. W. (1995). The solution-diffusion model: A review. *Journal of Membrane Science, 107*, 1–21.

Wojnar, L. (1999). *Image analysis: Applications in materials engineering*. Boca Ratón: CRC Press.

Zhu, Y., Xia, S., Liu, G., & Jin, W. (2010). Preparation of ceramic-supported poly(vinyl alcohol)–chitosan composite membranes and their applications in pervaporation dehydration of organic/water mixtures. *Journal of Membrane Science, 349*, 341–348.

Zhu, M., Qian, J., Zhao, Q., An, Q., Song, Y., & Zheng, Q. (2011). Polyelectrolyte complex (PEC) modified by poly(vinyl alcohol) and their blend membranes for pervaporation dehydration. *Journal of Membrane Science, 378*, 233–242.

Chapter 12
Travelling Waves Connected to Blood Flow and Motion of Arterial Walls

Zlatinka I. Dimitrova and Nikolay K. Vitanov

Abstract Blood contains large amount of water and it is very important for functioning of complex living organisms. Despite the fact that in the human body blood accounts only for approximately 8%–10% of its weight, the blood flow transports many ingredients which must be carried from one place to another in interior of the body. Our focus in this chapter will be on several mathematical results concerning traveling waves connected to arterial wall and blood flow in large arteries. In order to study these waves we use a method for obtaining exact solutions of nonlinear partial differential equations called Simple Equations method (SEsM). We present a brief summary of the method and apply it to obtain exact traveling wave solutions of nonlinear partial differential equations which model blood flow pulsations and nonlinearly affected motion of walls of large arteries.

Keywords Water as principal blood component · Viscosity · Blood flow · Blood flow along arterial walls · Travelling waves · Nonlinearities · Blood flow pulsations · Navier-Stokes equation · Mathematical modeling

1 Introduction

Human blood is a homogeneous liquid at the macroscopic level, but it has cellular and liquid components (McDonald 1974; Rodkiewicz 1983; Biswas 2000). The cellular component contains living blood cells, which are suspended in a nonliving fluid called blood plasma. The blood plasma is the less dense component of the blood and it is about 55% of the blood volume. The most dense component of the

Z. I. Dimitrova (✉)
Institute of Mechanics, Bulgarian Academy of Sciences, Sofia, Bulgaria
e-mail: dimizlati@gmail.com

N. K. Vitanov (✉)
Institute of Mechanics, Bulgarian Academy of Sciences, Sofia, Bulgaria
e-mail: vitanov@imbm.bas.bg

blood are the erythrocytes which are about 44% of the blood volume. The remaining about 1% of blood volume contains leukocytes and platelets. Blood is more dense than water and about five times more viscous, largely because of elements additional to blood plasma which consists mainly of water (about 90% of blood plasma volume). Blood performs a number of functions connected with distribution of substances. These functions include transport of oxygen from the lungs and nutrients from the digestive tract to all body cells as well as transport of metabolic waste products from cells to elimination sites (lungs, kidneys) or transport of hormones from endocrine organs to corresponding target organs.

In the human body blood flows in arteries and veins, which form a complex branching structure. Arterial human tree seems to have a fractal like structure. This structure is not scaled over an infinite range, but some features of fractality have been detected between the scales of the main artery. For example the fractal dimension of pulmonary arteries has been estimated around 2.7 (Mandelbrot 1983).

Blood is a fluid the flow of which obeys universal principles of conservation of mass, momentum and energy (Gustafson 1997; Landau and Lifshitz 1986; Batchelor 2000; Oertel 2004). Forces which drive blood flow are gravitation and pressure gradient force. What opposes the blood flow are the shear forces due to viscosity as well as turbulence (Freis and Heath 1964; Pedley 1980; Fung 1997). Turbulence is much studied phenomenon (Stehbens 1959, 1961; Tennekes and Lumley 1972; Lesieur 2008; Davidson 2015). It is important factor for arteries as artherosclerotic plaques are often found at sites of turbulence in blood vessels. In addition the velocity fluctuations connected to turbulence lead to pressure fluctuations.

The change of pressure in a fluid tube under steady conditions can be calculated by means of Bernoulli's equation from fluid mechanics (Granger 1995; Kundu et al. 2012) if the friction by viscous forces can be assumed to be negligible. But viscosity of blood is an important parameter for the blood flow. The importance of coefficient of viscosity is because the resistance to a flow in a circular cylindrical tube for the case of steady laminar flow is proportional to the coefficient of viscosity. In addition the volume flow rate of fluid in a long rigid circular cylindrical tube (i.e., volume of fluid flowing through the vessel in an unit time) is inversely proportional to the coefficient of viscosity. Larger coefficient of viscosity leads to smaller value of the volume flow rate. Blood viscosity is also important for the dimensionless parameter of the flow called Reynolds number which accounts for the relation between inertia force and viscous force connected to the flow. In several more words the Reynolds number has numerator and denominator. Characteristic parameter connected to inertia force is in the numerator (velocity of the flow multiplied by vessel's diameter). Characteristic parameter connected to viscous force is in the denominator and this is the kinematic viscosity of the fluid.

The blood in human body is moved by the heart which pumps blood through the arteries to peripheral organs. Usually, it is modeled by the set of pipes, which are connected into a hierarchical structure. In this structure, several pipes depart from a lower level pipe (Van Savage et al. 2008). The geometrical feature of the pipes mostly follows from Murray's law, which states that the internal diameters of the vessels take on such values that work for transport in minimized conditions (Sherman 1981).

Despite taking into account a number of theoretical and empirical assumptions in such models, they still do not represent all the properties of real systems. There is a need to study fluid flows in a single pipe, which assumed features better reflect observations of the real vessels.

Therefore, we make several remarks on some simple models connected to flows in cylindrical circular pipes which are of interest for the theory of blood flow in blood vessels. We consider first the flow through a circular cylindrical tube of radius a. The tube is oriented parallel to the x-axis and the flow has a velocity component (denoted by u) only in this direction. We assume that u does not depend on x, i.e., $u = u(y,z)$. The second law of Newton for the fluid motion (called also Navier-Stokes equation) in cylindrical polar coordinates (x,r,θ) is

$$(1/r)(\partial/\partial r)[r\partial u/\partial r]+(1/r^2)(\partial^2 u/\partial\theta^2)=(1/\mu)(dp/d\,)x \qquad (12.1)$$

where the term in the right-hand side of the equation accounts for the gradient of the pressure p and μ is the coefficient of viscosity. Assuming symmetry in the flow (i.e., there is no dependence on θ) we obtain

$$(1/r)(d/dr)[r\,du/dr]+(1/r^2)(d^2u/d\theta^2)=(1/\mu)(dp/d\,)x$$

which has the solution

$$u=\left[(r^2)/(4\mu)\right](dp/dx)+A\ln(r)+B.$$

The constants A and B can be determined by the (no-slip) boundary conditions, namely,

$$u=0, r=a,$$

$$du/dr=0, r=0,$$

and we obtain the velocity profile of the Hagen-Poiseuille flow

$$u=-\left[1/(4\mu)\right](a^2-r^2)dp/dx, \qquad (12.2)$$

and for the flow rate (2π multiplied by integral from 0 to a with respect to r) we obtain

$$Q=-\left[\pi a^4/(8\mu)\right]\Delta p/L, \qquad (12.3)$$

where Δp is pressure drop in a segment of blood vessel of length L.

As a complication the tube can be elastic and we can consider steady laminar flow in such a circular cylindrical tube. For flow maintained by pressure gradient we can write in analogy with (12.3)

$$dp/dx = -\left[8\mu/\left(\pi A^4\right)\right]Q. \qquad (12.4)$$

From this relationship we obtain the pressure – flow relation

$$p(x) = p(0) - \frac{8\mu}{\pi}Q\int_0^x d\tilde{x}\,\frac{1}{a^4(\tilde{x})} \qquad (12.5)$$

where $p(0)$ is the pressure at $x = 0$. We note that tube length is L and the tube radius a depends on the coordinate x. The dependence of the tube radius on the coordinate is connected to the pressure – radius dependence in the elastic tube. For a kind of blood vessels (pulmonary blood vessels) the pressure – radius relation can be approximated as

$$a = a_0 + \alpha\, p/2, \qquad (12.6)$$

where a_0 is the tube radius at $x = 0$ and α is a constant. From (12.6)

$$da/dx = (\alpha/2)dp/dx, \qquad (12.7)$$

and then from (12.4)

$$a^4 da/dx = -4(\mu\alpha/\pi)Q. \qquad (12.8)$$

From (12.8) we obtain

$$a^5(x) = -(20\,\mu\alpha/\pi)Q\,x\,, \qquad (12.9)$$

and using the boundary condition $a(x) = a(0)$ we obtain

$$(20\,\mu\alpha L/\pi)Q = [a(0)]^5 - [a(L)]^5. \qquad (12.10)$$

Thus if the tube radius $a(L)$ at the end of the tube is sufficiently smaller than the tube radius $a(0)$ at the entry of the tube it does not influence much (12.10). For an example if $a(L) = a(0)/2$ then $[a(L)]^5 = [a(0)]^5/32$.

In the sections below we shall discuss the wave propagation in blood vessels. The wave propagation is governed by a wave equation. The most simple (linear) version of this equation can be obtained as follows. We consider the flow of homogeneous, incompressible, non-viscous fluid in an infinitely long straight cylindrical elastic tube. The simplification is already very large (the blood is a viscous fluid) but let us consider in addition the propagation of waves of small amplitude in such a tube. Wave motion means that the tube is disturbed at some place and this disturbance propagates as waves of finite speed along the tube. Disturbance is assumed to be small compared to the radius of the tube. One further assumption is that the wave velocity has only longitudinal component u (other components are assumed to be negligible) and u depends only on the axial coordinate x and on the time t. Let p_i be the pressure in the tube and ρ be the density of the fluid. The equation of the motion of the fluid (second law of Newton or Navier-Stokes equation without presence of stress tensor and without external forces – Euler equation) is

$$\partial u / \partial t + u \partial u / \partial x + (1 / \rho) \partial p_i / \partial x = 0. \tag{12.11}$$

We consider small disturbances in stationary fluid-filled circular cylindrical tube. In this case u is small and the second term in (12.11) can be neglected. Differentiation of what remains with respect to x leads to

$$\partial^2 u / (\partial x \partial t) + (1 / \rho) \partial^2 p_i / \partial x^2 = 0. \tag{12.12}$$

We are going to obtain another equation which will be subtracted from (12.12) and the result of this subtraction will be the wave equation. This second equation comes from the conservation of the mass

$$\partial S / \partial t + \partial (u S) / \partial x = 0, \tag{12.13}$$

where $S(x,t)$ is the cross-sectional area of the tube. Next we shall account for the elasticity of the tube by means of equation similar to (12.6)

$$a = a_0 + (\alpha / 2) p_i. \tag{12.14}$$

From (12.14) we obtain

$$da = (\alpha / 2) dp_i. \tag{12.15}$$

Taking into an account that $S = \pi a^2$ we obtain from (12.13)

$$\partial u / \partial x + (\alpha / a) \partial p_i / \partial t = 0, \tag{12.16}$$

under the assumption $(\partial a / \partial x << 1)$. Differentiation of (12.16) with respect to t leads to

$$\partial^2 u / (\partial x \partial t) + (\alpha / a) \partial^2 p_i / \partial t^2 = 0. \tag{12.17}$$

Subtracting of (12.17) from (12.12) and setting

$$c = \left[a / (\alpha \rho) \right]^{1/2}, \tag{12.18}$$

as wave speed we obtain the wave equation

$$\partial^2 p_i / \partial x^2 - (1 / c^2) \partial^2 p_i / \partial t^2 = 0. \tag{12.19}$$

The solution of (12.19) is of form of a wave

$$p = p(x - ct), \tag{12.20}$$

propagating with velocity c in the direction of increasing x. Thus the pressure waves in our idealized case are governed by a linear wave equation. We note that the linear wave equation is obtained for the case of non-viscous fluid. The blood is a viscous fluid and because of this we have to consider more complicated models. This will be done in the following section.

2 More Complicated Models Connected to Blood Flow and Motion of Arterial Wall

Let us now consider more complicated models of blood flow and of motion of arterial walls which will lead us to models based on nonlinear partial differential equations. We shall start again from the Eqs. (12.11) and (12.13). But instead of (12.14) we shall adopt more complicated equation for the radial motion of artery wall under the forces caused by the fluid

$$\rho_w h \left(\partial^2 a / \partial t^2 \right) = p - p_e - (h/a)\sigma_\theta. \tag{12.21}$$

In (12.21) ρ_w is the density of the artery wall, h is the thickness of the tube of radius $a(x,t)$, p_e is the external pressure and σ_θ is the extending stress in the tangential direction. For external pressure we have the relationship

$$p_e = p_0 - (h_0 / R_0)\sigma_\theta^0, \tag{12.22}$$

obtained as equilibrium relation for artery of radius R_0 and wall thickness h_0 inflated at the diastolic pressure p_0. Then by means of relationships for the differential pressure

$$P = p - p_0, \tag{12.23}$$

and for the (small) radial elongation of the arterial wall

$$\gamma = (R - R_0)/ R_0, \tag{12.24}$$

we obtain the following relationship from (12.22)

$$P - \eta\sigma_\theta^* /(1 + \gamma)^2 = \left[\rho_w h_0 R_0 /(1 + \gamma) \right] \partial^2\gamma / \partial t^2, \tag{12.25}$$

where we assume wall incompressibility ($hR = h_0 R_0$), set $\eta = h_0/R_0$ and the stress component σ_θ^* is connected to the Young modulus E and to the coefficient of nonlinear elasticity α as follows

$$\sigma_\theta^* = \gamma E \left(1 + \alpha \gamma \right). \tag{12.26}$$

The motivation of employing Eq. (12.26) in a quadratic form with $\alpha > 0$ derives from applying a well-known structural-mechanical relation pertinent to non-Hookean behavior of hydrated elastin biomolecules of which the arterial walls, thus, important parts of the whole cardiovascular system, are composed. When inspecting thoroughly (Cocciolone et al. 2018), one may find out, see Fig. 4c and d therein, that the Hooke's law, if applicable to elastin arterial-wall systems (for a certain amount of the biomolecular component), extends toward a nonlinear viz. typically quadratic regime with $\alpha > 0$, as depicted clearly in the mentioned figures, and corroborated by the properly assumed relation (12.26) of the present study; for another stochastic approach, also revealing the peculiarities of transport in fractally designed blood vessels, see (West and Deering 1994; Weber and Pepłowski 2016).

In order to obtain a model equation we have to make additional assumptions. First assumption is that the rest radius R_0 depends only on the coordinate x and changes slowly along this coordinate – $R_0 = R_0(X)$. Here $X = \varepsilon x$ is a slow variable. The same assumption is made about the Young modulus – $E = E(X)$. Next assumption is that the velocity of the blood wave is much larger than the blood velocity u. Thus

$$u \left(\partial / \partial z \right) \propto \varepsilon \left(\partial / \partial t \right). \tag{12.27}$$

Assuming that γ and ε are small parameters we can make expansions of quantities in (12.11) and (12.13) in series containing powers of these small parameters. In (12.13) we keep terms of order up to ε^0 and γ^2 and obtain

$$\partial^2 u / \left(\partial x \partial t \right) = 2 \left(\partial \gamma / \partial t \right)^2 - 2 \left(1 - \gamma \right) \partial^2 \gamma / \partial t^2. \tag{12.28}$$

Next we differentiate twice (12.25) with respect to x and subtract from the obtained result (12.11) differentiated with respect to x. The result is an equation for the motion of the wall of the artery

$$\partial^2 \gamma / \partial t^2 - v_0^2 \partial^2 \gamma / \partial x^2 = \left(1/2 \right) \left[\partial^2 \left(\gamma^2 \right) / \partial t^2 \right] + s \partial^4 \gamma / \left(\partial x^2 \partial t^2 \right) + \beta v_0^2 \partial^2 \gamma / \partial x^2, \tag{12.29}$$

where v_0 is the phase velocity of the linear waves

$$v_0 \left(\varepsilon x \right) = \left[\eta E \left(\varepsilon x \right) / \left(2 \rho \right) \right]^{1/2}, \tag{12.30}$$

$\beta = \alpha - 2$ and

$$s \left(\varepsilon x \right) = \eta \rho_w R_0^2 \left(\varepsilon x \right) / \left(2 \rho \right). \tag{12.31}$$

(12.29) is a nonlinear partial differential equation accounting for the intrinsic nonlinearity of the fluid mechanics equations, for the nonlinear coefficient of elasticity of the artery wall and for dispersion connected to the inertial effects of the wall. Solution of (12.29) leads to complicated equation of kind (12.28) for the velocity of the waves in the blood and this complicated equation has to be solved numerically.

Other nonlinear partial differential equations can arise as models of motion of artery wall and of the motion of blood (Demiray 1992, 1996, 1997; Demiray and Antar 1997). We shall discuss in some detail solutions of two of them. First one (Demiray 2008; Dimitrova 2015) leads to the equation

$$\partial u / \partial \tau + \mu_1 u \partial u / \partial \xi + \mu_2 (\tau) \partial u / \partial \xi + \mu_3 \partial^3 u / \partial \xi^3 = 0 \qquad (12.32)$$

And the second one (Tay 2006; Tay et al. 2007; Tay and Demiray 2008; Demiray 2008) leads to the model equation

$$\partial u / \partial \tau + \mu_1 u \partial u / \partial \xi - \mu_2 \partial^2 u / \partial \xi^2 + \mu_3 \partial^3 u / \partial \xi^3 + \mu_4 (\tau) \partial u / \partial \xi + \mu(\tau) = 0. \quad (12.33)$$

Next we discuss how to obtain analytical traveling wave solutions of the model nonlinear partial differential Eqs. (12.29), (12.32) and (12.33) by means of a method called Simple Equations Method (SEsM) (Vitanov 2019a, 2020; Vitanov and Dimitrova 2019). Numerical solution of (12.28) will not be discussed in this text.

3   Simple Equations Method (SESM)

The Simple Equations Method (SEsM) was developed in order to be used for obtaining exact solutions of nonlinear PDEs of the models of natural and social systems (Ames 1965; Debnath 2012; Grimshaw 1993; Leung 1989; Murray 1977; Verhulst 1990; Vitanov 2016; Vitanov and Vitanov 2016, 2018, 2018a, 2019a, b). The exact solutions of these model equations are interesting as they lead to better understanding of states and processes in the modeled systems. In addition the exact solutions are used to test computer programs for numerical simulations. If these programs work properly they have to reproduce the corresponding exact solutions. In many problems, the model equations are nonlinear partial differential equations (PDEs) and in numerous cases these equations have traveling-wave solutions. The traveling wave solutions are studied intensively (Ablowitz et al. 1973; Holmes et al. 1996; Kudryashov 1990; Scott 1999; Tabor 1989; Vitanov 1996, 1998; Vitanov et al. 2009, 2009a) and complicated methodology has been developed for obtaining such exact solutions. Examples are the method of Inverse Scattering Transform or the method of Hirota (Hirota 2004; Ablowitz et al. 1974; Remoissenet 1993, Infeld and Rowlands 1990). The above methods perform very well for the case of integrable nonlinear PDEs. Other approaches for obtaining exact special solutions of nonlintegrable nonlinear PDEs have been developed in the recent years (see for examples Malfliet and Hereman 1996, Fan and Hon 2003, He and Wu 2006, Wazwaz 2004,

2009). Below we shall consider a method called Simple Equations Method – SEsM. This method is a generalization of the method of simplest equation and its version called Modified Method of Simplest Equation (Kudryashov 2005; Kudryashov and Loguinova 2008; Vitanov et al. 2010, 2015; Vitanov 2011, 2011a). The Method of Simplest Equation is based on a procedure analogous to the first step of the test for the Painleve property. In the version of the method called Modified Method of Simplest Equation this procedure is substituted by the concept for the balance equations. SEsM (Vitanov 2019a, b, c, 2020; Vitanov and Dimitrova 2019) is based on the possibility of the use of more than one simple equation for obtaining an exact analytical solution of the solved nonlinear partial differential equation. We do not use anymore the words simplest equation because the simple equations which solutions are used to construct the solution of the more complicated nonlinear partial differential equation may be quite complicated. The Modified Method of Simplest Equation is a particular case of SEsM when one uses one simple equation and one balance equation. The Modified Method of Simplest Equation has already numerous applications, e.g., obtaining exact traveling wave solutions of the generalized Kuramoto – Sivashinsky equation, reaction – telegraph equation, generalized Swift – Hohenberg equation and generalized Rayleigh equation, reaction – diffusion equation, generalized Fisher equation, generalized Huxley equation, b-equation, generalized Degasperis – Procesi equation, extended Korteweg-de Vries equation, etc. (Vitanov 2010, 2011a; Vitanov and Dimitrova 2010, 2014, 2018; Vitanov et al. 2011, 2013, 2013a, 2017).

Let us now describe the SEsM. We consider a nonlinear partial differential equation

$$DE(u,\ldots) = 0, \tag{12.34}$$

where $DE(u,\ldots)$ depends on the function $u(x,\ldots,t)$ and some of its derivatives (u can be a function of more than 1 spatial coordinate). The steps of the methodology of SEsM are as follows.

Step 1: Transformation of the Nonlinearity The goal here is to transform the nonlinearity of the solved equation to a more treatable kind of nonlinearity, e.g., polynomial nonlinearity or even to remove the nonlinearity (if possible). The transformation is

$$U(x,\ldots,t) = T\left[F(x,\ldots,t)\right]. \tag{12.35}$$

The transformation $T(F)$ is a function of another function F. In general the transformation $F(x,\ldots,t)$ is a function of the spatial variables as well as of the time. No general form of $T(F)$ is known nowadays. Several possible forms of $T(F)$ are.

- $u(x,t) = 4\,tan^{-1}F(x,t)$ for the case of 1+1-dimensional the sine – Gordon equation,

- $u(x,t) = 4\ tanh^{-1}F(x,t)$ for the case of 1+1-dimensional Poisson-Boltzmann equation (for applications of the last two transformations see, e.g. Martinov and Vitanov 1992, 1992a, 1994; Vitanov and Martinov 1996; Vitanov 1998).
- the Painleve expansion (Hirota 1971)

The form of $T(F)$ can also be different from the above forms. In many particular cases one may skip this step (then we have just $u(x,...,t) = F(x,...,t)$ and $T(F) = F$) but in numerous cases the step is necessary for obtaining a solution of the studied nonlinear PDE. The application of (12.35) to (12.34) leads in the most cases to a nonlinear PDE for the function $F(x,...,t)$ but in some cases the resulting equation for F may be a linear one.

Step 2: Choosing the Form of the Function F The function $F(x,...,t)$ is a function of other functions $f_1,...,f_N$. These functions are solutions of some differential equations, which can be partial or ordinary differential equations, that are more simple than (12.34). The possible values of N are $N = 1,2,...$ (also there can be an infinite number of functions f_i). No general form of the function $F(f_1,...,f_N)$ is known up to now. The forms of the function $F(f_1,...,f_N)$ can be different, e.g.,

$$F = \alpha + \sum_{N}^{i=1} \beta_i f_i + \sum_{N}^{i=1}\sum_{N}^{j=1} \gamma_{i,j} f_i f_j + ... + \sum_{N}^{i=1}...\sum_{N}^{n=1} \sigma_{i,j,...,n} f_i ... f_n, \qquad (12.36)$$

where $\alpha, \beta,...,\sigma$ are parameters. We note that $F(f_1,...,f_N)$ can have also another form which is different from (12.36). SEsM is very flexible with respect to the form of F. We shall use the form (12.36) in this text. We note that the relationship (12.36) for F contains the relationship used by Hirota (1971) as particular case.

Step 3: Choosing the Equations for $f_1,...,f_N$ In general the functions $f_1,...,f_N$ are solutions of partial differential equations (PDEs). These equations are simpler than the solved nonlinear PDE. There are two possibilities. First one can use solutions of the simple PDEs if such solutions are available. The second possibility is connected to transformation of the simpler PDEs by means of appropriate ansaetze (e.g., traveling-wave ansaetze such as $\xi = \alpha x + \beta t;\ \zeta = \mu y + \nu t$). Thus the solved differential equations for $f_1,...,f_N$ can be reduced to differential equations E_l, containing derivatives of one or several functions

$$E_l\left[a(\xi), a_\xi, a_{\xi\xi},...,b(\zeta), b_\zeta, b_{\zeta\zeta},...\right] = 0, l = 1,...,N \qquad (12.37)$$

In many cases (for example, if the equations for the functions $f_1,...,f_N$ are ordinary differential equations) one may skip this step but step 3 may be necessary if the equations for $f_1,...,f_N$ are complicated partial differential equations.

Step 4: The Form of the Functions $a(\xi)$, $b(\zeta)$,... The functions $a(\xi)$, $b(\zeta)$,... etc., can be functions of other functions, such as, $v(\xi)$, $w(\zeta)$,..., etc.. This means that we have

$$a(\xi) = A\left[v(\xi)\right]; b(\xi) = B\left[w(\xi)\right]. \tag{12.38}$$

Note that SEsM does not prescribe the forms of the functions $A,B,...$ Often one uses a finite-series relationship, e.g.,

$$a(\xi) = \sum_{\pi}^{\mu=-\nu} q_\mu A_\mu\left[v(\xi)\right]; b(\zeta) = \sum_{\rho}^{\mu=-\theta} r_\mu B_\mu\left[w(\zeta)\right], \tag{12.39}$$

where $q, r,...$, are coefficients. However other kinds of relationships, and more complicated ones, can be used too.

Step 5: The Simple Equations The functions $v(\xi)$, $w(\zeta)$, ... are solutions of simple ordinary differential equations. We note that the number of the simple equations can be larger than 1. For about 10 years, we have used a particular case of the described methodology that was based on the use of one simple equation. This simple equation was called simplest equation, and the methodology based on one equation was called Modified Method of Simplest Equation. SEsM contains the Modified Method of Simplest Equation as one of its numerous particular cases (for more particular cases see Vitanov and Dimitrova 2019) and we had to change the name of the methodology as the used simple equations may be not the simplest possible ones.

Step 6: Balance Equations The application of steps 1–5 to (12.34) transforms the left-hand side of this equation. We consider the case when the result of this transformation is a function which is a sum of terms containing some function multiplied by a coefficient. This coefficient is a relationship which includes some of the parameters of the solved equation, some of the parameters of the solution, and some of the parameters of the simple equations. In many cases, a balance procedure must be applied to ensure that the above mentioned relationships for the coefficients contain more than one term (e.g., if the result of the transformation is a polynomial then the balance procedure has to ensure that the coefficient of each term of the polynomial is a relationship that contains at least two terms). In some cases the balance procedure is not needed and such an example will be presented for the case of Korteweg – de Vries equation. The balance procedure may lead to one or more additional relationships among the parameters of the solved equation and parameters of the solution. The last relationships are called balance equations. The satisfaction of the balance equations is closely connected to the obtaining solutions of the solved nonlinear partial differential equation.

Step 7: Solution of the System of Nonlinear Algebraic Equations We may obtain a nontrivial solution of (12.34) if all coefficients mentioned in Step 6. are set to zero. This leads to a system of nonlinear algebraic equations for the coefficients of: (i) the solved nonlinear PDE, (ii) the coefficients of the solution, and (iii) the coefficients of the simple equations. Any nontrivial solution of this algebraic system leads to a solution the studied nonlinear partial differential equation. Usually the above system of algebraic equations contains many equations that have to be solved with the help of a computer algebra system. The algebraic system can be quite complicated and then it is not possible to solve it even by means of a computer algebra system. In this case, we cannot obtain an exact solution of the solved nonlinear partial differential equation by SEsM but we can try to solve the algebraic system numerically, and if this can lead to a numerical solution to the solved nonlinear partial differential equation.

4 Use of SESM for Obtaining Exact Solutions of Model Equations Connected to Blood Flow

Let us now obtain analytical traveling wave solutions of some of the nonlinear model equations derived above. The simplest of these equations is (12.32). We can transform this equation by means of the new coordinate

$$\eta = \xi + \tau - \int_{\tau}^{0} ds \mu_2(s). \tag{12.40}$$

The result is

$$du / d\eta + \mu_1 u \, du / d\eta + \mu_3 d^3 u / d\eta^3 = 0, \tag{12.41}$$

and it can be obtained from the classical form of the Korteweg – de Vries equation

$$\partial U / \partial t + \alpha U \partial U / \partial x + \beta \partial^3 U / \partial x^3 = 0, \tag{12.42}$$

by means of transformation

$$\alpha U / 6 \rightarrow U; x \beta^{1/2} \rightarrow x; t / \beta^{1/2} \rightarrow t. \tag{12.43}$$

The result is

$$\partial U / \partial t + 6U \partial U / \partial x + \partial^3 U / \partial x^3 = 0, \tag{12.44}$$

and the search for travelling wave solutions of the kind $U(\xi) = U(x - vt)$ (where v is velocity of the wave) leads to (12.41) when $\mu_1 = -6/v$; $\mu_3 = -1/v$.

Equation (12.42) is particular case (when $p = 1$) of the equation

$$\partial U / \partial t + \alpha U^p \partial U / \partial x + \beta \partial^3 U / \partial x^3 = 0. \tag{12.45}$$

The solution of (12.45) by means of SEsM is as follows (for more details see Vitanov et al. 2015). We search for travelling wave solutions of the kind $U(\xi) = U(\mu x + \nu t)$ (where μ and ν) are parameters. We will not use transformation of the nonlinearity (step 1 of SEsM). Next we have to choose the form of function F (step 2 of SEsM). The choice is particular case of (12.36)

$$U = b_0 + b_1 g(\xi), \tag{12.46}$$

where b_0, b_1 are parameters and $g(\xi)$ is solution of the equation (steps 3–5 of SEsM)

$$\left(\frac{dg}{d\xi}\right)^2 = \sum_{2+p}^{j=0} a_j g^j. \tag{12.47}$$

Note that in our case $p = 1$ and thus $2 + p = 3$. The substitution of (12.46) and (12.47) in (12.45) leads to the balance (step 6 in SEsM) because of the choice $j_{max} = 2 + p$ in (12.47). The substitution of (12.46) and (12.47) in (12.45) leads also to a system of algebraic equations which has the solution

$$a_0 = a_1 = 0; a_2 = -v / \mu^3; a_3 = \ldots = a_{p-1} = 0;$$
$$a_p = 2\alpha b_1^p / \left[\mu^2 \alpha (p+1)(p+2)\right], b_0 = 0. \tag{12.48}$$

Substitution of (12.48) in (12.47) gives the specific form of the simple equation. The solution of this simple equation leads to the following solution of (12.45)

$$U(\xi) = \Omega b_1 / \cosh^{2/p}(\xi), \tag{12.49}$$

where Ω is a parameter. This solution allows us to obtain the following solitary wave solution of (12.41)

$$u(\eta) = (v/2)/\cosh^2\left[v^{1/2}\eta / 2\right], \tag{12.50}$$

which is a solitary wave.

Let us now consider the more complicated model Eq. (12.33). We introduce new dependent variable

$$V(\xi, \tau) = u(\xi, \tau) - \int d\tau \mu(\tau). \tag{12.51}$$

In addition we introduce the coordinate transformation:

$$\tau' = \tau; \xi' = \xi - \int d\tau \left[\mu_4(\tau) - \mu_1 \int d\tau \mu(\tau)\right]. \tag{12.52}$$

Transformations (12.51) and (12.52) reduce (12.33) to the generalized Korteweg – deVries equation:

$$\partial V / \partial \tau' + \mu_1 V \partial V / \partial \xi' - \mu_2 \partial^2 V / \partial \xi'^2 + \mu_3 \partial^3 V / \partial \xi'^3 = 0. \tag{12.53}$$

We search for traveling wave solution of (12.53) and introduce new coordinate $\zeta = \xi' - v^* \tau'$ where v^* is the velocity of the traveling wave. We will not transform the nonlinearity in (12.53) (step 1 of SEsM). The form of solution is particular case of (12.36) (step 2 of SEsM) namely

$$V = a_0 + a_1 g + a_2 g^2, \tag{12.54}$$

where $g(\zeta)$ is solution of the simple equation (steps 3–5 of SEsM)

$$dg / d\zeta = b_0 + b_1 g + b_2 g^2, \tag{12.55}$$

and $a_0, a_1, a_2, b_0, b_1, b_2$ are parameters. (12.54) and (12.55) are chosen in such a way that the balance equation for (12.53) is satisfied (step 6 of SEsM). The substitution of (12.54) and (12.55) in (12.53) leads to a system of algebraic equations for the parameters of the solution and parameters of the equation (step 7 of SEsM). The solution of this system leads to the following solution of (12.55):

$$g(\zeta) = -b_1 / (2b_2) - [\Delta / (2b_2)] \tanh[\Delta(\zeta + \zeta_0)/2] + \exp[\Delta(\zeta + \zeta_0)/2]$$
$$/\{2\cosh[\Delta(\zeta + \zeta_0)/2] b_2 / \Delta + 2C^* \exp[\Delta(\zeta + \zeta_0)/2] \cosh[\Delta(\zeta + \zeta_0)/2]\}, \tag{12.56}$$

and the solution of (12.53) is as follows

$$V(\zeta) = -(1/25)\left[-3\mu_2^2 - 30\mu_1\mu_3 b_1 + 75\mu_3^2 b_1^2 + 25\nu\mu_3\right] / (\mu_1\mu_3)$$
$$-12\left[b_2(5\mu_3 b_1 - \mu_2)/(5\mu_1)\right] g(\zeta) - 12\left[\mu_3 b_2^2 / \mu_1\right] g^2(\zeta). \tag{12.57}$$

This solution describes a solitary wave.

Finally let us obtain soliton solutions of (12.29). By means of the reductive perturbation method (Taniuti and Wei 1968) and by means of appropriate rescaling of coordinates (Paquerot and Remoissenet 1994) (12.29) can be reduced to the Korteweg –de Vries equation

$$\partial U / \partial t + \sigma U \partial U / \partial x + \partial^3 U / \partial x^3 = 0, \tag{12.58}$$

where σ is a parameter. Eq. (12.58) has a two-soliton solution which can be obtained by SEsM as follows. We make a transformation of the nonlinearity in (12.58) by means of substitution $U = \partial p/\partial x$ and by means of the transformation $p = (12/\sigma) \partial(\ln F)/\partial x$ (step 1 of SEsM). Step 2 of SEsM is to determine the form of the function F. We take the following particular case of (12.36)

$$F(x,t) = 1 + f_1(x,t) + f_2(x,t) + cf_1(x,t)f_2(x,t), \tag{12.59}$$

where f_1, f_2 are functions and c is a parameter (step 2 of SEsM). The substitution of $U = \partial p/\partial x$, $p = (12/\sigma) \partial(\ln F)/\partial x$, and (12.59) in (12.58) leads to a large equation containing the functions f_1, f_2. Step 3 of SEsM requires to set relationships for the functions f_1, f_2. These relationships are

$$\partial f_i / \partial x = \alpha_i f_i; \partial f_i / \partial t = \beta_i f_i, i = 1, 2. \tag{12.60}$$

In addition we assume

$$f_1(x,t) = a(\xi), f_2(x,t) = b(\zeta), \xi = \alpha_1 x + \beta_1 t + \gamma_1, \zeta = \alpha_2 x + \beta_2 t + \gamma_2, \quad (12.61)$$

where $\alpha_i, \beta_i, \gamma_i, i = 1,2$, are parameters. Step 4 of SEsM determines the functions a and b. We use particular case of (12.38)

$$a(\xi) = q_1 v(\xi), b(\zeta) = r_1 w(\zeta), \quad (12.62)$$

where v and w are functions which are determined at step 5 of SEsM. The form of these function here is simple

$$dv / d\xi = v; dw / d\zeta = w, \quad (12.63)$$

and the corresponding solutions of (12.63) are

$$v(\xi) = w_1 \exp(\xi); w(\zeta) = w_2 \exp(\zeta). \quad (12.64)$$

We assume that parameters w_1, w_2 are included in parameters q_1 and r_1 and q_1 and r_1 can be included in ξ and ζ.

It is interesting here that we do not need to perform the balance procedure and to obtain a balance equation as the substitution of all above in the Korteweg – deVries equation leads to a relationship containing sum of exponential functions each of which is multiplied by a coefficient which contains parameters of the equation and parameters of the searched solution. We need just to set there coefficients equal to zero and this leads us to the system of nonlinear algebraic equations from the step 7 of SEsM. The nontrivial solution of this system of algebraic equations leads to the two-soliton solution of the Korteweg – se Vries equation

$$U(x,t) = (12/\sigma)\partial^2 / \partial x^2 \{1 + \exp(\alpha_1 x - \alpha_1^3 t + \gamma_1) + \exp(\alpha_2 x - \alpha_2^3 t + \gamma_2) +$$
$$\left[(\alpha_1 - \alpha_2)^2 / (\alpha_1 + \alpha_2)^2\right] \exp\left[(\alpha_1 + \alpha_2)x - (\alpha_1^3 - \alpha_2^3)t + \gamma_1 + \gamma_2\right]\}. \quad (12.65)$$

With respect to notations used in (12.29) this two-soliton solution can be written as

$$\gamma(x,t) = 8\left[\mu_1^2 f_1 + \mu_2^2 f_2 + 2(\mu_1 - \mu_2)f_1 f_2 + (\mu_1^2 f_1 f_2^2 + \mu_2^2 f_1^2 f_2)(\mu_1 - \mu_2)^2 / (\mu_1 + \mu_2)^2\right] /$$
$$\left[1 + f_1 + f_2 + f_1 f_2 (\mu_1 - \mu_2)^2 / (\mu_1 + \mu_2)^2\right]^2, \quad (12.66)$$

where $i = 1,2$,

$$f_i = \exp\{[(2\mu_i x - 8\mu_i^3 t]\}; \mu_i = \left[(6/(b+\beta))(v_i / v_0 - 1)\right]^{1/2} / 2, \quad (12.67)$$

and v_i are initial velocities of the two pulses of the two-soliton solution.

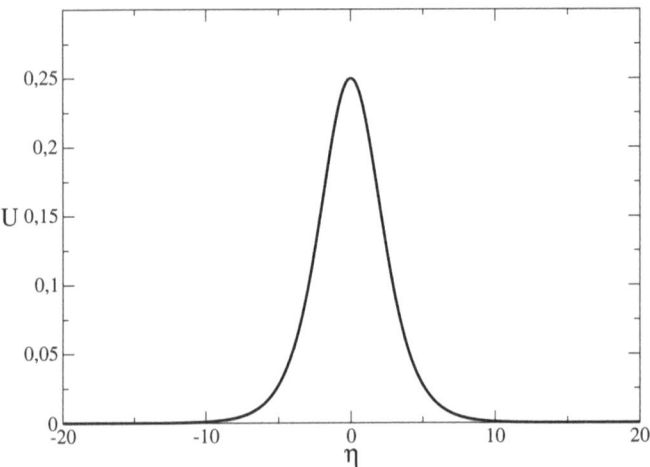

Fig. 12.1 Solution $U(\eta)$ (12.50) of Eq. (12.48). $\nu = 0.5$

5 Discussion

Blood is water-based liquid which is very important for functioning of the human and animal bodies. Blood flows in complex system of arteries and veins and this flow is maintained by the pulsations of heart. Arteries and veins are elastic tubes and as a consequence of the above pulsations waves propagate in the arterial walls and in the blood. Research on these waves leads to interesting mathematical problems connected to obtaining solutions of the model nonlinear partial differential equations. We obtain exact solutions of kinds of solitary wave and two-soliton solutions of some of the model equations by means of method called Simple equations Method (SEsM). Several of these solutions are shown in Figs. 12.1, 12.2 and 12.3.

What is of large interest for the area of description of motion of blood in elastic arteries are solutions of the model equation which describe pulsations. In the model equations one assumes that the amplitude of the pulsation is zero when the spatial coordinate has large absolute value. Such pulsations are described by solitary wave solutions of the model equations. We see from Figs. 12.1, 12.2 and 12.3 that the different model equations lead to analytical relationship for the pulsation of the arterial wall. The blood flow associated with this pulsation has to be calculated numerically and this is out of the goal of this chapter which is devoted to analytical results and on the ways of obtaining of such results.

Finally we note that several more applications of this method for obtaining exact solutions of blood flow problems and other cases are for an example (Dimitrova 2012a, b; Nikolova et al. 2017, 2018; Nikolova 2018). SEsM is a useful and robust method for obtaining exact analytical solutions of nonlinear partial differential equations used to model different processes in Nature and society.

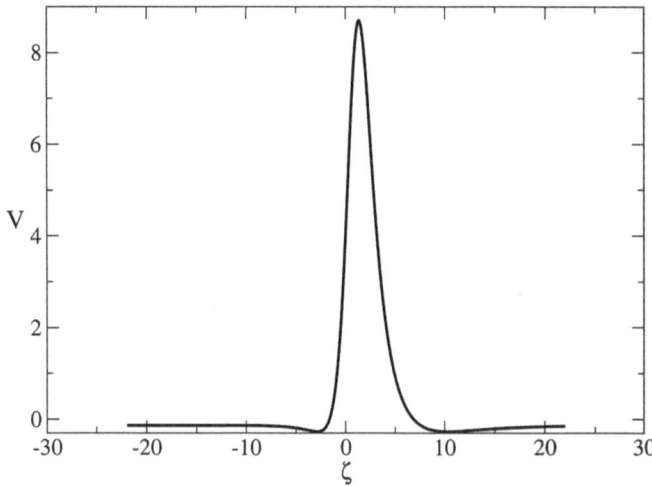

Fig. 12.2 Solution $V(\zeta)$ – (12.57) of Eq.(12.53). The values of the parameters are as follows: $\nu = 0.5$, $b_1 = 0, b_2 = -2, d = 0.5, c = 2, \mu_1 = 0.1, \mu_2 = 0.15, \mu_3 = -0.2$

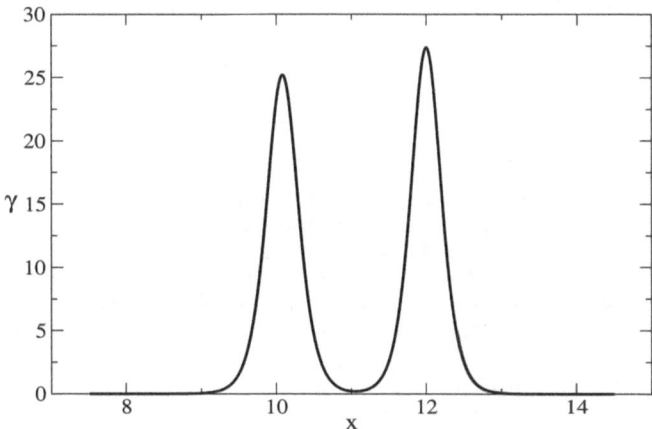

Fig. 12.3 Solution $\gamma(x, t)$ – (12.66) of Eq. (12.58). $t = 0.2$. The values of parameters are as follows: $\mu_1 = 3.7$, $\mu_2 = 3.55$

Acknowledgements We acknowledge the partial support by the project BG05M2OP001-1.001-0008 "National Center for Mechatronics and Clean Technologies", funded by the Operating Program "Science and Education for Intelligent Growth" of Republic of Bulgaria and by the National Scientific Program "Information and Communication Technologies for a Single Digital Market in Science, Education and Security (ICTinSES)", contract No D01205/23.11.2018, financed by the Ministry of Education and Science of Republic of Bulgaria.

The work for assessing the final stage of the manuscript by Dr. Piotr Weber (Gdańsk University of Technology) is appreciated.

References

Ablowitz, M. J., Kaup, D. J., & Newell, A. C. (1973). Nonlinear evolution equations of physical significance. *Physical Review Letters, 31*, 125–127.

Ablowitz, M. J., Kaup, D. J., Newell, A. C., & Segur, H. (1974). Inverse scattering transform – Fourier analysis for nonlinear problems. *Studies in Applied Mathematics, 53*, 249–315.

Ames, W. F. (1965). *Nonlinear partial differential equations in engineering.* New York: Academic.

Batchelor, K. G. (2000). *An introduction to fluid dynamics.* Cambrige: Cambridge University Press.

Biswas, D. (2000). *Blood flow modes – A comparative study.* New Delhi: Mittal Publications.

Cocciolone, A. J., Hawes, J. Z., Staiculescu, M. C., Johnson, E. O., Murshed, M., & Wagenseil, J. E. (2018). Elastin, arterial mechanics, and cardiovascular disease. *American Journal of Physiology. Heart and Circulatory Physiology, 315*, H189–H205.

Davidson, P. (2015). *Turbulence. An introduction for scientists and engineers.* Oxford: Oxford University Press.

Debnath, L. (2012). *Nonlinear partial differential equations for scientists and engineers.* New York: Springer.

Demiray, H. (1992). Wave propagation through a viscosed fluid contained in a prestressed thin elastic tube. *International Journal of Engineering Science, 30*, 1607–1620.

Demiray, H. (1996). Solitary waves in prestressed elatic tubes. *Bulletin of Mathematical Biology, 58*(5), 939–955.

Demiray, H. (1997). Solitary waves in initially stressed thin elastic tubes. *International Journal of Non-Linear Mechanics, 334*(3), 571–588.

Demiray, H. (2008). Non-linear waves in a fluid filled inhomogeneous elastic tube with variable radius. *International Journal of Nonlinear Mechanics, 43*, 241–245.

Demiray, H., & Antar, N. (1997). Nonlinear waves in an inviscid fluid contained in prestressed viscoelastic thin tube. *Zeitschrift für angewandte Mathematik und Physik ZAMP, 48*(2), 325–340.

Dimitrova, Z. (2012a). On traveling waves of lattices: The case of Riccati lattices. *Journal of Theoretical and Applied Mechanics, 42*(3), 3–22.

Dimitrova, Z. (2012b). Relation between G'/G-expansion method and the modified method of simplest equation. *Comptes Rendus de L'Academie Bulgare des Sciences, 65*, 1513–1520.

Dimitrova, Z. I. (2015). Numerical investigation of nonlinear waves connected to blood flow in an elastic tube with variable radius. *Journal of Theoretical and Applied Mechanics, 45*(4), 79–92.

Fan, E., & Hon, Y. C. (2003). A series of travelling wave solutions for two variant Boussinesq equations in shallow water waves. *Chaos, Solitons & Fractals, 15*, 559–566.

Freis, E. D., & Heath, W. C. (1964). Hydrodynamics of aortic blood flow. *Curculation Research, 14*, 105–116.

Fung, J. C. (1997). *Biomechanics. Circulation.* New York: Springer.

Granger, R. (1995). *Fluid mechanics.* New York: Dover.

Grimshaw, R. (1993). *Nonlinear ordinary differential equations.* Boca Raton: CRC Press.

Gustafson, K. (1997). *Lectures on computational fluid dynamics, mathematical physics, and linear algebra.* Singapore: World Scientific.

He, J.-H., & Wu, X.-H. (2006). Exp-function method for nonlinear wave equations. *Chaos, Solitons & Fractals, 30*, 700–708.

Hirota, R. (1971). Exact solution of Korteweg-de Vries equation for multiple collisions of solitons. *Physical Review Letters, 27*, 1192–1194.

Hirota, R. (2004). *The direct method in soliton theory.* Cambridge: Cambridge University Press.

Holmes, P., Lumley, J. L., & Berkooz, G. (1996). *Turbulence, coherent structures, dynamical systems and symmetry.* Cambridge: Cambridge University Press.

Infeld, E., & Rowlands, G. (1990). *Nonlinear waves, solitons and chaos.* Cambridge: Cambridge University Press.

Kudryashov, N. A. (1990). Exact solutions of the generalized Kuramoto -Sivashinsky equation. *Physics Letters A, 147*, 287–291.

Kudryashov, N. A. (2005). Simplest equation method to look for exact solutions of nonlinear differential equations. *Chaos, Solitons & Fractals, 24,* 1217–1231.

Kudryashov, N. A., & Loguinova, N. B. (2008). Extended simplest equation method for nonlinear differential equations. *Applied Mathematics and Computation, 205,* 396–402.

Kundu, P. K., Cohen, I. M., & Dowling, D. R. (2012). *Fluid mechanics.* Amsterdam: Elsevier.

Landau, L. D., & Lifshitz, E. M. (1986). *Fluid mechanics.* Oxford: Pergamon Press.

Lesieur, M. (2008). *Turbulence in fluids.* Dordrecht: Springer.

Leung, A. W. (1989). *Systems of nonlinear partial differential equations. Applications to biology and engineering.* Dordrecht: Kluwer.

Malfliet, W., & Hereman, W. (1996). The tanh method: I. Exact solutions of nonlinear evolution and wave equations. *Physica Scripta, 54,* 563–568.

Mandelbrot, B. (1983). *Fractal geometry of nature.* New York: W. H. Freeman.

Martinov, N., & Vitanov, N. (1992). Running wave solutions of the two-dimensional sine-Gordon equation. *Journal of Physics A: Mathematical and General, 25,* 3609–3613.

Martinov, N., & Vitanov, N. (1992a). On some solutions of the two-dimensional sine-Gordon equation. *Journal of Physics A: Mathematical and General, 25,* L419–L426.

Martinov, N. K., & Vitanov, N. K. (1994). New class of running-wave solutions of the (2+1)-dimensional sine-Gordon equation. *Journal of Physics A: Mathematical and General, 27,* 4611–4618.

McDonald, D. A. (1974). *Blood flow in arteries.* Philadelphia: Williams & Wilkins.

Murray, J. D. (1977). *Lectures on nonlinear differential equation models in biology.* Oxford: Oxford University Press.

Nikolova, E. V. (2018). On nonlinear waves in a blood-filled artery with aneurism. *AIP Conference Proceedings, 1978,* 470050.

Nikolova, E. V., Jordanov, I. P., Dimitrova, Z. I., & Vitanov, N. K. (2017). Evolution of nonlinear waves in a blood-filled artery with aneurism. *AIP Conference Proceedings, 1895,* 070002.

Nikolova, E. V., Jordanov, I. P., Dimitrova, Z. I., & Vitanov, N. K. (2018). Nonlinear evolution equation for propagation of waves in an artery with aneurism: An exact solution obtained by the modified method of simplest equation. In *Advanced computing in industrial mathematics* (pp. 141–144). Cham: Springer.

Oertel, H. (2004). *Prandtl's essentials of fluid mechanics.* New York: Springer.

Paquerot, J.-F., & Remoissenet, M. (1994). Dynamics of nonlinear pressure waves in large artheries. *Physics Letters, 194,* 77–82.

Pedley, T. J. (1980). *The fluid mechanics of large blood vessels.* Cambridge: Cambridge University Press.

Remoissenet, M. (1993). *Waves called solitons.* Berlin: Springer.

Rodkiewicz, C. M. (Ed.). (1983). *Arteries ans arterial blood flow.* Wien: Springer.

Scott, A. C. (1999). *Nonlinear science. Emergence and dynamics of coherent structures.* Oxford: Oxford University Press.

Sherman, T. F. (1981). On connecting large vessels to small: The meaning of Murray's law. *The Journal of General Physiology, 78,* 431–453.

Stehbens, W. E. (1959). Turbulence of blood flow. *Quarterly Journal of Experimental Physiology, 44,* 110–117.

Stehbens, W. E. (1961). Discussion on vascular flow and turbulence. *Neurology, 11,* 66–67.

Tabor, M. (1989). *Chaos and integrability in dynamical systems.* New York: Wiley.

Taniuti, T., & Wei, C. C. (1968). Reductive perturbation method in nonlinear wave propagation. *Journal of the Physical Society of Japan, 21,* 209–212.

Tay, K. G. (2006). Forced Korteweg – de Vries equation in an elastic tube filled with inviscid fluid. *International Journal of Engineering Science, 44,* 621–632.

Tay, K. G., & Demiray, H. (2008). Forced Korteweg – deVries – Burgers equation in an elastic tube filled with a variable viscosity fluid. *Chaos, Solitons & Fractals, 38,* 1134–1145.

Tay, K. G., Ong, C. T., & Mohamad, M. N. (2007). Forced perturbed Korteweg – de Vries equation in an elastic tube filled with a viscous fluid. *International Journal of Engineering Science, 45,* 339–349.

Tennekes, H., & Lumley, J. L. (1972). *A first course in turbulence*. Cambridge, MA: The MIT Press.

Van Savage, M., Deeds, E. J., & Fontana, W. (2008). Sizing up allometric scaling theory. *PLoS Computational Biology, 4*(9), e1000171.

Verhulst, F. (1990). *Nonlinear differential equations and dynamical systems*. Berlin: Springer.

Vitanov, N. K. (1996). On travelling waves and double-periodic structures in two-dimensional sine - Gordon systems. *Journal of Physics A: Mathematical and General, 29*, 5195–5207.

Vitanov, N. K. (1998). Breather and soliton wave families for the sine–Gordon equation. *Proceedings of the Royal Society of London A, 454*, 2409–2423.

Vitanov, N. K. (2010). Application of simplest equations of Bernoulli and Riccati kind for obtaining exact traveling wave solutions for a class of PDEs with polynomial nonlinearity. *Communicatons in Nonlinear Science and Numerical Simulation, 15*, 2050–2060.

Vitanov, N. K. (2011). Modified method of simplest equation: Powerful tool for obtaining exact and approximate traveling-wave solutions of nonlinear PDEs. *Communications in Nonlinear Science and Numerical Simulation, 16*, 1176–1185.

Vitanov, N. K. (2011a). On modified method of simplest equation for obtaining exact and approximate solutions of nonlinear PDEs: The role of the simplest equation. *Communications in Nonlinear Science and Numerical Simulation, 16*(11), 4215–4231.

Vitanov, N. K. (2016). *Science dynamics and research production. Indicators, indexes, statistical laws and mathematical models*. Cham: Springer.

Vitanov, N. K. (2019). Modified method of simplest equation for obtaining exact solutions of nonlinear partial differential equations: History, recent developments of the methodology and studied classes of equations. *Journal of Theoretical and Applied Mechanics, 49*, 107–122.

Vitanov, N. K. (2019a). The simple equations method (SEsM) for obtaining exact solutions of nonlinear PDEs: Opportunities connected to the exponential functions. *AIP Conference Proceedings, 2159*, 030038.

Vitanov, N. K. (2019b). Recent developments of the methodology of the modified method of simplest equation with application. *Pliska Studia Mathematica, 30*, 29–42.

Vitanov, N. K. (2020). *Schroedinger equation and nonlinear waves* (pp. 37–92) in Simpao, V. A., & Little, H. C. *Understanding the Schroedinger equation*. New York: Nova Science Publishers.

Vitanov, N. K., & Dimitrova, Z. I. (2010). Application of the method of simplest equation for obtaining exact traveling-wave solutions for two classes of model PDEs from ecology and population dynamics. *Communications in Nonlinear Science and Numerical Simulation, 15*, 2836–2845.

Vitanov, N. K., & Dimitrova, Z. I. (2014). Solitary wave solutions for nonlinear partial differential equations that contain monomials of odd and even grades with respect to participating derivatives. *Applied Mathematics and Computation, 247*, 213–217.

Vitanov, N. K., & Dimitrova, Z. I. (2018). Modified method of simplest equation applied to the nonlinear Schröodinger equation. *Journal of Theoretical and Applied Mechanics, 48*, 59–68.

Vitanov, N. K., & Dimitrova, Z. I. (2019). Simple equations method (SEsM) and other direct methods for obtaining exact solutions of nonlinear PDEs. *AIP Conference Proceedings, 2159*, 030039.

Vitanov, N. K., & Martinov, N. K. (1996). On the solitary waves in the sine-Gordon model of the two-dimensional Josephson junction. *Zeitschrift fuer Physik B, 100*, 129–135.

Vitanov, N. K., & Vitanov, K. N. (2016). Box model of migration channels. *Mathematical Social Sciences, 80*, 108–114.

Vitanov, N. K., & Vitanov, K. N. (2018). Discrete-time model for a motion of substance in a channel of a network with application to channels of human migration. *Physica A, 509*, 635–650.

Vitanov, N. K., & Vitanov, K. N. (2018a). On the motion of substance in a channel of a network and human migration. *Physica A, 490*, 1277–1294.

Vitanov, N. K., & Vitanov, K. N. (2019a). Statistical distributions connected to motion of substance in a channel of a network. *Physica A, 527*, 121174.

Vitanov, N. K., Dimitrova, Z. I., & Ivanova, T. I. (2017). On solitary wave solutions of a class of nonlinear partial differential equations based on the function 1/cosh $(\alpha x + \beta t)$. *Applied Mathematics and Computation, 315*, 372–380.

Vitanov, N. K., Dimitrova, Z. I., & Kantz, H. (2010). Modified method of simplest equation and its application to nonlinear PDEs. *Applied Mathematics and Computation, 216*, 2587–2595.

Vitanov, N. K., Dimitrova, Z. I., & Kantz, H. (2013). Application of the method of simplest equation for obtaining exact traveling-wave solutions for the extended Korteweg-de Vries equation and generalized Camassa-Holm equation. *Applied Mathematics and Computation, 219*, 7480–7492.

Vitanov, N. K., Dimitrova, Z. I., & Vitanov, K. N. (2011). On the class of nonlinear PDEs that can be treated by the modified method of simplest equation. Application to generalized Degasperis – Processi equation and b-equation. *Communications in Nonlinear Science and Numerical Simulation, 16*, 3033–3044.

Vitanov, N. K., Dimitrova, Z. I., & Vitanov, K. N. (2013a). Traveling waves and statistical distributions connected to systems of interacting populations. *Computers & Mathematics with Applications, 66*, 1666–1684.

Vitanov, N. K., Dimitrova, Z. I., & Vitanov, K. N. (2015). Modified method of simplest equation for obtaining exact analytical solutions of nonlinear partial differential equations: Further development of the methodology with applications. *Applied Mathematics and Computation, 269*, 363–378.

Vitanov, N. K., Jordanov, I. P., & Dimitrova, Z. I. (2009). On nonlinear population waves. *Applied Mathematics and Computation, 215*, 2950–2964.

Vitanov, N. K., Jordanov, I. P., & Dimitrova, Z. I. (2009a). On nonlinear dynamics of interacting populations: Coupled kink waves in a system of two populations. *Communications in Nonlinear Science and Numerical Simulation, 14*, 2379–2388.

Wazwaz, A.-M. (2004). The tanh method for traveling wave solutions of nonlinear equations. *Applied Mathematics and Computation, 154*, 713–723.

Wazwaz, A.-M. (2009). *Partial differential equations and solitary waves theory*. Dordrecht: Springer.

Weber, P., & Pepłowski, P. (2016). Gaussian diffusion interrupted by Lévy walk. *Journal of Statistical Mechanics: Theory and Experiment, 2016*(10), 103202.

West, B. J., & Deering, W. (1994). Fractal physiology for physicists: Lévy statistics. *Physics Reports, 246*(1–2), 1–100.

Chapter 13
Fractal Properties of Flocs, Filtration Cakes and Biofilms in Water and Wastewater Treatment Processess

Beata Gorczyca

Abstract Physical properties of aggregates and their deposits are determined by the internal arrangement of particles inside the aggregate/deposit, i.e. the structure. The particular aggregate structure, which is determined by the mechanism of its formation, can be well described in terms of fractal geometry. Although the concepts of fractal geometry have been demonstrated to be suitable for characterization of aggregates and deposits formed in water and wastewater treatment, the determination of fractal dimensions has been proven to be challenging. Different measurements methods yield different fractal dimensions. A spectrum of fractal dimensions, rather that one dimension may be required to characterize these materials. The relationship between fractal dimensions and the aggregates/deposit properties is also not straight forward. There is definitely a need for more research in this area, which can have numerous applications in studies of general material structures.

This paper presents summary of major developments and knowledge gaps regarding fractal properties of particle aggregates, deposits (cakes) and biofilms formed in the water and wastewater treatment processes. Similarities and differences between water treatment and wastewater treatment flocs, deposits and biofilms are emphasized.

Keywords Particle aggregates · Filtration cakes · Biofilms · Physical properties · Fractal dimensions

B. Gorczyca (✉)
Department of Civil Engineering, University of Manitoba, Winnipeg, Canada
e-mail: Beata.Gorczyca@umanitoba.ca

© The Author(s), under exclusive license to Springer Nature
Switzerland AG 2021
A. Gadomski (ed.), *Water in Biomechanical and Related Systems*,
Biologically-Inspired Systems 17, https://doi.org/10.1007/978-3-030-67227-0_13

1 Introduction

The area of particle aggregate analyses have attracted researchers from a wide range of disciplines. Some focus on more theoretical concepts relating aggregates geometry to its formation mechanism (Meakin 1988). Others observe relationship between the aggregate mechanism of formation and resulting geometric properties in highly controlled laboratory experiments using plastic beads or pure kaolin clay (Elimelech et al. 2013).

There are also reports on physical properties of particle aggregates formed in their natural environments, for example natural water sediments, marine snow, or aggregates formed in some industrial processes. This paper falls in this last category.

The objective of this paper is to provide summary of most important discoveries in the area of characterization of particle aggregates, filtration cakes and biofilms formed in water treatment processes, with focus on use of fractal geometry for coherent conception of hydrodynamic behaviour of these biomaterials.

2 Flocs

The importance of physical properties of particle aggregates (flocs) formed in water/wastewater for the efficiency of treatment processes have been known for many decades (Ganczarczyk and Kosarewicz 1961). Significant advancement in this field had been made in the seventies (Tambo and Watanabe 1978, 1979). The concept of characterizing flocs and other biomaterials in water/wastewater using fractal geometry was coined by Li and Ganczarczyk in 1986.

The area has been continuously advanced until about 2015 (Vahedi and Gorczyca 2014; Maggi 2015). Although this research area continues to attract attention (Moruzzi et al. 2020) the author is not aware of any major and ground-breaking developments in this field in the past 5 years.

Floc Size Size of particle aggregates formed in water and wastewater treatment affects settling, filtration and many other unit processes. The floc sizes have typically been determined directly using microscopes or indirectly with particle counters. Particle counters calculate floc diameter indirectly from light scattering or other properties of aggregates (Elimelech et al. 2013). Floc equivalent circular diameter can be calculated from the projected area of the floc image using a microscope coupled with image analysis system (Tambo and Watanabe 1978, 1979; Li and Ganczarczyk 1991; Gorczyca and Ganczarczyk 1996;Elimelech et al. 2013). Equivalent circular diameters reported for alum coagulation floc range from about 5–300 μm. The size range of biological aggregates in wastewater treatment have been reported to be typically several times larger than that for chemical aggregates (Gorczyca and Ganczarczyk 2001), however, floc size range determinations are strongly depended on the method and equipment used (Wheatland et al. 2020).

Floc Projected Area and Perimeter Two dimensional (2D) floc projected area and perimeter are often used to obtain information about three dimensions (3D) floc surface area and volume. Floc projected areas have typically been analysed on resin embedded aggregates' thin sections with optical microscope (Ganczarczyk et al. 1992) or directly using confocal microscope (Thill et al. 1998). These floc properties determine many qualities of sludge's, such as sludge volume, dewatering and adsorptive properties (Gorczyca and Ganczarczyk 2002).

Floc Surface Area, Volume 3D floc surface area and volume can be determined as summation of 2-Dfloc sections perimeters and cross-sectional areas respectively. Vahedi and Gorczyca has successfully reconstructed 3D images of lime softening flocs from 2D projections, using an optical microscope equipped with a motorized stage. The microscope has inverted lenses that allowed direct imaging of the flocs in suspension while preserving the floc structure. It is equipped with a temperature and moisture controlled chamber. Software used for 3D reconstruction of NMRI images, was used for reconstruction of 3D images of lime softening flocs with diameters up to 235 μm (Vahedi and Gorczyca 2011).

Floc surface properties are central to aggregates' hydrodynamic behavior, yet the surface geometry has often been described by a single parameter, ie. floc projection boundary perimeter. The external surface characteristics of flocs, filtration cakes and biofilms have been characterized using projected area perimeter which could be related to the drag force acting on these biomasses (Jha et al. 2018; Maggi et al. 2006; Zahid and Ganczarczyk 1994a, b, c).

Floc Mass Floc mass determines floc settling rate as well as many important properties of sludges. It has typically been determined indirectly from measurement of the floc settling rate by calculation of floc density from Stokes' Law (Tambo and Watanabe 1978; Gorczyca and Ganczarczyk 1996; Vahedi and Gorczyca 2011; Trofa and D'Avino 2020).

Floc Density, Porosity and Permeability Floc density, porosity and permeability are determined by the internal arrangement of particles or structure inside the aggregate. The "chaotic" distribution of particles inside the floc is difficult to describe using concepts of conventional geometry; however, it can be well described in terms of fractal geometry (Mandelbrot 1983; Kaye 1989). Knowledge of floc structure is essential to develop understanding of floc behaviour during separation from water. The vast majority of papers describe porosity of a floc by one average value. However, the porosity of flocs cannot be described by one number because the flocs have multilevel structures, i.e. primary particles aggregate into flocculi, flocculi aggregate into microflocs and microflocs form the aggregate (Fig. 13.1). Each of these levels of aggregation (flocculi, microfloc and aggregate) has their own porosity value. Therefore, the floc porosity is non-homogeneous and can only be discussed in the context of floc structure (Beata Gorczyca and Ganczarczyk 1999; Logan 1999; Moruzzi et al. 2020).

Fig. 13.1 Multilevel
structure of flocs

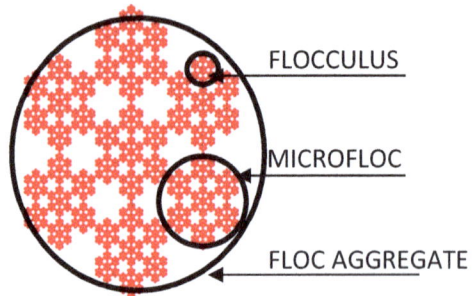

The flow of water inside a particle aggregate is determined by floc permeability, compressibility and the drag force. All of these properties are extremely difficult to determine experimentally. However, with the knowledge of the floc internal structure it is possible to calculate floc porosity and permeability. Porosity and permeability of flocs have been successfully described using concepts of fractal dimensions (Li and Ganczarczyk 1988; Li and Ganczarczyk 1989; Logan 1999; Gorczyca and Ganczarczyk 2002; Maggi et al. 2006; Maggi and Tang 2015).

Floc Properties and Floc Composition Floc properties are affected by the properties of primary particles forming the aggregates (Gorczyca and Ganczarczyk 1996). The properties of biological aggregates are affected by many more factors than the properties of primarily mineral flocs and several differences between these aggregates have been reported (Gorczyca and Ganczarczyk 2001, 2002; Maggi and Tang 2015; Tang and Maggi 2016).

One important reason for these differences is that microbiomes, that are vital components of biological aggregates, have ability to modify floc structure to control transport of electron donors and acceptors, nutrients and many other elements required for living organisms (Costerton et al. 1995; Cousin and Ganczarczyk 1998; Bruno et al. 2018; Li et al. 2020).

Fractal Dimensions 2 Dimensional fractal dimension (2D) of floc projected area or area perimeter have been determined directly using box counting method using image analysis system. In this method, the image of the floc perimeter or projected area is covered by N_r elements (pixels) of size r (Kaye 1989):

$$D = \lim_{r \to 0} \frac{\log\left(N_r\right)}{\log\left(1/r\right)} \tag{13.1}$$

Floc surface area and volume can be characterized by 3 dimensional fractal dimensions (3D). Floc surface area and volume are extremely difficult to determine experimentally, therefore, several methods have been proposed to calculate 3D fractal dimensions from measured 2D fractal dimensions (Maggi et al. 2006; Maggi and Tang 2015; Mandelbrot 1983; Tang and Maggi 2015).

3D floc mass fractal dimension have often been determined indirectly from other properties of flocs such as settling rates or light scattering properties (Gorczyca and Ganczarczyk 1996; Li and Ganczarczyk 1987; Vahedi and Gorczyca 2011; Wu et al. 2002; Guérin et al. 2019).

A. Vahedi compared floc fractal dimensions determined using different techniques (Vahedi and Gorczyca 2011). A significant discrepancy between fractal dimensions determined directly on floc images and those determined indirectly from floc settling velocity was reported. This difference is most likely due to a multitude of assumptions involved in the indirect determination of fractal dimensions, especially related to the estimate of the drag acting on the settling floc (Maggi 2013, 2015; Mola et al. 2020). Most of these assumptions stem from the use of Euclidean geometry and do not take into consideration the fractal nature of the flocs. Light scattering spectra can only be used for determination of fractal dimension of flocs that fall in a particular size range (Wu et al. 2002).

Multiple Fractal Dimension Floc fractal dimension is inherently linked to the mechanism of aggregate formation, and flocs in water treatment are formed as a result of several mechanisms (Gorczyca and Ganczarczyk 1996; Vahedi and Gorczyca 2012). Therefore, one may need multiple fractal dimensions to describe the geometry and hydrodynamic behaviour of these floc. A. Vahedi successfully predicted settling velocities of a population of lime softening flocs by incorporating a variable 3D floc fractal volume dimension into the modified Stokes' Law (Vahedi and Gorczyca 2012; Vahedi and Gorczyca 2014) . A comprehensive information on the use of the multifractal spectrum and its application is available elsewhere (Maggi 2005).

3 Filtration Cakes

Separation of flocs from water determines success of many water/wastewater treatment processes. Filtration involve application of pressure to aggregates suspensions, whereby water is extracted and the aggregates start to from a network, i.e. the cake (Fig. 13.2).

Filtration cake analyses typically consist of: quantification of foulant dry density in mg/cm^2, determination of foulant composition, microbiological testing, and microscopy to investigate the foulant structure. Most of these analyses are conducted on dried fouling layers, yet, natural water and wastewater filtration cakes, are often composed of organic materials with extremely low density that can occupy significant volume, such as soluble microbial products (Wiesner and Laine 1996). Structures of such materials change during drying, therefore, it can only be investigated through non-destructive, wet state analysis; such as Confocal Scanning Laser Microscope (CSLM) or Small Angle Neutron Scattering (SANS) methods (Mendret et al. 2006, 2007; Lee et al. 2010).

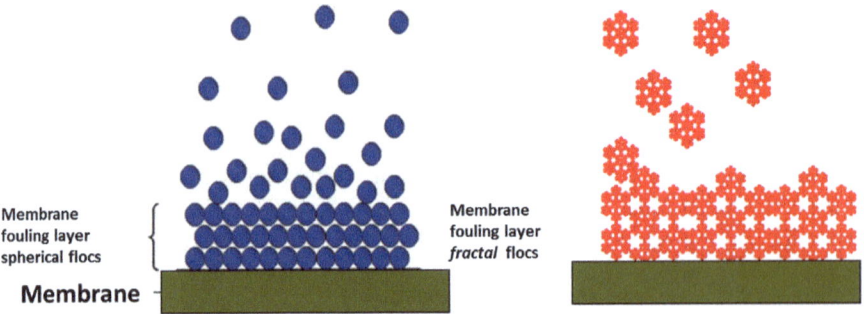

Fig. 13.2 Possible relationship between the structure of flocs and structure of filtration cakes

Much research has focused on the effect of the size of the floc on filtration resistance, although the floc size may change under pressure imposed during filtration (Mendret et al. 2006).

Several structural similarities between the flocs and filtration cakes have been reported (Antelmi et al. 2001; Madeline et al. 2006, 2007; Parneix et al. 2009). Behaviour of some sludge deposits (cakes) that have been formed under pressure have been successfully linked to properties of flocs using concepts of consolidation (compression) theory (Gorczyca and Ganczarczyk 2002).

Suspension chemistry and related to it floc structure can certainly affect permeability of the cake formed on the surface of membrane filters, yet few studies relate the structure of the flocs to filter fouling (Tarabara et al. 2002). Some suggested that porosity of membrane filter fouling cake is determined by the structure of flocs in the pre-treatment process (Mendret et al. 2010). Also, the porous cake formed on the membrane filter surface may serve as an additional filter capturing dissolved organic compounds. Madeline et al. studied changes in floc structures under pressure filtration using the Small Angle Neutron Scattering (SANS) technique. General compression behaviour of the flocs was proposed: as the pressure increases, the largest voids inside the aggregate collapse first, followed by compression of the smaller voids (Fig. 13.1) (Madeline et al. 2006, 2007).

Estimation of filter cake permeability is typically conducted using classical filtration theory, i.e. the Darcy's or Karman-Cozeny equation. The assumptions of this theory, especially homogeneous media porosity, limit its application for the estimation of permeability of the flocs, cakes and biofilms (Adler 1992; Gorczyca and Ganczarczyk 2002; Renslow et al. 2012). The permeability of flocs has been shown to be estimated quite well assuming a heterogeneous distribution of pores, similar to that of the Sierpinski fractal carpet (Zahid and Ganczarczyk 1994a, b; Gorczyca and Ganczarczyk 2002).

Some studies have considered the effects of floc fractal dimensions on membrane filtration resistance (Barbot et al. 2008). Due to many variables affecting the value of the determined fractal dimension discussed above, it is not possible to make many comparisons or conclusions from these studies.

4 Biofilms

Biofilms accumulate at all surfaces immersed in water, especially wastewater. They represent a very complex ecosystem that has been studied by many. Engineers primarily focus on the aspects of transport in biofilms (Li and Ganczarczyk 1988; Wanner and Reichert 1996; Lewandowski and Beyenal 2005; Wanner et al. 2006; Renslow et al. 2012), while microbiologist investigate the microbiomes (Pinto et al. 2012; Bruno et al. 2018). This paper focuses primarily on some major development in understanding transport phenomena in biofilms.

Biofilm structure affects its activity, in term of removal contaminants from water as well as biofilm sloughing. Several similarities between flocs and biofilms have been reported in the literature and it has been suggested that biofilm can be represented as a group of compressed particle aggregates (flocs) immobilized on a substratum. Biofilms are composed of microcolonies, which can be analogical to flocculus in the floc model (Fig. 13.1) (Ganczarczyk 1996; Wilking et al. 2011).

Characteristics of particular microbiome forming the biofilm as well as transport of electron donors and acceptors for biological activity are the paramount factors affecting biofilm structure. The mass transfer through a biofilm is accomplished both through diffusion and advection. Both strongly dependent on the biofilm's structure and porosity which has been successfully modelled also using fractal models (Wanner and Reichert 1996; Zahid and Ganczarczyk 1994a, b, c).

External surface area of the biofilm is important, as it creates particular hydrodynamic conditions at the water biofilm interface where the uptake of nutrients, as well as some contaminants occurs. Fractal dimensions have been demonstrated to provide the best description of development of very complex surfaces, including that of the biofilm (Zahid and Ganczarczyk 1994a, b).

While biofilms differ from flocs and filtration cakes similar techniques applied to study flocs and cakes can be applied. One relevant difference between the filtration cakes and biofilms is that flocs and biofilms are formed primarily with microorganism under atmospheric pressure.

5 Summary

Properties of particle aggregates, filtration cakes and biofilms formed in water/ wastewater determine success of most treatment processes. Description of properties of these materials and linking it to the performance of the treatment unit is very challenging. Concepts of fractal geometry have been demonstrated to be suitable for characterization of such materials, however, determination of particular fractal dimensions has been proven to be challenging, depending on method of determination as well as the type of material. A spectrum of fractal dimensions, rather that one number may be required to characterize these complex biomassess. There is definitely a need for more research in this area, which can have numerous applications in studies of general material structures.

References

Adler, P. (1992). *Porous media: Geometry and transport*. Stoneham: Butterworth Henemann.

Antelmi, D., Cabane, B., Meireles, M., & Aimar, P. (2001). Cake collapse in pressure filtration. *Langmuir, 17*(22), 7137–7144. https://doi.org/10.1021/la0104471.

Barbot, E., Moustier, S., Bottero, J. Y., & Moulin, P. (2008). Coagulation and ultrafiltration: Understanding of the key parameters of the hybrid process. *Journal of Membrane Science, 325*(2), 520–527. https://doi.org/10.1016/j.memsci.2008.07.054.

Bruno, A., Sandionigi, A., Bernasconi, M., Panio, A., Labra, M., & Casiraghi, M. (2018). Changes in the drinking water microbiome: Effects of water treatments along the flow of two drinking water treatment plants in a urbanized area, Milan (Italy). *Frontiers in Microbiology, 9*(October), 1–12. https://doi.org/10.3389/fmicb.2018.02557.

Costerton, J. W., Lewandowski, Z., Caldwell, D. E., Korber, D. R., Sn, S., & Lappin-scott, H. M. (1995). Microbial biofilms. *Annual Review of Microbiology, 49*, 711–745.

Cousin, C. P., & Ganczarczyk, J. J. (1998). Effects of salinity on physical characteristics of activated sludge flocs. *Water Quality Research Journal of Canada, 33*(4), 565–587. https://doi.org/10.2166/wqrj.1998.032.

Elimelech, M., Gregory, J., & Jia, X. (2013). *Particle deposition and aggregation*. Stoneham: Butterworth Henemann.

Ganczarczyk, J. J. (1996, August 28). Microbial films and microbial flocs some similarities and differences. *3rd International IAWQ Special Conference on Biofilm Systems*, Copenhagen.

Ganczarczyk, J., & Kosarewicz, O. (1961). Dispersion of activated sludge flocs. *Gaz Woda i Technika Sanitaria (Poland), 35*, 19–21.

Ganczarczyk, J. J., Zahid, W. M., & Li, D.-H. (1992). Physical stabilization and embedding of microbial aggregates for light microscopy studies. *Water Research, 26*(12), 1695–1699. https://doi.org/10.1016/0043-1354(92)90170-9.

Gorczyca, B., & Ganczarczyk, J. J. (1996). Image analysis of alum coagulated mineral suspensions. *Environmental Technology, 17*, 1361–1369. https://doi.org/10.1080/09593331708616505.

Gorczyca, B., & Ganczarczyk, J. J. (1999). Structure and porosity of alum coagulation flocs. *Water Quality Research Journal of Canada, 34*(4), 653–666.

Gorczyca, B., & Ganczarczyk, J. J. (2001). Fractal analysis of pore distributions in alum coagulation and activated sludge flocs. *Water Quality Research Journal of Canada, 36*(4), 687–700. Retrieved from http://scholar.google.com/scholar?hl=en&btnG=Search&q=intitle:Fractal+Analysis+of+Pore+Distributions+in+Alum+Coagulation+and+Activated+Sludge+Flocs#0.

Gorczyca, B., & Ganczarczyk, J. J. (2002). Flow through alum coagulation and activated sludge flocs. *Water Quality Research Journal of Canada, 37*(2), 389–398.

Guérin, L., Frances, C., Liné, A., & Coufort-Saudejaud, C. (2019). Fractal dimensions and morphological characteristics of aggregates formed in different physico-chemical and mechanical flocculation environments. *Colloids and Surfaces A: Physicochemical and Engineering Aspects, 560*(September 2018), 213–222. https://doi.org/10.1016/j.colsurfa.2018.10.017.

Jha, N., Kiss, Z. L., & Gorczyca, B. (2018). Fouling mechanisms in nanofiltration membranes for the treatment of high DOC and varying hardness water. *Desalination and Water Treatment, 127*, 197–212. https://doi.org/10.5004/dwt.2018.22830.

Kaye, B. (1989). *A random walk through fractal dimensions*. New York: VCH Publishers.

Lee, W. N., Yeon, K. M., Hwang, B. K., Lee, C. H.., & Chang, I. S. (2010). Effect of PAC Addition on the Physicochemical Characteristics of Bio-Cake in a Membrane Bioreactor Effect of PAC Addition on the Physicochemical Characteristics of Bio-Cake in a Membrane Bioreactor. *Separation Science and Technology, 45*, 896–903.

Lewandowski, Z., & Beyenal, H. (2005). Biofilms: Their structure, activity, and effect on membrane filtration. *Water Science & Technology, 51*, 181–192.

Li, D.-H., & Ganczarczyk, J. J. (1987). Stroboscopic determination of settling velocity, size and porosity of activated sludge flocs. *Water Research, 21*(3), 257–262. https://doi.org/10.1016/0043-1354(87)90203-X.

Li, D.-H., & Ganczarczyk, J. (1988). Flow through activated sludge flocs. *Water Research, 22*(6), 789–792. https://doi.org/10.1016/0043-1354(88)90192-3.

Li, D.-H., & Ganczarczyk, J. (1989). Fractal geometry of particle aggregates generated in water and wastewater treatment processes. *Environmental Science and Technology, 23*(11), 1385–1389. https://doi.org/10.1021/es00069a009.

Li, D., & Ganczarczyk, J. J. (1991). Distribution of activated sludge flocs. *Journal of Water Pollution Control Federation, 63*(5), 806–814.

Li, Z. H., Guo, Y., Hang, Z. Y., Zhang, T. Y., & Yu, H. Q. (2020). Simultaneous evaluation of bioactivity and settleability of activated sludge using fractal dimension as an intermediate variable. *Water Research, 178*, 115834. https://doi.org/10.1016/j.watres.2020.115834.

Logan, B. E. (1999). *Environmental transport processes*. New York: Wiley.

Madeline, J. B., Meireless, M., Botet, R., & Cabane, B. (2006). The role of interparticle forces in colloidal aggregates: Local investigations and modelling of restructuring during filtration. *Water Science and Technology, 53*(7), 25–32. https://doi.org/10.2166/wst.2006.204.

Madeline, J. B., Meireles, M., Bourgerette, C., Botet, R., Schweins, R., & Cabane, B. (2007). Restructuring of colloidal cakes during dewatering. *Langmuir, 23*(4), 1645–1658. https://doi.org/10.1021/la062520z.

Maggi, F. (2005). *Flocculation dynamics of cohesive sediment* (Vol. 5). Retrieved from https://repository.tudelft.nl/islandora/object/uuid:0dd37043-d40c-44c3-a87b-741caa10b85e?collection=research

Maggi, F. (2013). The settling velocity of mineral, biomineral, and biological particles and aggregates in water. *Journal of Geophysical Research: Oceans, 118*(4), 2118–2132. https://doi.org/10.1002/jgrc.20086.

Maggi, F. (2015). Experimental evidence of how the fractal structure controls the hydrodynamic resistance on granular aggregates moving through water. *Journal of Hydrology, 528*(August), 694–702. https://doi.org/10.1016/j.jhydrol.2015.07.002.

Maggi, F., & Tang, F. H. M. (2015). Analysis of the effect of organic matter content on the architecture and sinking of sediment aggregates. *Marine Geology, 363*, 102–111. https://doi.org/10.1016/j.margeo.2015.01.017.

Maggi, F., Manning, A. J., & Winterwerp, J. C. (2006). Image separation and geometric characterisation of mud flocs. *Journal of Hydrology, 326*(1–4), 325–348. https://doi.org/10.1016/j.jhydrol.2005.11.005.

Mandelbrot, B. (1983). *The fractal geometry of nature*. New York: W.H.Freeman and Company.

Meakin, P. (1988). Models for colloidal aggregation. *Annual Review of Physical Chemistry, 39*(1), 237–267. https://doi.org/10.1146/annurev.physchem.39.1.237.

Mendret, J., Guigui, C., Cabassud, C., Doubrovine, N., Schmitz, P., Duru, P., et al. (2006). Development and comparison of optical and acoustic methods for in situ characterisation of particle fouling. *Desalination, 199*(1–3), 373–375. https://doi.org/10.1016/j.desal.2006.03.212.

Mendret, J., Guigui, C., Schmitz, P., Cabassud, C., & Duru, P. (2007). An optical method for in situ characterization of fouling during filtration. *AICHE Journal, 53*(9), 2265–2274. https://doi.org/10.1002/aic.

Mendret, J., Guigui, C., Cabassud, C., & Schmitz, P. (2010). Numerical investigations of the effect of non-uniform membrane permeability on deposit formation and filtration process. *Desalination, 263*(1–3), 122–132. https://doi.org/10.1016/j.desal.2010.06.048.

Mola, I. A., Fawell, P. D., & Small, M. (2020). Particle-resolved direct numerical simulation of drag force on permeable, non-spherical aggregates. *Chemical Engineering Science, 218,* 115582. https://doi.org/10.1016/j.ces.2020.115582.

Moruzzi, R. B., Bridgeman, J., & Silva, P. A. G. (2020). A combined experimental and numerical approach to the assessment of floc settling velocity using fractal geometry. *Water Science and Technology: a Journal of the International Association on Water Pollution Research, 81*(5), 915–924. https://doi.org/10.2166/wst.2020.171.

Parneix, C., Persello, J., Schweins, R., & Cabane, B. (2009). How do colloidal aggregates yield to compressive stress? *Langmuir, 25*(8), 4692–4707. https://doi.org/10.1021/la803627z.

Pinto, A. J., Xi, C., & Raskin, L. (2012). Bacterial community structure in the drinking water microbiome is governed by filtration processes. *Environmental Science and Technology, 46*(16), 8851–8859. https://doi.org/10.1021/es302042t.

Renslow, R., Lewandowski, Z., & Beyenal, H. (2012). Biofilm image reconstruction for assessing. *Biotechnology and Bioengineering, 108*(6), 1383–1394. https://doi.org/10.1002/bit.23060. BIOFILM.

Tambo, N., & Watanabe, Y. (1978). Physical characteristics of flocs-I. The floc density function and aluminium floc. *Water Research, 13,* 409–419.

Tambo, N., & Watanabe, Y. (1979). Physical aspect of flocculation process-I: Fundamental treatise. *Water Research, 13*(5), 429–439. https://doi.org/10.1016/0043-1354(79)90035-6.

Tang, F. H. M., & Maggi, F. (2015). Reconstructing the fractal dimension of granular aggregates from light intensity spectra. *Soft Matter, 11*(47), 9150–9159. https://doi.org/10.1039/c5sm01885d.

Tang, F. H. M., & Maggi, F. (2016). A microcosm experiment of suspended particulate matter dynamics in nutrient- and biomass-affected waters highlights: The increase of nutrient availability and biomass growth enhanced SPM flocculation. Biomass-affected aggregates had size 60% greater th. *Water Research, 89,* 76–86.

Tarabara, V. V., Pierrisnard, F., Parron, C., Bottero, J. Y., & Wiesner, M. R. (2002). Morphology of deposits formed from chemically heterogeneous suspensions: Application to membrane filtration. *Journal of Colloid and Interface Science, 256*(2), 367–377. https://doi.org/10.1006/jcis.2002.8617.

Thill, A., Veerapaneni, S., Simon, B., Wiesner, M., Bottero, J. Y., & Snidaro, D. (1998). Determination of structure of aggregates by confocal scanning laser microscopy. *Journal of Colloid and Interface Science, 204*(2), 357–362. https://doi.org/10.1006/jcis.1998.5570.

Trofa, M., & D'Avino, G. (2020). Sedimentation of fractal aggregates in shear-thinning fluids. *Applied Sciences (Switzerland), 10*(9), 1–20. https://doi.org/10.3390/app10093267.

Vahedi, A., & Gorczyca, B. (2011). Application of fractal dimensions to study the structure of flocs formed in lime softening process. *Water Research, 45*(2), 545–556.

Vahedi, A., & Gorczyca, B. (2012). Predicting the settling velocity of flocs formed in water treatment using multiple fractal dimensions. *Water Research, 46*(13), 4188–4194.

Vahedi, A., & Gorczyca, B. (2014). Settling velocities of multifractal flocs formed inchemical coagulation process. *Water Research, 53,* 322–328.

Wanner, O., & Reichert, P. (1996). Mathematical modeling of mixed-Cu. *Biotechnology and Bioengineering, 49,* 172–184.

Wanner, O., Eberl, H., Morgenroth, E., Noguera, D., Picioreanu, C., Rittmann, B., & van Loosdrecht, M. (2006). *Mathematical modeling of biofilms* (IWA scientific and technical report) (Vol. 18). London: IWA Publications. https://doi.org/10.2166/9781780402482.

Wheatland, J. A. T., Spencer, K. L., Droppo, I. G., Carr, S. J., & Bushby, A. J. (2020). Development of novel 2D and 3D correlative microscopy to characterise the composition and multiscale structure of suspended sediment aggregates. *Continental Shelf Research, 200*(April), 104112. https://doi.org/10.1016/j.csr.2020.104112.

Wiesner, M. R., & Laine, J. M. (1996). *Water treatment membrane processes.* New York: McGraw-Hill.

Wilking, J., Angelini, T., Seminara, A., Brenner, M., & Weitz, D. (2011). Biofilms as complex fluids. *MRS Bulletin, 36*(5), 385–391. https://doi.org/10.1557/mrs.2011.71.

Wu, R. M., Lee, D. J., Waite, T. D., & Guan, J. (2002). Multilevel structure of sludge flocs. *Journal of Colloid and Interface Science, 252*(2), 383–392. https://doi.org/10.1006/jcis.2002.8494.

Zahid, W., & Ganczarczyk, J. (1994a). Fractal properties of RBC biofilm structure. *Water Science and Technology, 29,* 271–279.

Zahid, W., & Ganczarczyk, J. (1994b). A technique for a characterization of RBC biofilm surface. *Water Research, 28*(10), 2229–2231. https://doi.org/10.1016/0043-1354(94)90036-1.

Zahid, W. M., & Ganczarczyk, J. J. (1994c). Structure of RBC biofilms. *Water Environment Research, 66*(2), 100–106. https://doi.org/10.2175/wer.66.2.2.

Chapter 14
Soil Hydrology

Zoltan Futo and Karoly Bodnar

Abstract The general introduction of hydrological cycle is followed by the roles of water in plant tissues and physiology. Water management has a predominant role in soil structure and fertility. Water acts as a solvent, as a reagent and as a transport medium. It participates in the physical, chemical and biological processes of the soil. Soils can be characterized as having different water balance depending on structure, location and environmental factors which gives information regarding the water supply of the area and the quantity of water provided for plants. Certain elements of cultivation practice are all important factors of improving the effectiveness of water consumption. The water uptake mechanism through the roots is discussed. Modern irrigation systems are used to prevent water shortage (water stress) to avoid yield loss, and it is advisable to start irrigation before the onset of the visible symptoms of water deficiency, before the moisture content of the soil falls below 50% of its water capacity. Calculation methods are given to determine the irrigation water requirement. A case study on effects of an up-to-date irrigation system on maize yields is presented. A tape drip irrigation method was tested on the level of yields and yielding elements of maize. Irrigation satisfying the 100% water requirement of the crop was supplemented with complex water-soluble fertilizer (N-P-K). The results show that the yields of sweet corn could be significantly increased in the very favourable water supply.

Keywords Hydrological cycle · Water management · Soil hydrology · Irrigation · Plant cultivation · Water influence on plant physiology · Groundwater · Water in cells · Hydrogen bonds

Z. Futo (✉) · K. Bodnar (✉)
Institut of Irrigation and Water Management, Szent Istvan University, Szarvas, Hungary
e-mail: Futo.Zoltan@szie.hu

© The Author(s), under exclusive license to Springer Nature
Switzerland AG 2021
A. Gadomski (ed.), *Water in Biomechanical and Related Systems*,
Biologically-Inspired Systems 17, https://doi.org/10.1007/978-3-030-67227-0_14

1 Introduction

Water is a colorless, odorless, tasteless liquid. The pleasant taste of drinking water is due to the dissolved substances. Water is the only substance on Earth which can be found in all three physical states. Water density is the highest at +4 °C. In the winter only the top layer of rivers and lakes freezes over, so wildlife remains unharmed under the ice. In ice water molecules form crystals, i.e. a molecular lattice. Water is a good solvent. One liter of water forms approx. 1750 l of steam. As a result of exposure to high temperatures (e.g. molten metal), decomposition occurs, i.e. water is decomposed into hydrogen and oxygen – the mixture of which is the highly explosive oxyhydrogen. The density of water is maximal (1000 kg/m^3) on 4 °C, 998.2 kg/m^3 on 20 °C.

The oxygen and hydrogen found in water molecules have different electronegativies, thus oxygen attracts bonding electron pairs more towards itself. The resulting charge shift causes the polarity of the bond and the molecule as well. As a result, the water molecule has a negative and a positive pole, which greatly affects the behavior of water molecules in the soil as well as in plants. Water molecules are polar molecules, and as such, it is a good solvent of many ionic compounds, like table salt (NaCl, KCl, etc.). In the water molecule, oxygen atoms are negatively, while hydrogen atoms have poisitvely charged, which is caused by the charge shift described above. Due to their polar nature, hydrogen bonds are formed between water molecules, which give exceptional physical properties to water (Fig. 14.1). Due to the hydrogen bonds, water has a quasi-crystalline structure, which is continuously created, decomposed and reassembled. The colder the water, the stronger the hydrogen bonds are between water molecules and the harder it is for a water molecule to be removed from this aggregate.

Hydrology deals with the manifestations of water on Earth, the laws of the water cycle, and certain elements of the cycle. All of that water forms constitute the "water shell" of our planet, the so-called hydrosphere. In the hydrosphere, water occurs in all three states in a varied distribution in space and time. The hydrosphere overlaps with other terrestrial spheres: the Earth's solid rocks layer the lithosphere (up to a depth of about 5 km), the atmosphere, and the biosphere, respectively (Gombos 2011).

The water cycle, also known as the hydrological cycle, is the continuous and natural cycle of water in the Earth's hydrosphere (Vermes 1997), maintained by energy from solar radiation. Surface and groundwater, as well as atmospheric and groundwater content participate in the cycle (Fig. 14.2). Physical-meteorological factors of the terrestrial water cycle: solar radiation, temperature, air pressure, humidity and wind.

The initial phase of the cycle is evaporation. The water of rivers, lakes and oceans is constantly evaporating, but so are living organisms. Then the light mist rises, precipitates high, and forms clouds. Precipitation falls from the clouds, which can be liquid (rain) or solid (snow, ice). The fallen raindrops are absorbed by the soil on the one hand, and utilized by living beings in the course of their life. Water that infiltrated into the soil flows under the surface and collects. The water flowing

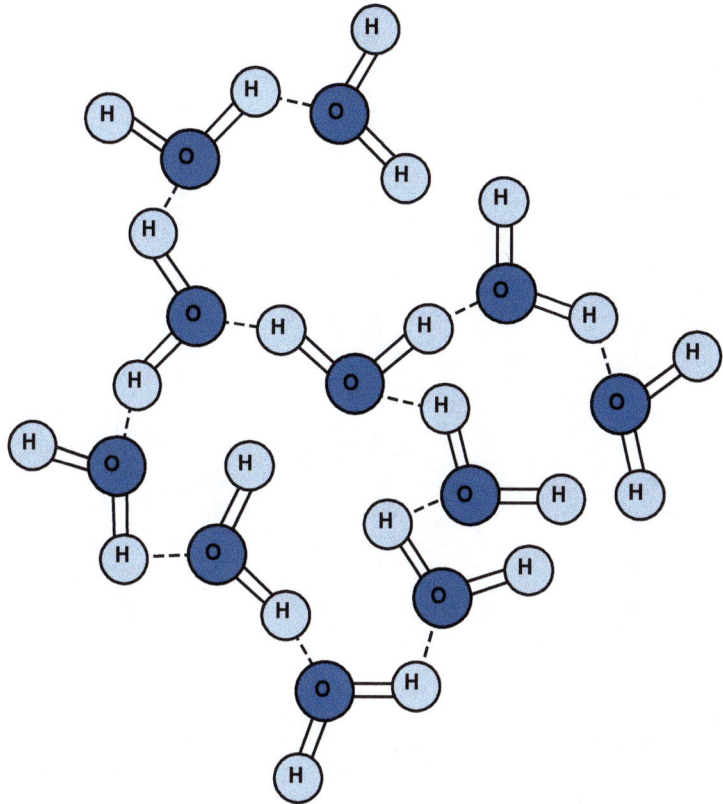

Fig. 14.1 Hydrogen bonds forming between water molecules

beneath the surface reappears on the surface in springs. Eventually, it swells into a small stream and then a river, and it flows into the sea, while its water is constantly evaporating. The volume of water is constant. Approximately 1386 million cubic kilometers of water are in constant motion. The steps of the permanent circulation of water are therefore: evaporation of surface waters and plants, cloud and precipitation, the leakage of precipitation into the soil, the uptake of water by living organisms, the emergence of groundwater, the evaporation of surface and transpiration of living organisms (evapotranspiration).

In the recent chapter the forms and movements of water in the soils and its roles in crop production are discussed with a special focus on the agricultural importance of these phenomena.

The water below the surface of the ground is the groundwater. Within this, we can distinguish several forms based on their location. The water in the top layer of the soil is the soil moisture. Above the first watertight layer, the water that fills the pores of the soil in a coherent manner is groundwater. Below (in water-bearing layers), the water between the watertight layers is called stratified water. Karst water is a form of water found in the fissures and passages of solid-textured rocks. Most of

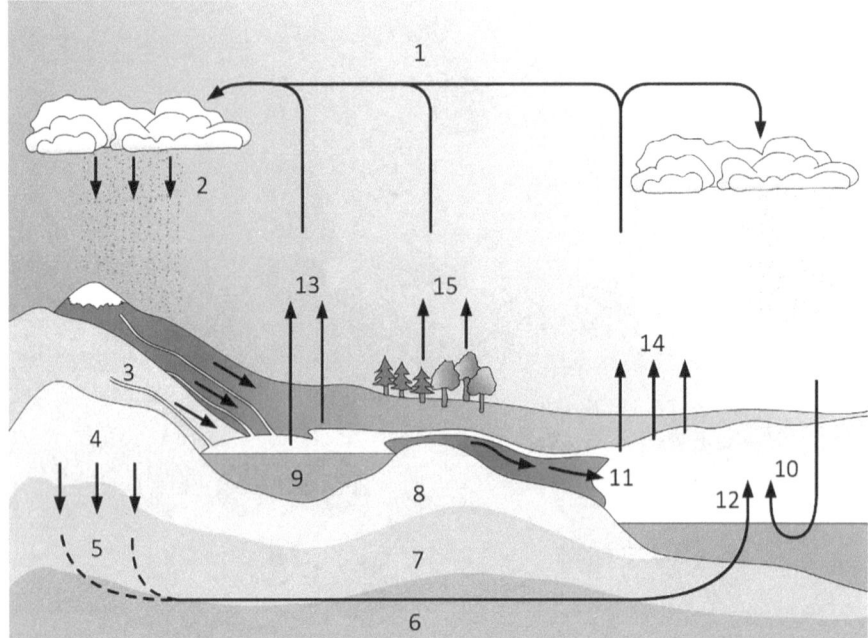

Fig. 14.2 The hydrologic cycle. 1 – Water storage in atmosphere and condensation, 2 – Precipitation, 3 – Surface runoff, 4 – Infiltration, 5 – Percolation, 6 – Groundwater, 7 – Soil moisture, 8 – Land, 9 – Lakes and rivers, 10 – Ocean, 11 – Surface outflow, 12 – Subsurface outflow, 13 – Evaporation from soil and water surface, 14 – Evaporation from oceans, 15 – Transpiration (Based on tudasbazis 2020)

the groundwater is generated by the leakage of rainwater into the soil. These are called infiltrating, otherwise known as infiltration waters. Between deeper layers the movement of water is called percolation.

The precipitating (condensing) water is formed by the cooling or precipitation of water vapor in the pores of the soil, mainly near the soil surface. From a water management point of view, the amount is not significant.

2   Role of Water Cycle in Plant Cultivation Area

2.1   The Role of Water in Plants

The evaporation of live vegetation, seas and land areas on Earth, and the continuous changes in the physical state of water vapor in the atmosphere maintain a constant water cycle, as evaporation is always followed by condensation. Plant vegetation contributes to this circulation with its own water turnover and management.

The water turnover of plants consists of the plant taking up the water, utilizing it during its physiological processes for the synthesis and construction of its own organism, and then evaporating it into the environment. Water management of plants is therefore a very complex, coordinated and highly regulated process consisting of water uptake and water loss. The complexity of the water management of plants is increased also by the fact that plants belonging to different ecological groups have considerably different mechanisms for it.

Transpiration is the evaporation of water through the stomata of a plant. Not only transpiration enables water uptake and continuous solute flow in the plant's transport system, but it also provides for the cooling of the plant. Due to its large heat capacity, water reduces the rate of temperature change in plants. Growing plant tissues are composed of 80–95% water. Plant seeds are the driest; they contain 5–15% water. The water content of chloroplasts and mitochondria is around 50%, while vacuoles might have 98%.

2.2 Water Cycle and its Importance on Plant Physiology

Water uptake and water loss in plants is continuous; under ideal circumstances, plants manage the available water. The water management of plants consists primarily of the available and disposable water resources, and the water built into the tissues and utilized by the plant during its life processes, then evaporated through the leaves.

Depending on their life stage, plant tissues contain large amounts of water in, but in general it can be stated that only a very small portion, 1–2% of the absorbed water is built into the plant's body components. A significant part of the water taken up exits the plant's body by its evaporation activity. However, all of this has a very important role throughout the plant's nutrient uptake, thermal management and organic buildup processes. Water absorbed by the plant and evaporated through the leaves is ideally in a state of equilibrium.

By comparing the two sides, a value characteristic of the plant's water management is obtained, the so-called water balance. Water balance can only be maintained securely if water absorption and water release are in accordance with each other. Water balance becomes negative when water supply does not cover the amount of evaporated water.

In agriculture, this often occurs in two situations. The first case is soil drought, when there is no adequate amount of available water in the soil, while the other is atmospheric drought, when the vapor and water extraction effect of hot, dry air is greater than what can be recovered by the plants through the roots. In either case, wilting and negative water balance can be experienced We define short- and long-term changes in water balance. During the day, the plant's water balance is negative (due to increased transpiration), which is restored during the night. This lasts for a rather short time. During dry periods, however, the water balance of the plant becomes increasingly negative due to the soil's diminishing water resources, which can only be restored in the next rainy period, or by irrigation.

The most important roles and functions of water in plant physiology:

- Water is the most important cellular component of terrestrial and aquatic plant organisms, since all cell components are in a dissolved state, therefore a significant portion of the living plant consists of water.
- Water is the solvent of mineral nutrients, as well as the transport medium of nutrient ions and dissolved assimilates (eg. sugars) in plants.
- It has a fundamental role in all biochemical conversion processes, since it is the reaction medium, as well as a component of organic material (sugars, carbohydrates, proteins, etc.) synthesis.
- Water is essential for sustaining the viability of leaves playing a central role in photosynthesis and carrying out assimilation activities, and for maintaining the specific water saturation degree of tissues.
- For the functioning of plant cells, an appropriate internal pressure (turgidity) is required, and the fundamental and essential basic component of this process is water.
- The temperature of the plant's body is strongly dependent on external temperature. Plants only have limited means for temperature control; one such example is the regulation of high temperatures by means of evaporation and transpiration, of which a main component is again water.
- Significant amounts of water are needed to achieve high yields through unimpeded photosynthetic processes. In the world of plants, it is very common that up to 450–850 l of water are required for the production of 1 kg of dry material.
- The clearest example of water release and uptake can be observed in the life of seeds: the maturation of seeds includes water loss (seed dormancy), while intense uptake is characteristic of germination (swelling and germination).

Water plays a special role in the life and organizational structure of plants. According to the physical laws of osmosis, the concentration difference between the plant cell and the environment determines the direction of water flow.

If the external medium is more diluted, water flows into the cell, among the plant's organelles, the vacuole and the cytoplasm are filled up with water, their volume increases, and the cell becomes swollen. The cytoplasm is pressed against the cell wall, which resist the cell's volume increase depending on its constant of elasticity. Therefore, pressure builds up between the cytoplasm and the cell wall, which is called turgidity. On the organizational level, the sum of each cell's turgidity composes the hydrostatic skeleton of the plant, which is essential for supporting its life. This plant hydrostatic skeleton collapses in a spectacular way if a significant water loss (wilting) event occurs.

If the external medium is more concentrated than the cell, or the plant loses significant amounts of water due to other reasons, water moves out of the plant, the vacuole and the cytoplasm shrink, and plasmolysis occurs. In this case, the turgidity value is zero, the first stage is reversible, but in the case of steady water shortage, the plant enters an irreversible wilting state. This state can lead to the death of the cell, but very often to the death of the whole plant.

The water turnover of arable crops is very complex. For the above-ground plant shoots, the direct source of water is air humidity or precipitation, while the indirect source is the water transported from the roots in the direction of the shoots. For the roots and subsurface shoots, groundwater is the direct water source, the availability of which, however, depends largely on the quality of the soil, its actual water content, and the water's motion in the soil as well. Therefore, water transport from the root to the shoot, and the development of tissues in the shoots that regulate water-loss are of great importance for terrestrial plants.

In addition to the aforementioned functions, water plays a fundamental role in the transport of minerals in plants as well. For the uptake and transport of minerals, that is, the mineral nutrition of plants is associated with water uptake. In the case of arable crops, minerals are taken up mostly, but not exclusively by the roots. Aqueous solutions absorbed by the roots have to be transported to all living cells of the plant. This transport process may occur in two ways:

- it can take place in the cell wall (apoplastic transport);
- the solution enters the root parenchyma cells (symplastic transport), and from there, the transport of the aqueous solution takes place in the differentiated transport vascular bundles (vessel elements).

This transport process assumes the presence of water as a transport medium in all cases, therefore this role of water is indispensable in plant nutrition.

2.3 Water Cycle of Soils, Water Management

Water management has a predominant role in soil fertility. Water acts as a solvent, as a reagent and as a transport medium. It participates in the (physical, chemical and biological) weathering, formation and degradation processes of the soil.

Water management of soils can be characterized by:

- the amount of water in the soil,
- its movement, and
- its spatial and temporal changes.

In terms of plant production, the fertility of soils is fundamentally affected by:

- the soil moisture content available for plants,
- water movement in the soil (influx and runoff, etc.) and
- the chemical composition of water (i.e. water valuable for crop production, or high salinity groundwater).

Soils can be classified into different groups in terms of water transport (Boorman et al. 1995). Soils can be categorized into four main types in terms of water transport, which are as follows (Fig. 14.3):

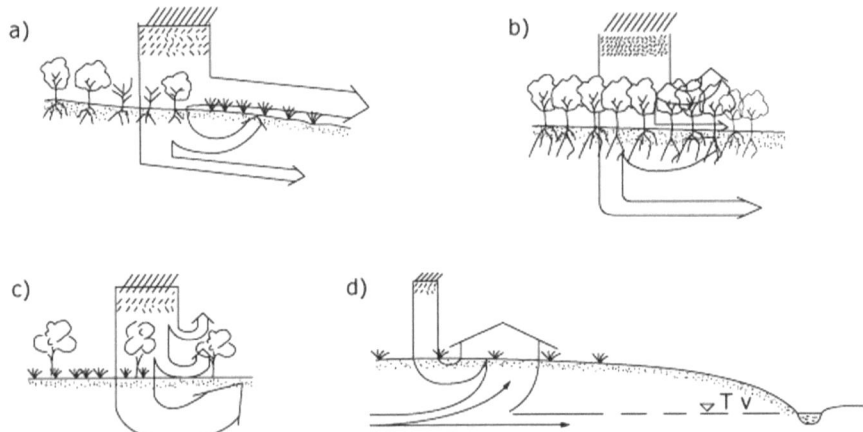

Fig. 14.3 Major types of water transport in soils (*Tv*: level of groundwater)

(a) Strong surface runoff (sloping soil surface, surface runoff, erosion damage) e.g. rocky skeletal soil.
(b) Leaching water flow, strong downward water movement (the greater proportion of the large amount of precipitation trickles into the soil, excessive leaching) e.g. luvisol.
(c) Equilibrium water balance (annual water flow balance of upward and downward water movements, alternations within the year; periodic motion of materials are typical, e.g. calcic chernozem soil.
(d) Evaporative water flow (predominantly upward water movement, low-lying areas, the impact of groundwater, low groundwater salinity → no salinization; effect of stagnant saline groundwaters → salinization) e.g. typical meadow soil, solonchak soils

Water capacity is the amount of water that can be absorbed by a given soil under certain circumstances, or retained against gravitational force.

Soil water capacity provides information about the soil's water content, the amount of water retained against gravitational force and the water supplying capacity of the soil. These soil properties help us make decisions about crop water and nutrient management.

Main types of water capacity (Fig. 14.4):

1. Field water capacity (FWC): The quantity of water that can be withheld by the soil against gravity under natural circumstances.
2. Maximum water capacity (WC_{max}): The quantity of water at which the soil pores are 100% saturated with water. This includes the large gravitational pores as well.
3. Minimum water capacity (WC_{min}): The quantity of water withheld against gravity in case the effect of soil water is negligible. FWC ≈ MWC (under natural circumstances, the values of field capacity and minimum water capacity are not significantly different from each other.)

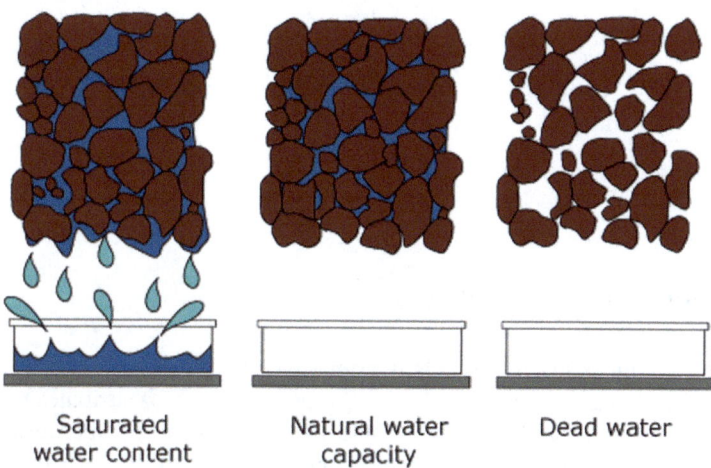

Saturated   Natural water   Dead water
water content   capacity

Fig. 14.4 Different forms of water capacity and the pattern of dead water content

4. Capillary water capacity (CWC): moisture content of the soil layer saturated by capillary action (in a 10 cm tall column).

It is essential to clarify two concepts regarding plant cultivation. These are: the dead water content of the soil, and the concept of water resources that can be taken up (available water). These concepts are necessary to clarify so that the total water content of the soil, which may vary depending on the soils, might not confuse the irrigation plan and crop production technology decisions.

Dead water is defined as the water content that is very tightly bound to soil colloids and in the smallest capillaries (smaller than 2 microns). The compaction force is greater than 15 atm (bar).

Determination of dead water (DW) content:

- using the wilting experiment: the vessels of test plants are allowed to dry out until only the dead water content remains, then we can infer the amount of dead water based on the wilting point;
- steel walled pF device: The pF device is capable of measuring the quantity of water that is still bound at 15 bar pressure.
- Calculated from the "Hy" value: The weight percent amount of dead water content can be inferred from the Hy values appearing in the soil test report. The method of calculation: DWc% = 4 Hy.

Available Water Content (AW) is the amount of water that is bounded by less than 15 atm (bar) of force. Available water is the form of water utilizable by plants, since it is bound to soil particles by less than 15 bars of pressure, so the suction pressure of roots can exceed it. Its calculation:

$$AWmax = AWCmin - DW$$

Table 14.1 The effect of soil types on the available and dead water content

Texture	FWC	AW	(DW)	AW	(DW)
	Volume %			In WC%	
Sand	10	8	2	80	20
Loam	31	16	15	51	49
Clay	46	13.	33	28	72

2.4 Water Capacity of Soils

The water binding capacity of soils is primarily the interaction between the solid phase (soil colloids, clay particles, etc.) and water, and it can be attributed to adsorption (adhesion) and capillary forces (Table 14.1). Adhesion moisture forms a thin layer of film on the surface of soil particles, but the impact of forces decreases rapidly farther away from the surface. The first layer of water molecules near the surface of soil colloids is bound with a great deal of force, ~50 bar which decreases with distance, and reaches the value below 15 bar that already available.

The water retention and lifting capacity of capillaries can be interpreted as an effect of adhesion forces and the force of attraction between water molecules (cohesion). A portion of the water entering the soil as precipitation of irrigation water is bound in the soil caverns, and this amount is called field water capacity. Water capacity is strongly influenced by soil structure, e.g. loose, coarse-grained soils like sand store less water (approx. 8–10%), and clayey, fine-grained soils can bind more water (max. 31%). Therefore, better structured soils can absorb and store larger amounts of water. In compacted soils containing a lot of clay, the bulk of the large amount of water (more than 70% of the water capacity) is in a firmly bound form, which increases the dead water content, and the amount of water available for plants becomes less.

2.5 Water Balance of Soils

Soils can be characterized as having different water balance depending on structure, location and environmental factors (Fig. 14.5). Water balance gives a good clue regarding the water supply of the area and the quantity of water provided for plants:

$$W_p + \left(W_i\right) + W_{cr} + W_{inf} = W_{ep} + W_{tr} + W_{min} + W_{eff} \pm \Delta W$$

Where the factors are the following:

W_p = recharge from atmospheric precipitation
W_i = amount of irrigation water
W_{cr} = rate of capillary rise
W_{inf} = recharge from influent waters

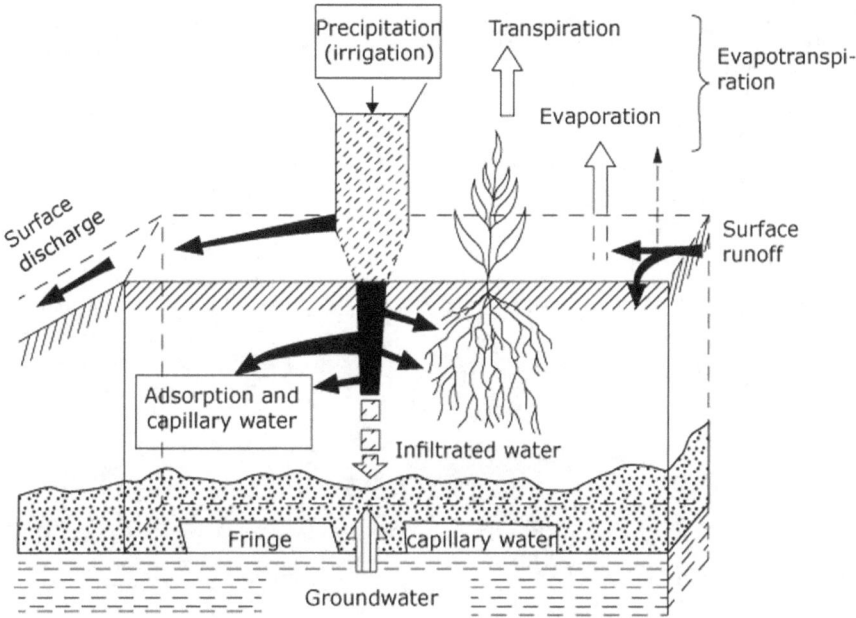

Fig. 14.5 Major factors of water circulation in soils

W_{ep} = evaporation loss of the area
W_{tr} = transpiration loss of the area
W_{min} = amount of percolating ground water
W_{eff} = amount of effluent water
ΔW = change in water storage of the area

When calculating water balance, excess precipitation and water losses of the area are also taken into account. Different soil types have different characteristics. In the case of loose, sandy soils, the primary loss factor in the formula is W_{perc}, the amount of percolating water, while this factor is of less importance in the case of well-structured loamy soil types, and it is almost minimal in the case of soils with clay-type physical structure. Therefore, water balance data give good clues not only for soil water balance or changes in water storage, but also for the components of water management parameters and the properties of the soil (Hunt et al. 2020). Water consumption of plants in field experiments can be well measured using lysimeters (Fig. 14.6). Lysimeters measure evaporation, transpiration, infiltrated water, adsorption and capillary water. The amount of water used for plant biomass can also be measured in lysimeters.

Fig. 14.6 Ploughland lysimeter, in a corn experiment

2.6 Basics of Plant Water Management

One of the most important conditions for the normal vital functions for higher plants is that their water balance – that is the ratio of water absorption and water consumption – shall be ensured for a long time without any prominent deficit.

The water balance of plants, however, is often in the negative range. Although the reason for that may appear to be very simple, still it is the result of an extremely complex phytophysiological process. Plants the functions of life of plants indicate that they evaporate more water than they can absorb.

Given the climatological conditions of Hungary this evaporation is so fast that the entire water content of the plant is replaced within a single day. It means that the water balance of the plant is very dynamic, but it is often fluctuating and extremely unstable. The amount of water absorbed by the plant depends on the extent of evaporation, the dimensions and effectiveness of the root system, and the availability of water in the soil. Water balance is regulated by two factors. One of them is the soil factor, the other one is the plant factor, the physiological status of the plant.

Scarcity of water modifies a number of physiological processes, e.g. it affects breathing (it stimulates the breathing of leaves, while reduces that of seeds due to their receding water content). It generally affects internal biochemical processes (e.g. it reduces the sugar content in tobacco leaves, and increases the quantity of nitrogenous compounds and nicotine in the cells). During the course of evolution plants have developed drought resistance to a certain extent, which depends on the species, and in particular on the given tribe (Nemeskeri and Helyes 2019). Nevertheless, even draught resistant plants are unable to fend off extensive water shortage lasting for a long time. Finally the plant dies in the end after an irreversible process of withering.

The amount of water "flowing through" the plant is incredibly large: during the entire growing season a sunflower or corn evaporates 200–250, or even more litres of water. The water requirement if fruit trees flucuates during the period of vegetation. Their water requirement is larger when their shoots are growing in the spring, and also when fruits are growing. In order to classify these periods, then the one with the biggest water requirement would be shoot growth taking place during the spring months. The second most water-consuming period is flowering and fruiting (the period of pollination and fertilisation), which generally takes place in April–May in our climatic zone. The third most water-consuming period is the one of intensive fruit-growth, the first phase of ripening, which, depending on the relevant fruit type, may last from July until even September. In the case of winter apple and winter pear this period takes place in August in our climatic zone. Of course, the earlier ripening summer fruits need a more abundant water supply earlier in order to be able to grow their fruits to the required size.

The peak of water requirement can be clearly identified in the lives of plants as well. Usually the most water-consuming periods are germination (in this case not the amount of water is important, but the presence of water required for the swelling of seeds is critical), the period around flowering, while the third most water-consuming period is that of intensive crop growth and grain filling.

2.7 Water Consumption of Plant Species and Plant Varieties

Certain elements of cultivation practice, such as soil cultivation, nutrition, crop rotation and plant protection are all important factors of improving the effectiveness of water consumption. The role of plant varieties should not be forgotten, either. Selecting the appropriate variety plays a tremendously important role in improving the efficiency of the available water quantity. In Hungary a sufficient number of varieties of the main arable crops are available for farmers.

There may be significant differences between the varieties or hybrids in terms of water requirement, water consumption and, in this context, drought tolerance. From amongst the characteristics of the varieties the length of the growing season affects water consumption the most. The length of the growing season determines the dynamics and also the quantity of water consumption. Generally speaking, shorter growing season results in less water consumption, therefore better drought tolerance, although obviously there are exceptions. In the case of winter wheat varieties a difference of 2 weeks means 50–70 mm difference in water requirement. As for corn, if the growing season is longer with a month, it will result an extra 100 mm water requirement.

Arable crops also have different water requirements. Paprika, for example, requires a lot of water. On hot summer days (30–35 °C) the plant requires even as much as 3 l of water for optimal growth and crop yield. According to the results of precise measurements, 20–25% of the plant's daily water consumption takes place during the hot and dry period between 12–14 p.m.

The growth rate of the root systems, root mass, and, in this context, the effectiveness of absorbing water and nutrients from the soil may be different in the case of each variety. Water-stress may affect the quantity and quality of the crop yield in a significantly different way, depending on the life stage of the plant it occurs (before or after entering the generative phase). Water-stress occurring during the early period, in the stage of vegetative growth significantly decreases the reproduction potential of the plant. The vegetative growth stage of variables with shorter growing season is shorter, therefore water shortage causing serious damage is less likely to occur.

Varieties with better production potential utilise water more efficiently. Amongst irrigated circumstances, in case the ecological conditions make it possible, it is advisable to choose a variety with longer growing season in order to exploit better production potential.

Evapotranspiration, the extent of water vapour released by the soil and the vegetation mainly depends on the features of vegetation. The connection between the surface of the vegetation and the extent of vapour release is not linear, as the amount of energy per surface unit is inversely proportional to the extension of the surface. The results of lysimeter studies demonstrate that among identical circumstances the water requirement related to the leaf area of the varieties of the same plant species with different leaf areas increases at a decreasing rate. This correlation may be characterized with a saturation curve (Szalai 1989).

During irrigation-related researches the water consumption of plants can be precisely quantifiable and measurable. Based on the research of Sándor Szalóki (1991) the water consumption of main plant species can be determined, which were measured on the lysimeter-plant of IRI (Irrigation Research Institute) during water management research projects (Table 14.2).

2.8   Water Uptake of Plants

Higher terrestrial plants take up water with their cuticle-free roots, the emission of which is reduced by the cuticle and the regulated movement of the stoma. In the case of higher plant species water uptake via the parts above the ground is minimal (Berry et al. 2018).

On the root itself water is absorbed by the root-hairs. The root-hairs are located on the root tip, close to the dividing, young cells. Root-hairs are in close connection with the particles of the soil, and absorb nutritive salts from it in the form of of an aqueous solution. Roots keep growing in the soil until they reach the depth where water supply is sufficient.

Table 14.2 The water requirement of plants and the main characteristics of irrigation in dry years occurring with a frequency of 20%

Designation	Critical period	Rooting depth	Moisture requirement	Water requirement (mm)	Irrigation water (mm)
Sugar beet	June-August	D	M	550–600	180–250
Maize (medium-late season)	July-August	M	M	400–550	150–200
Fertilisation of soy bean	July-August	M	MB	400–500	120–180
Alfalfa	June-August	VD	A	600–700	200–300
Intensive lawn	May-September	SH	VB	600–700	300–400
Table grapes	June-July	VD	M	570–670	150–200
Apple, pear (dwarf variety)	July-August	M-D	B	500–600	150–250

S short, *M* medium, *B* big, *VB* very big, *SH* shallow, *D* deep
Based on Szalóki (1991)

The quantity of utilisable water resources in different soil types may extremely vary. In sandy loam and loam soils the ratio of useful and dead water is ideal. Clay soils contain too much colloid material enclosing a larger quantity of dead water, which is water bonded with such a force that it becomes unabsorbable for the plant.

The water uptake of roots is also affected by the temperature of the soil. At lower temperatures the resistance of roots to water flow increases, the intensity of vegetable metabolism decreases, root growth, and, consequently, the pace of water uptake and water transportation slows down because as a result of cooling down the inner substance of cells becomes denser.

Extremely high temperature (35 °C) also affects water uptake – usually it decreases absorption.

Plants are able to absorb both the gravitational water moving freely amongst the soil particles and the weakly bonded water in the capillary columns of the soil. The water of hydration strongly bonded on the surface of soil particles is unabsorbable for plants. Water can only be absorbed if the water potential of the root-hairs is more negative than that of the soil. The active root surface of cultivated plants is very big. The most active root-hairs are around the renewing root tip.

The roots growing in the soil can absorb water in case their own water potential is lower than that of the soil. It can be demonstrated that the changes in the water potential of the root system are directly proportional to that of the soil. The difference varies between 0.2 és 0.8 MPa depending on the water supply. The root system is able to actively reduce its water potential. It takes up the ions required for its growth, and, of course, for water absorption, and stockpiles them in the cell walls and the vacuoles of the cells. Increased salt concentration reduces water potential value, as a result of which the water potential of the root system decreases as compared to that of the soil. This allows the absorption of water from the soil zone located next to the root system, which, in turn, begins to desiccate. Therefore the water potential at the soil part neighbouring the plant also decreases. The desiccation caused by the root system also affects soil zones being farther away, thus a water potential gradient pointing towards the roots are established, which causes water flow towards the roots. This effect applies in a distance of approximately 8–10 cm from the surface of the roots.

In dry weather, when there is no replacement of precipitation for a long time, the soil gets more and more desiccated. Its water potential reduces to such an extent that the root system becomes unable to cause a water potential difference by uptaking ions, thus the water potential of the root system and the soil equalises. In such a case the plant is unable to absorb water fron the soil, but it still loses water via its stomas. Finally the turgor pressure decreases and the plant wilts.

The drier the soil is the more negative its water potential becomes to which plants adapted in several different ways. By increasing the concentration of the solution in the vacuoles, they decrease their osmotic pressure, thus their water potential becomes more negative than their environment. Plant root hairs are able to grow towards soil zones with higher water content, which is required because the movement of water in the soil is slow. The roots may grow asymmetrically, which means that the root hairs remaining dry decay while others vigorously grow. In case the soil desiccates to such an extent that its root system is unable to absorb any water from it, then the plan starts to wilt. It is generally accepted that in case the water potential value of the soil reaches −15 Bars, then it is called permanent wilting point. At this stage cultivated plants begin to perish in case they do not get any water, which means their irrigation shall be started immediately.

2.9 Excess of Water and its Effects

Conditions indicating extreme water balance situations are particularly important for horticultural professionals. The system of conditions leading to harmful water abundance, its reasons and prevention, the possibilities of mitigating the related risks are all important topics with great relevance.

The general objective of agricultural water management is not to let water become a limit of effectiveness during the process of production. Its fundamental aim is to prevent or even cease water management conditions being detrimental to agricultural production.

During the course of crop production water management interventions enable to prevent the occurrence of unfavourable water management situations (drought, harmful excess water), and to cease any existing harmful condition as soon as possible.

Over the last decades drought in Hungary has become increasingly frequent, although certain regions of the country the extent of this change may be different. Therefore in connection with agriculture it is expected that the desired amount of crop yield shall be achieved with a decreasing quantity of water. As far as finding a cure to the problem is concerned, settling for irrigation can only mean a partial solution, as both our surface- and groundwater resources are limited. The decreasing quantity of fresh water reserves is a global problem, the significance of which has not yet been sufficiently recognised by humanity.

Due to the water saturation in the pore space volume, constant oxygen excess is formed. The diffusion of O_2 is approximately 10,000 times smaller than in the air, therefore its replacement from the air practically ceases to exist.

In an oxygen-deprived environment the facultative and obligate anaerobic microorganisms become dominating. Anaerobic decomposing processes come to the fore, therefore, instead of the mineralisation of organic materials, they are reduced and hydrocarbons are formed. Under certain conditions anaerobic microorganisms reduce H^+ ions partially into H_2 molecules, while a part of organic materials are reduced into methane (CH_4). During the reduction processes, for example, NO_3^- ions are transformed into NO_2^-, and through the intermediary of N_2O, they form N_2, while as a result of further reduction they are transformed into NH_4^+ ions, Mn^{4+} and Mn^{3+} are transformed into Mn^{2+}, and Fe^{3+} turns into Fe^{2+}. Reduction of SO_4^{2-} ions into H_2S may also commence; that is why such soils have a characteristic sulphuric acid smell. As a result of long-lasting anaerobic conditions a large amount of reduced compounds and ions may accumulate in the soil, which, on the one hand, results in the deterioration of the soil structure, while on the other hand it is being extremely unfavourable concerning the nutrition of plants. The reduction potential of soils is primarily determined by their airiness, while temperature and pH-value play a modifying role.

Parallel with the progress of processes taking place in the soil as a result of the constant abundance of water, plant communities preferring, or tolerating shallow water coverage or high ground waters are quickly formed.

2.10   Determination of the Soil Water Content

According to traditional views, irrigation was started when the crop was already showing the signs of water deficiency. Research has shown that in the assimilation activity of a plant under water stress conditions is restored only days after the improvement of water supply. When symptoms are visible, such changes have occured in the plant that are still reversible for a while, but can be restored only in a long period of time.

Modern irrigation is used to prevent water shortage. In order to avoid yield loss, it is advisable to start irrigation before the onset of the visible symptoms of water deficiency, when the moisture content of the soil falls below 50% of its water capacity.

At this time, the plants are not yet in a state of water stress, but the moisture content of the soil is unable to cover their water requirements. At less than 30% of water capacity, plants start showing the signs of wilting, and they reach the condition of water stress.

Based on the water balance equation, the estimated current soil moisture content can be determined by daily calculation. The actual values reduce evaporation, precipitation increases the initial set humidity. Using the results, we can determine the starting date for irrigation. Of course, if the area was irrigated, the applied quantity of water is added to the precipitation. Naturally, the current soil moisture content may also be determined by measurements, e.g. using tensiometers that can be installed in various depths. Tensiometers are used for the measurement of soil moisture suction, from which the moisture content of a given layer can be inferred. A cup-shaped object made of a variety of porous materials (plaster, ceramics, plastic) is sunk into the ground, where it is connected to a vacuum monometer by a tube.

The theoretical basis for conductivity based measurement is the significant difference between the conductivity of solid soil and water. Electrodes are inside a gypsum block, their life span is limited, about a year, but there can be significant differences. Resistance increases in proportion with the decreasing moisture content. Soil moisture content can be measured with these tools more accurately in lower ranges, although as moisture content increases, measurement errors increase as well. The measurement is affected by the salinity of soil water content as well.

Methods based on capacity measurement provide rapid result, and they continuously supply data after installation. The theoretical basis for this is the difference between the dielectric constant of different materials. It is a widely used method due to its ease of use.

The method based on the measurement of neutron scattering is fast, and is therefore also suitable for frequent measuerments; its disadvantage is that the device is expensive (Vereecken et al. 2015), probe pipes have to be installed, and in the top 30–40 cm layer where the soil is disturbed, the outcome is uncertain.

Preliminary results show the NASA Land Information System (LIS) model is suitable for estimating percolation indirectly from the equation of soil water balance. Outcomes of deep percolation rates are connected with precipitation meanwhile coupled with the topography. The relationship between surface and ground-water flow is taken into consideration in the LIS model (Elbana et al. 2019).

The current soil moisture supply of the given day is determined by measurement or calculation, then using the daily evaporation values, the number of days for which soil moisture content will be enough for the plants can be estimated. The result of the calculation can be regarded as estimation even with the utmost care, because many other factors are impossible to take into account. For example, precipitate is not utilized in 100% usually – and water moves laterally in the soil as well, and the upper layers may be moistened by the water rising from the lower layers, even if ground water is deeply situated.

Water requirement (wr) is identical with the actual water consumption of stocks of plants with optimal water supply. Calculation of the water requirement of plants:

- by measuring the water balance of lysimeters (cultivation tanks lowered into the soil), by applying different types of lysimeters: floating, weighing and compensating lysimeters;

- by measuring field water balance: one of its preconditions is optimal water supply (satisfying statical water requirements), which enables unlimited water absorption of plants;
- with simulators: meters measuring surface-water evaporation where the surface of evaporated water can be modified, thus imitating the changes in the leaf-surface of plants, therefore its evaporating procedure;
- with calculation, modelling.

Determining water requirement (wr) with simple lysimeters (Zsembeli et al. 2018; Sołtysiak and Rakoczy 2019): this method is the most reliable, the water balance of cultivation tanks lowered into the soil can be easily measured and traced. Its greatest advantage is that there is no surface onflow and leakage, and that the quantity and quality of water leaking and flowing through the soil can be measured. While implementing the method, however, attention needs to be paid to avoid any occurrent edge- or oasis-effect, together with the accuracy of irrigation.

2.11 Calculation of the Irrigation Water Requiretments

The amount of irrigation water can be calculated knowing the water requirements of the plant, the amount of precipitation and the moisture of the soil. The starting available moisture of the soil and the amount of the precipitation are subtracted from the water demand.

$$IW_r = W_r - W_0 - P\left[mm\right]$$

- IW_r – irrigation water requirement (mm)
- W_r – water requirement of the plant (mm)
- W0 – the starting available water content of the soil (mm)
- P – amount of precipitation (mm)

The amount of precipitation is not previously known, and the fault of prognostication is very high. This can be calculated with the area-specific average precipitation data, therefore irrigation water requirements can be determined in a certain level of probability. If the precipitation is measured, the calculation based on the averages can be corrected by the actual measurements. This way, the calculation becomes more accurate.

The amount of the irrigation water is usually measured in mm or m^3/ha. Conversion is simple: 1 mm of moisture content means 10 m^3 water on 1 ha. It is advisable to carry out irrigation in a way that the dispersed water amount is 70–80% of the total water capacity of the soil. Dispersion above 80% or by the total water capacity can have several negative effects and bear a great risk:

- Moisture content above the static water demand prevents the root system to take up the required oxygen amount, therefore the water and nutrient uptake is being impeded, and damages can occur in the root fibers.
- The microbiological life of the soil changes due top the excessive moisture content, denitrification becomes dominant, and nitrogen loss can occur.
- Chemical processes in the soil shift toward negative reduction, and poisonous reduced ions can develop.
- The structure of the soil is damaged, proliferation can occur.
- On slope areas, run-off water and erosion problems can occur.

To calculate irrigation water needs, the losses during application must be considered, besides the irrigation water needs. Losses originate from evaporation, leakage and run-off water. Losses change depending on several factors. Evaporation losses, for example, primarily depend on the temperature and humidity of the air, droplet size and wind conditions. Losses caused by run-off water are affected by the irrigation intensity, the water capacity of the soil and the slope of the fields.

$$IW_n = IW_r + E + L + R\,[mm]$$

- IW_r = irrigation water requirement
- E – evaporation loss (mm)
- L – leakage loss (mm)
- R – run-off water (mm)

Irrigation water requirement can be calculated by the dispersion efficiency as well: [mm]

$$IW_r = IW_d\,/\,\eta$$

- IW_r = irrigation water requirement
- η = dispersion efficiency

Efficiency depends on the irrigation conditions, usually varies between 0.7 and 0.95.

Traditional irrigation practice disperses a higher amount of water at once (even 60–80 mm). This is disadvantageous from several viewpoints. This method does not consider the requirements of the plant, and water excess can negatively affect the vegetation, the microbiological life of the soil, the chemical processes and the soil structure. Some damages can be permanent, for example the structure damage of the soil. Its advantage is to decrease the number of irrigation turns, and due to less relocation, treading causes less damage and it requires less manpower.

The spreading practice of modern irrigation uses lesser water standard with more frequent irrigation. A single dose is 15–30 mm. This submits to the requirements of the plants. The method can be started earlier and can be finished later, therefore it lengths the irrigation season. This means better utilization of machines and best irrigation practices (Chartzoulakis and Bertaki 2015). Due to lesser water amounts, this enables irrigation of soils that did not allow irrigation due to low water capacity.

Its disadvantage is the frequent relocation, although with modern irrigation equipment (linear system, center pivot) this is not a problem. Lesser water amount means increased evaporation loss.

3 Some Results of Irrigation Experiments (Case Study)

In 2016 and 2017 the effect of tape drip irrigation was tested on the change of corn yields and yielding elements of maize (Futo and Bencze 2018). In the study the Aqua Traxx tape drip system was used, sold by Metra Company. During the experiment, non-irrigated (control) plots were used, then plots satisfying 75% and 100% tape drip irrigation parcels in the maize water requirement and finally, irrigation satisfying the 100% water requirement was supplemented with complex water-soluble fertilizer (N-P-K) in the fourth treatment. In 2017 a humic acid treatment was used instead of water-soluble fertilizer. In the experiment, we investigated a leading Pioneer hybrid, a leading Monsanto hybrid, and a hybrid sweet corn of Martonvasar (Hungary).

The soil of the experiment is characterized by the fact that its physical characteristics is clay, as for its acidity it is acidic and slightly acidic, the cultivated layer does not contain $CaCO_3$, and the N-content of the soil is medium ranged based on humus content. The water management of the soil is characterized by poor water flow capability and high water retention capacity. The cultivated level is compressed, its porosity, and within that, the ratio of gravity pores is smaller.

The results show that water supply increased the relative chlorophyll content of maize only at 100% water demand (Table 14.3). The 75% water supply this year did not differ significantly from the results of the control irrigation plots, due to the excellent precipitation distribution.

First of all, the average yield of sweet corn was examined, which occured only in the 2016 study. During the measurement of the averge yields, we compared the yields of plots without irrigation (control), the irrigated plots to 75% water demand, the irrigated plots to 100% water demand and the irrigated plots with fertilization.

Table 14.3 The relative chlorophyll content of maize in 2016–2017. (SPAD value)

	Control	75% water-based irrigation	100% water-based irrigation	100% water-base and nutrient solution
Sweet corn	41.7	41.6	46.1	46.6
P9903	43.2	43.5	46.7	46.8
DKC4541	43.0	43.6	46.6	46.8
Average 2016	42.63	42.90	46.46	46.73
P9903	44.1	43.7	45.9	46.4
DKC4541	43.2	43.9	45.8	46.7
Average 2017	42.80	43.17	45.67	46.33

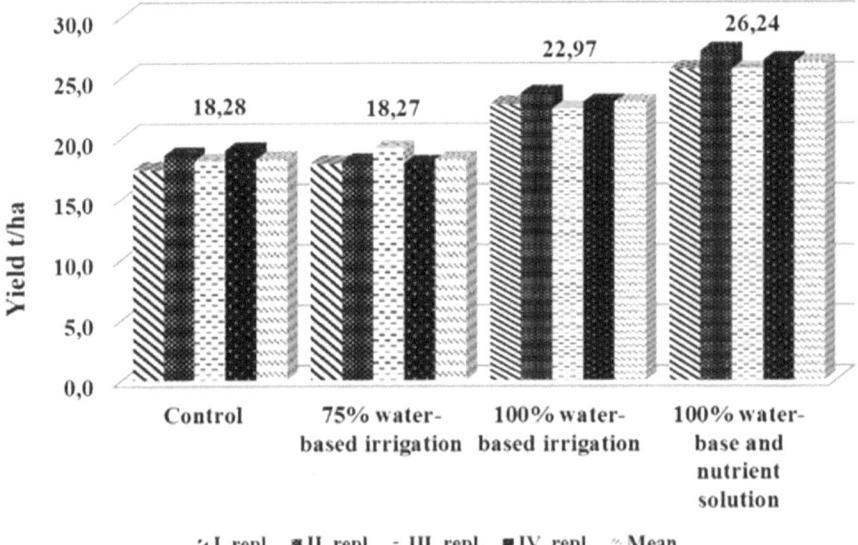

Fig. 14.7 The average yield of sweet corn in a tape-drip irrigation experiment 2016 (Szarvas, Hungary)

The yields of sweet corn were expressed by a higher (~60–70%) moisture content than the average and a cob harvest weight.

From the results, it can be seen that the yields of sweet corn could be significantly increased in the very favorable year 2016 by the use of tape drip irrigation technology (Fig. 14.7). Due to the favorable precipitation, there was no difference between the yield of parcels without irrigation (control) and those with irrigation of 75% water demand in the experiment. However, the crop-enhancing effect of irrigation, which satisfies the entire water demand of the plant, was very significant even in this year's favorable water supply. The yield of sweet corn reached 22.97 t/ha, which is very favorable. This yield in the experiment was only surpassed by the yields of the plots with nutrient supply, the yield reached 26.24 t/ha.

In the next group of studies, we examined two feedstuff maize hybrids whose crops were monitored during the test. The analysis of the yields of feedstock corn showed similar results in 2016 than sweet corn hybrid (Figs. 14.8, 14.9, 14.10 and 14.11). There was no difference between the yields of no irrigation and the yields with irrigation that meet the 75% water demand. This was due to the satisfying amount and distribution of precipitation. However, the total water demand of the plant could not be covered by the naturally falling rainfall even in this favorable year, which meant that the yield could be increased by satisfying the 100% of water demand of corn in 2016. The yields increased by 22.3–24.5% compared to the yields of control plots.

In 2017 precipitation was much less favorable, and precipitation in the growing season did not reach the 30-year average values. Therefore, average yields were significantly lagging behind the results of the previous year.

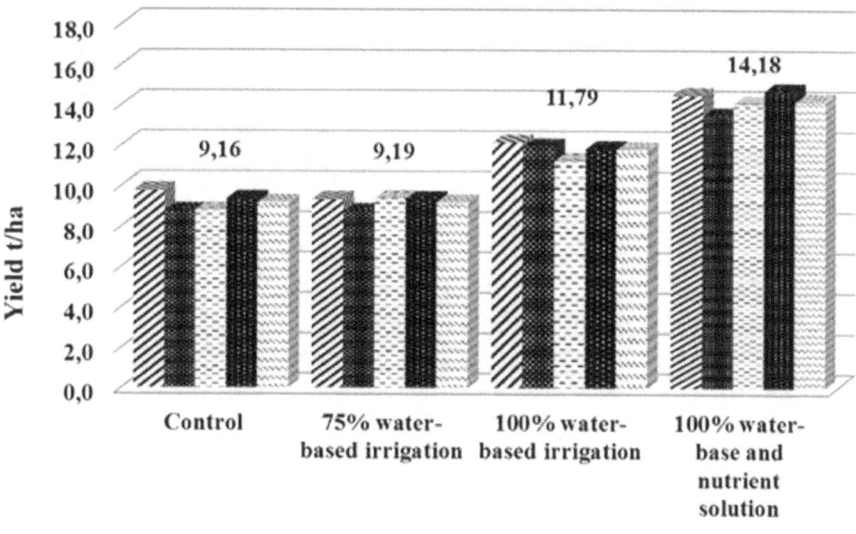

Fig. 14.8 Evolution of the yield of P9903 hybrid in a tape drip irrigation experiment 2016 (Szarvas)

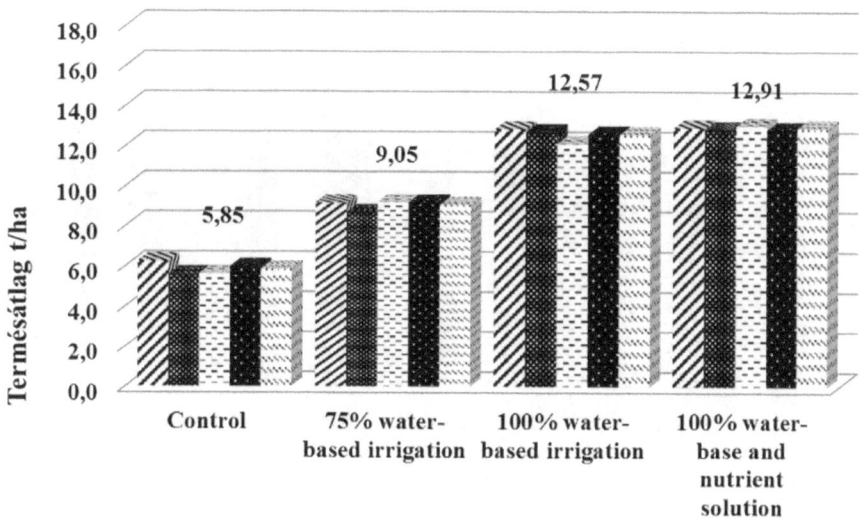

Fig. 14.9 Evolution of P9903 hybrid crop average in a tape drip irrigation experiment 2017 (Szarvas)

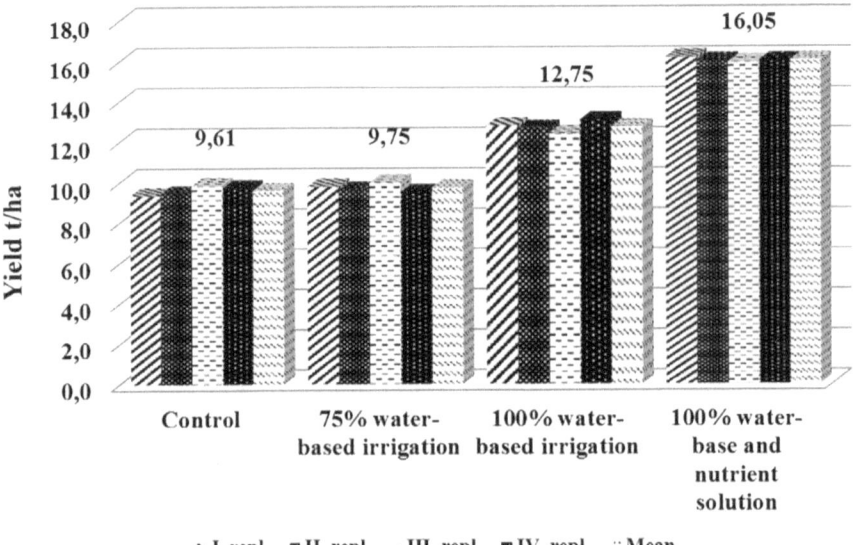

Fig. 14.10 Evolution of the DKC4541 hybrid yield ratio in a tape drip irrigation experiment 2016 (Szarvas)

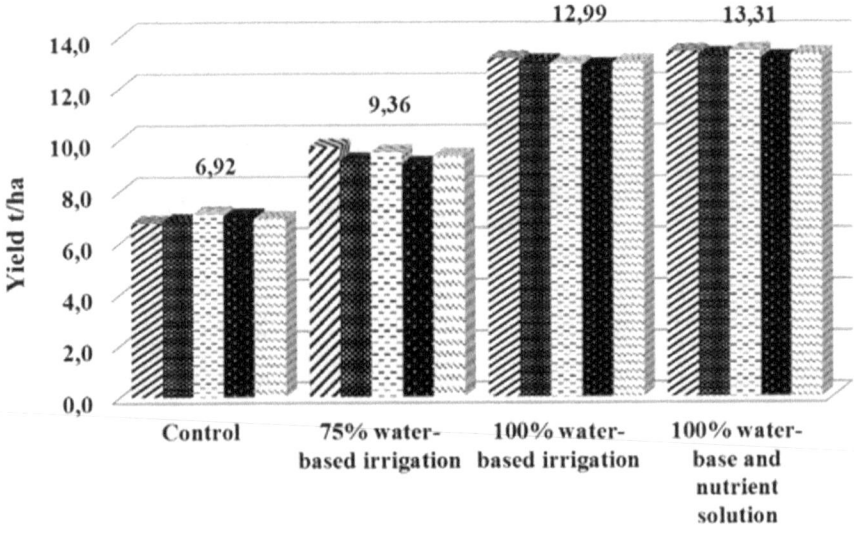

Fig. 14.11 Evolution of the DKC4541 hybrid yield average in a tape drip irrigation experiment 2017 (Szarvas)

The largest drop in yields compared to the previous years was on control plot of no irrigation. This is mainly due to the low precipitation period of maize water supply (July, August). With irrigation of satisfying the 75% of the water demand of the plant, the yields have increased very strongly, for both tested hybrids it exceeded 9 t/ha (9.05 and 9.36 t/ha).

If the water requirement of the plant was 100% satisfied during the irrigation, yields of 12–13 tons of hybrids were formed. The favorable water supply resulting from irrigation in hybrids shows that the decreasing effect of ever-increasing dry periods in climate change is significant, yields decreased in the control plots with bad water supply by 6.72 t/ha (P9903) and 6.07 t/ha (DKC4541).

In 2016, the nutrient parcels could further increase this, which is primarily due to the favorable phytophysiological condition of the plant that the plant immediately comes to the dissolved form of nutrient in the root hair zone of the root. This also refers to the important fact that optimal nutrient supply is possible only in the presence of sufficient amounts of water in the form of being available for the plant. The average yield of the plots with nutrient supply ranged from 14.18 to 16.05 t/ha, which reached the limit of the economically favorable production and profitability by the results of the experiment.

In 2017, instead of the conventional nutrient solution, a humic acid treatment was tested by applying tape drip irrigation. As a result of the treatments, similarly to year 2016, we could measure a further increase in crop average. The yield increase in 2017 was 340 kg/ha (P9903 hybrid) and 320 kg/ha (DKC4541 hybrid).

4 Conclusions

Overall, it was found that tape drip irrigation of maize is of very low water-use, energy-efficient and generally efficient irrigation technology, which can be a major domestic technical innovation for maize irrigation in future for intensive farming.

The yield of maize can be significantly increased by improving the water supply of the plant. In many areas, only little water is available for irrigation. The effect of drip irrigation in our experiment was investigated for corn yields in 2016 and 2017. In 2016, yields increased by 22.3–24.5% compared to the yields of control plots, while the yield gains in the drier year of 2017 reached 46.73–53.46%. In our experiment, the growth of the average yield was economically measurable.

Acknowledgements This chapter was supported by EFOP-3.6.1-16-2016-00016 project: The specialization of the research and training profile of SZIU Campus of Szarvas in the issues of water management, hydroculture, precision mechanical engineering and alternative crop production.

References

Berry, Z. C., Emery, N. C., Gotsch, S. G., & Goldsmith, G. R. (2018). Foliar water uptake: Processes, pathways, and integration into plant water budgets. *Plant, Cell & Environment, 42*, 1–14.

Boorman, D. B., Hollis, J. M., & Lilly, A. (1995). *Hydrology of soil types: A hydrologically-based classification of the soils of the United Kingdom* (IH report no. 126). Wallingford: Institute of Hydrology. 146 p.

Chartzoulakis, K., & Bertaki, M. (2015). Sustainable water management in agriculture under climate change. *Agriculture and Agricultural Science Procedia, 4*, 88–98.

Elbana, M., Refaie, K., Elshirbeny, M. A., AbdelRahman, M. A. E., Abdellatif, B., Elgendy, R., & Attia, W. (2019). Indirect estimation of deep percolation using soil water balance equation and NASA land simulation model (LIS) for more sustainable water management. *Egyptian Journal of Soil Science, 59*(4), 363–383.

Futo, Z., & Bencze, G. (2018). The results of the modern drip irrigation of maize in Szarvas. *Research Journal of Agricultural Science, 50*(3), 58–68.

Gombos, B. (2011). *Hidrológia – hidraulika: digitális tankönyv.* (In Hungarian) https://regi.tankonyvtar.hu/hu/tartalom/tamop412A/2010-0019_hidrologia-hidraulika/index.html. Accessed 6 Apr 2020.

Hunt, A., Faybishenko, B., Ghanbarian, B., Egli, M., & Yu, F. (2020). Predicting water cycle characteristics from percolation theory and observational data. *International Journal of Environmental Research and Public Health, 17*(3), 734–753.

Nemeskeri, E., & Helyes, L. (2019). Physiological responses of selected vegetable crop species to water stress. *Agronomy, 9*(8), 447–466.

Sołtysiak, M., & Rakoczy, M. (2019). An overview of the experimental research use of lysimeters. *Environmental & Socio-economic Studies, 7*(2), 49–56.

Szalai, Gy. (edit.) (1989) *Az öntözés gyakorlati kézikönyve.* Mezőgazdasági Kiadó, Budapest, 473 p. (In Hungarian).

Szalóki, S. (1991). A növények vízigénye és öntözésigényessége. In J. Lelkes & F. Ligetvari (Eds.), *Öntözés a kisgazdaságokban* (pp. 21–42). Budapest: Fólium Könyvkiadó Kft. (In Hungarian).

Tudasbazis. https://tudasbazis.sulinet.hu/hu/szakkepzes/mezogazdasag/a-mezogazdasagi-termeles-fobb-okologiai-tenyezoi/az-elettelen-kornyezeti-tenyezok/hidrologiai-adottsagok. Accessed 18 Mar 2020.

Vereecken, H., Huisman, J. A., Hendricks Franssen, H. J., Bruggemann, N., Bogena, H. R., Kollet, S., Javaux, M., van der Kruk, J., & Vanderborght, J. (2015). Soil hydrology: Recent methodological advances, challenges, and perspectives. *Water Resources Research, 51*, 2616–2633.

Vermes, L. (1997). *Vízgazdálkodás.* Budapest: Mezőgazdasági Szaktudás Kiadó. 462 p. (In Hungarian).

Zsembeli, J., Czellér, K., Sinka, L., Kovács, G., & Tuba, G. (2018). New techniques in agricultural water management. In P. Machal (Ed.), *Creating a platform to address the techniques used in creation and protection of environment and in economic management of water in the soil* (pp. 69–79). Brno: Visegrad Grant, Mendel University.

Chapter 15
External Solicitations, Pollution and Patterns of Water Stock: Remarks and some Modeling Proposals

Roy Cerqueti

Abstract This Chapter aims at outlining some quantitative models for the analysis of two crucial themes associated to the lifecycle of the water. By one side, we discuss the pollution of water generated by the impact of human activities on biological occurrences; by the other side, we present some arguments on the stock of available water, by including also some details on the causes of its evolution and its possible exhaustion.

Keywords Water lifecycle · Water biopollution · Water stocks · Shallow lakes · Patterned water stocks · Mathematical modeling · Markov chain · Stochastic approach · Sustainability · Environmental protection

1 Introduction

In this period of climate change and overconsumption of natural resources, the theme of sustainability is at the center of the debate. In particular, a growing interest is paid on how to manage pollution and to avoid the exhaustion of the irreplenishable natural resources. In this context, a crucial role is played by water.

This contribution deals with the analysis of the evolution of the stock of water, with the precise scope of discussing how the external events are able to modify it.

We present two types of contexts, leading to two different characterizations of the term 'stock': by one side, we consider the issue of the water pollution –so that, 'stock of water' has to be intended as 'clean water'; by the other side, we discuss the

R. Cerqueti (✉)
Department of Social and Economic Sciences, Sapienza University of Rome, Rome, Italy

School of Business, London South Bank University, London, UK
e-mail: roy.cerqueti@uniroma1.it

© The Author(s), under exclusive license to Springer Nature
Switzerland AG 2021
A. Gadomski (ed.), *Water in Biomechanical and Related Systems*,
Biologically-Inspired Systems 17, https://doi.org/10.1007/978-3-030-67227-0_15

303

theme of depletion of water –so that, 'stock of water' means 'physical amount of water'.

In the former setting, we provide some notes on the so-called 'shallow lakes', which are remarkably sensitive to the pollution created by human activities and agriculture; in the latter one, we introduce and discuss some stochastic models for describing the evolution of the amount of water, to face the relevant theme of avoiding its depletion.

Needless to say, the physical stock of water has to be intended in a specific region, and when we deal with depletion, we have in mind the exhaustion of water in a limited area. Thus, we face the relevant task of green areas becoming arid.

Moreover, the shallow lake setting has also biological ground, being the pollution of a shallow lake generated by the action of the vegetation of the lake. In general, the patterns of the stock of water is remarkably affected by human activities and physical interventions on the environment. In all the considered meanings of stock of water, a not irrelevant role is played by solicitations of mechanical nature. Examples can be found in the fractured soil conditions, which are driven by a growth of the temperature or the level of humidity. Under this perspective, one can experience a sudden increase of the level of pollution of a lake or a collapse of the available stock of water.

Furthermore, the mediums where water is naturally stored might lead to a quick decrease of its level, mainly in presence of porous medium effect.

Water scarcity and water quality represent two crucial themes for policymakers and scientists.

To deal with such tasks, a policymaker must strike a balance between the economic reasoning -which pushes consumption- and conservation of water reserves. In this context, human actions and natural phenomena represent sources of water pollution and they are able to modify the quantities of available stock of water. This said, the evolution of water stock and water pollution is surrounded by uncertainty, which calls for randomness.

A large strand of literature has focused on the identification of the factors affecting the level of pollution in the water or its availability. In the context of human activities and climate changes, refer to Loukas et al. (2002), Jones and Post (2004), Tamerius et al. (2006), Wang et al. (2006), Zhang et al. (2010), Yu et al. (2011), Cazcarro et al. (2016), and Martin (2019).

Let us enter some detail on the specific contexts we deal with.

Shallow lakes have been extensively studied in Kiseleva and Wagener (2010) and Wagener (2013) through deterministic dynamical systems theory (see e.g. the survey in Perko 2013). We present a critical view of these studies, by enlarging also the perspective to a stochastic model of Markov chain type for assessing the regimes of the dynamics of pollution in the shallow lakes (see e.g. Anatolyev and Vasnev 2002; Huisman and Mahieu 2003 and, more recently, Cerqueti et al. 2013, 2015, 2017a, b).

It is important to point out that the pollution of the water and the case of the shallow lakes can be discussed also through the methods of the colloidal science, for which the employment of porous membranes and the exploitation of the hydrostatic osmosis might modify the natural pattern of the level of phosphorus in the water. In this respect, the Markovian assumption can be violated, and the evolutive model can be suitably grounded on processes of Poisson type. Under the same perspective, one can also discuss how the fractality-dependent decrease of suspension mass is associated to the departure from the Markovian setting. Such relevant themes are not developed in the present report, but they represent hints for future studies.

For what concerns the physical stock of water, we refer to the literature on stochastic growth models and its applications (see e.g. Ausloos and Kowalski (1992), Gadomski (1996, 2003), Ausloos and Vandewalle (1996), Gadomski and Ausloos (2006), and Vandewalle and Ausloos (1996a, b, 1997). In particular, we follow the approach followed by Mitra and Roy (2007) and propose some stochastic models based on Markov chains (see e.g. Norris 1998 for a panoramic view of this mathematical instrument). As a further analysis, we deviate from the Markovian context by outlining a model based on point process for the evolution of the available resource, as in Cerqueti (2014). The main assumption on this framework is that the stock of water is generated by impulsive events. In so doing, we are in line with stochastic growth model of non-Markovian type generated by impulsive events occurring at random times. In this respect, we point the attention of the reader to the relevant contribution of Łuczka et al. (1995), where Poisson processes are efficiently employed for modeling purposes. We are inspired by Łuczka et al. (1995), and we propose a stochastic growth model for stock of water whose distributional assumption is of Spatial Mixed Poisson Process type (see e.g. Grandell 1997).

The proposed models deserve a deep research activity. Nevertheless, the considered settings allow gaining some conclusions on the need of taking care of the quality of water and on its availability.

2 Bio-Polluted Water: The Case of the Shallow Lakes

We here give a quantitative perspective on the dynamics of pollution in the so-called shallow lakes, which are particularly sensitive to the human actions and to the activities that might bring polluting substances in the water.

This theme is of paramount relevance: indeed, the Legislator needs to pay attention to two competing objectives: by one side, the economic development of a non-urban area, to be understood as the exploitation of agricultural resources; by the other side, the need to preserve the natural environment and ecosystems in the region.

At a first glance, the problem seems to be of purely economic nature. However, as we will see below, it has also a biological root.

Let us deal with economic reasoning. It is evident that intensive cultivation is associated with high profits, and is therefore encouraged by agricultural operators; however, they also have a strong environmental impact. In this sense, it is well-known that the devastation of ecosystems also arises as an economic problem, due to the consequent depression of the tourism industry of the region involved and the huge costs associated with climate change generated by pollution.

In the context we are analyzing, shallow lakes represent a paradigmatic case. A lake is said to be 'shallow' when its maximum depth is 3 meters. Starting from this definition and unlike different lake systems, both the uniform growth of large aquatic plants and the resuspension of pollutants previously sedimented on the seabed are possible in a shallow lake.

The vegetation level of the lake is determined by the amount of nutrients present in its waters. When the nutrient level is low, lake vegetation consists of small algae. The increase in nutrients is strongly influenced by human intervention: in fact, the chemical fertilizers used for agricultural activities can be transported through the rain from the shores to the water of the lake system, generating an increase in the nutrient level.

The phenomenon of the increase in nutrients in the water of a shallow lake generates a chain of events.

The first occurrence is the growth of large algae. From this, it derives an increase in phytoplankton biomass, the direct consequence of which is a cloudiness of the water of the lake.

A layer of sediment then covers the leaves of the aquatic plants, and this helps to reduce the amount of light that reaches the bottom of the lake. At this point, therefore, the algae begin to die. The disappearance of aquatic plants is associated to the disappearance of the hiding places of the zooplankton, which takes refuge during the day among the submerged leaves. In this way, the fish easily manage to prey on zooplankton, whose population then reduces drastically.

Since zooplankton is the largest natural predator of phytoplankton, the population of phytoplankton increases, and this makes the water of the lake increasingly turbid. This effect is also amplified by the lack of algae in their function of anchoring sediments on the seabed.

We intend to provide some brief details on the irreversibility of the process described above. More in detail, we want to establish whether, by reducing the level of nutrients in the murky water of a shallow lake through appropriate agricultural policies, it is possible to obtain clear water again.

2.1 A Quantitative Perspective on the Pollution of a Shallow Lake

The most appropriate quantitative tool for analysis of this type is represented by the theory of dynamic systems. The founding principle of the study lies, in fact, in the formalization of an evolutionary phenomenon (or system) through an appropriate differential equation, with which there are associated balances that describe the states of the system. In our specific context, the system is given by the dynamics of the phosphorus concentration - which represents a proxy of the nutrient level - in the lake water. The balances represent the state of clarity of the lake system.

For a proper writing of the dynamics, one has to consider the way the phosphorus level evolves and the variables which are connected to it. In particular, the increase in the level of nutrients is directly related to the amount of fertilizers used in crops and inversely related to the amount of phosphorus already present in the lake.

In addition, it has been experimentally observed that the inverse relation between level of nutrients and phosphorus level depends also on a factor, denoted by b, which describes the sedimentation rate of the phosphorus.

Factor b is a number between 0 and 1, and by convention we have b = 0 when the phosphorus does not settle on the bottom and remains permanently in suspension, while b = 1 is the case in which the phosphorus precipitates immediately and completely at the bottom of the lake.

Interestingly, the reversibility of the process depends on the sedimentation rate b, so that there exists a critical threshold b* such that b > b* implies that the lake becomes permanently turbid (see Kiseleva and Wagener 2010; Wagener 2013).

However, it is important to give credit to technological procedures leading to a forceful reversibility of the pollution pattern. Indeed, the successful employment of porous means and membranes might lead to a reverse hydrostatic osmosis process, which is able to depurate water and reduce the phosphorus level in the shallow lakes. In this respect, one is authorized to model the sedimentation rate b as a function of such types of depuration devices.

2.2 An Optimal Regimes-Based Markov Chain Model for the Evolution of the Pollution of the Shallow Lakes

Section 2.1 discusses the impact of the chemical fertilizers in determining the pollution of the water of a shallow lake.

Starting from this premise, we now aim at outlining a stochastic model for making a proper prediction of the future evolution of the level of chemical fertilizers in the water of the lake.

The ground of the model lies in the randomness of what will be the agricultural activity on the lakesides in the future. It is clear that human interventions are able to rule the economic activities, so that the impact on them on the environment can be efficiently controlled. However, we propose a model based on the knowledge of past activities and without imposing constraints, to have an unbiased view of what will be the future evolution of the agricultural activities; in so doing, we wish to provide a device to be used by the policymakers for planning pollution control strategies.

Take now in consideration a sample of N consecutive properly measured observations of the level of the chemical fertilizers in the water of a shallow lake.

We assume that the amount of chemical fertilizers in the lake evolves accordingly to a Markov chain $\{X(t)\}_{t \geq 0}$. The states of the Markov chain are empirically identified by the available distinct observations, while the transition probabilities are obtained by looking at the empirical transitions from an observation to the subsequent one, on the basis of the observed sample. We denote the empirical transition probability of the Markov chain by **P**.

The Markovian assumption is one of the possible distributional conditions on the dynamics of pollution; however, it is quite suitable in the context we deal with. Indeed, one can argue that the level of fertilizers in a shallow lake increases/decreases with respect to the previous value, disregarding the previous history. This is due to the additive nature of the fertilizers on the lake, so that their level at a given time $t + 1$ is the one observed at time t with the addition of the fertilizers added in/removed from the lake at time $t + 1$. This is totally in line with the Markov chain hypothesis.

Generally, the number of distinct observations of the level of chemical fertilizers is so high that the resulting Markov chain $\{X(t)\}_{t \geq 0}$ becomes trivial for forecasting purposes. Indeed, one has empirically a few transitions from a state to another one – mostly, only one transition- so that transtition matrix **P** is filled by a large number of 0's and 1's.

To gain in meaningfulness, states have to be properly lumped together. Let us enter the details.

We assume that the interval $A = [u,v]$ is the variation range of the empirical observations.

Moreover, we build a partition π of A in J non-overlapping intervals $a_1, \ldots, a_J$, such that if x belongs to a_i and y belongs in a_{i+1} then $x < y$.

Of course, the observed sequence of observations leads to a corresponding sequence of intervals a's. Since more than one of the observations could hypothetically belong to the same interval, the sequence of observations leads to the identification of $H=H(N) \leq N$ different intervals a's, which will be denoted hereafter simply as $a_1, \ldots, a_H$.

Starting from partition π, we can identify a new Markov chain $\{X_\pi(t)\}_{t \geq 0}$ whose states are $a_1, \ldots, a_H$ and the transition probability matrix empirically derived and denoted by $\mathbf{P}_\pi$.

It is possible to select the optimal partition π^* such that the Markov chain $\{X_{\pi^*}(t)\}_{t \geq 0}$ is the 'best' approximation of the original Markov chain $\{X(t)\}_{t \geq 0}$, where 'best' has to be intended in the twofold sense of leading to the highest statistical similarity while avoiding perfect reproduction of the original stochastic process. Basically, this means that π^* is the solution of a minimization problem of the distance between **P** and $\mathbf{P}_\pi$, under the constraint that a diversity measure between **P** and $\mathbf{P}_\pi$ below a pre-fixed threshold is appropriately penalized.

The presence of a critical threshold for the sedimentation rate leading to the irreversibility of the pollution has to be modelled as a further constraint on being some states of $\{X_\pi(t)\}_{t \geq 0}$ of absorbing type.

The elements of the optimal partition are said to be the regimes of the dynamics of the amount of chemical fertilizers in the water of the shallow lake.

3 Discussing the Patterns of the Physical Stock of Water and Its Exhaustion: Some Stochastic Models

The debate on the depletion of non-renewable resources is particularly heated today. This interest is not based only on the assessment of the obvious impact that the exhaustion of natural resources would have on our existence, but also on recent results regarding the residual life time of non-renewable sources. More precisely, it is noted that the extraordinary economic-demographic expansion of the highly growing countries (like e.g. India and China) and the unstoppable consumption level of the West of the World are jointly responsible for the forecasts of exhaustion of some important resources in very short times.

The main purpose of this section is to provide some models for the evolution of the physical stock of water, by including in the discussion also the possibility of its extinction. This issue is addressed through the construction and study of appropriate quantitative models, in which the formal description of the evolution of the stock of water and of the event linked to its extinction, interpenetrates with a mathematical tractability that allows obtaining results of real practical interest. In essence, formalism and scientific rigor are maintained, but with the declared purpose of avoiding a mere philosophical speculation and, rather, arrive at the writing of policies to be proposed to the decision maker.

The models proposed for our study rest on two theories, i.e. Markovian models and point process models. The former context is explored here by means of two specific settings:

(A) Markov chains;
(B) Disturbed Markov chains.

The latter one relies to a specific class of Poisson processes, and it will be treated separately.

3.1 Markovian Stochastic Models

Before going into details, some preliminary considerations are appropriate.

Model (B) provides, as we will see, reasonable and truly feasible results, while model (A) offers answers of little practical interest. This dissonance is because Markov chains do not actually describe the problem we intend to address in a realistic way. However, their introduction is necessary because it is a prelude to the construction of the model (B), of which they constitute a trivial sub-case.

Models (A) and (B) share, as it is reasonable, the main characteristic of the phenomenon we are dealing with: they are stochastic models. In fact, it is not imaginable to describe the future evolution of the available quantity of a resource and the event linked to its extinction through deterministic models.

We denote with R_t the quantity of water available at time $t > 0$. By conventional agreement, it is assumed that $t = 0$ represents today (i.e. the starting time of our analysis), and therefore R_0 is the initial endowment of the considered natural resource.

It is assumed that R_0 is a nonnegative number and it is known, so that it is a deterministic quantity. Differently, having fixed a certain value $t > 0$, the value R_t is projected into the future, and therefore it is a stochastic term. From a mathematical perspective, this means that $\{R_t\}_{t>0}$, is a stochastic process.

We now present the stochastic models used for the development of the problem we deal with.

(A) Markov chain model

Suppose that the quantity R_t of water available at time t is a Markov chain. The number 0 must necessarily be one of the states of the chain, because it corresponds to the case in which the quantity of water is zero. In this type of model, we assume that if quantity is 0, then water has become depleted. This condition means that we do not admit that it will be possible to discover a new reservoir of water after it is declared to be exhausted.

Under a purely mathematical perspective, this assumption means that 0 represents an absorbing state of the chain, that is:

$$R_t = 0 \text{ implies that } R_{t+1} = 0, \text{ for each } t > 0.$$

Given this premise, the problem of extinction of water can be addressed by studying the probability of absorption of the chain, or the probability that there exists $t^* > 0$ such that $R_{t^*} = 0$. Clearly, given the the condition that 0 is an absorbing state, the problem aims at identifying the smallest value $t^* > 0$ such that $R_{t^*} = 0$.

To avoid unnecessary complications, therefore, we will in fact look for the probability that $R_{t^*} = 0$, knowing that before t^* the chain had never reached the absorbent state 0. This probability is expressed as follows:

$$P\left(R_{t^*} = 0 | R_{t^*-1} \neq 0, R_{t^*-2} \neq 0, \dots, R_0 \neq 0\right).$$

An important mathematical result of immediate intuition states that the probability expressed above depends on the initial endowment of resource R_0 and on the transition matrix of the Markov chain. We can therefore conclude that, in the case of Markov chain, the problem of the avoidance of extinction of water can find a solution only if the policymaker pursue the target of controlling the transition probabilities from a state to another one. This cannot be given for granted in a simple Markov chain model, since transition probabilities can be exogenously determined.

Moreover, the initial endowment is an exogenous datum, that the decision maker cannot manage, and therefore no intervention can be put in force in this respect to avoid the exhaustion of water.

However, the contribution of this type of models remains valid when the objective is adequately describing some features of the stock of water and its evolution.

In the next subsection, we overcome the limitations of the Markov chain approach presented above.

(B) Disturbed Markov chain model

The introduction of Disturbed Markov Chains (CMDs) responds to the need of building a model that includes real control by the decision maker. As we observed in the previous section, in fact, the Markov chains provide a purely descriptive analysis of the dynamics linked to the quantity of water.

A CMD is a stochastic process $\{R_t\}_{t \geq 0}$, the starting point is established at $t = 0$ which takes values in the elements of a discrete set called state space and whose elements are called states. Differently from a Markov chain, a CMD is endowed with a stochastic process $\{s_t\}_{t \geq 0}$, called disturbance, such that

$$P\left(R_t = j \mid R_{t-1} = i\right) = P\left(R_t = j \mid R_{t-1} = i; s_{t-1}; ...; s_0\right).$$

Above formula has a simple interpretation: the passage from time to state from one state to another also depends on the previous realizations of process $\{s_t\}_{t \geq 0}$.

The disturbance process $\{s_t\}_{t \geq 0}$ represents the quantitative translation of all events that can occur on a certain date and affect the process $\{R_t\}_{t \geq 0}$.

The variation range of the process $\{s_t\}_{t \geq 0}$ can be set to the entire real line or to an interval centered in zero. In so doing, by conventional agreement, we assume that a positive realization of s_t is a "good news", and it is able to enlarge the probability of increasing the amount of water. Similarly, a negative realization is a "bad news", and induces a large probability of a decrease in the stock of water. The case $s_t = 0$ means that, at time t, absolutely nothing has happened with an impact on the dynamics of the quantity water.

The introduction of the disturbance process precludes the possibility that both the memory loss property and the invariance of transition probabilities are satisfied; hence, process $\{s_t\}_{t \geq 0}$ formalizes the difference between the standard Markov chains models and the CMD. In this respect, notice that a CMD is not a Markov chain in general. By the opposite perspective, the converse statement holds true: a Markov chain is also a particular CMD with disturbance $s_t = 0$ with probability 1, for each $t \geq 0$.

Let us now assume that the quantity R_t of water available at time t is modeled through a CMD with disturbance $\{s_t\}_{t \geq 0}$.

Similarly to what was elaborated in the previous case, also in this situation we propose that the number 0 is one of the states, and it is associated to the extinction of the stock of water. In essence, 0 is an absorbent state also in the CMD model.

The probability that water achieves its absorbing state at time t* -and we assume that it does it for the first time- is

$$P\left(R_{t*} = 0 \mid R_{t*-1} \neq 0, R_{t*-2} \neq 0, ..., R_0 \neq 0; s_{t*-1}; ...; s_0\right).$$

It can be demonstrated that the probability expressed above depends on the initial endowment of resource R_0 and on all the realizations of the disturbance process before t*. This means that the decision maker is able, through regulatory policies or by investing in research and development, to implement actions able to reduce consumption of water, in order to minimize the probability of extinction expressed above or to make the absorption time t* very remote. The entities of the implemented actions are captured by the disturbance process $\{s_t\}_{t \geq 0}$.

3.2 Point Process Models

In this section, we offer a different view of the evolution of the stock of water. Specifically, we focus on impulsive events of random nature, which generate the dynamics of the stock. Thus, we pay attention on the jump component of the amount of available water.

We introduce a point process

$$S = \left\{ \left(\tau_i, \alpha_i \right) \right\}_{i=1,2,\dots}$$

representing the couples of random times τ's in which events with entities α's occurring in the surrounding environment appear –being clear that such events have an impact on the available stock of water. The index i is a counter, so that the event in τ_i appears before the one in τ_{i+1}.

As preannounced above, an event translates in a jump in the stock of water. We denote the point process of the jumps in the stock of water by

$$U = \left\{ \left(\gamma_i, \beta_i \right) \right\}_{i=1,2,\dots}$$

being γ_i and β_i given by the time and the entity of the jump in the stock of water generated by the event (τ_i, α_i).

Evidently, (γ_i, β_i) is obtained as a transformation of the terms in (τ_i, α_i).

We reasonably assume that γ_i is given by τ_i with the addition of a random delay. Such a condition captures the evidence that the effect of an event on the stock of water is not necessarily simultaneous, and the identification of such a delay can be driven by uncertainty.

The mark β_i is assumed to depend on the mark of the corresponding event α_i but also on time τ_i, so that the same event generates different effects on the water when measured at different times.

By summing the marks of process U, one is able to derive the time-dependent available amount of water on the basis of the occurrences captured by the events in S.

In this framework, the exhaustion of water can be declared in correspondence of a critical mark for the events such that the resulting (negative) jump in process U leads to null aggregated stock of water.

The estimation of the expect value of the stock of water can be performed over a prefixed time-interval J; this might lead to useful insights on the depletion of such a precious resource.

At this aim, some technical assumptions can be stated on the nature of the process S and on its components.

The main assumption on S it that is is a Spatial Mixed Poisson Process. This assumption allows to use an invariance property of this type of stochastic processes, so that mild conditions on the components of S and on the operator transforming S in U guarantee that also U is a Spatial Mixed Poisson Process.

The estimation procedure is of Bayesian type. The idea is to take a testing interval I. The estimation of the expected stock of water accumulated in J can be performed by knowing; the number of events occurring during I, the number of jumps in the stock of water measured in J and the number of events measured in I leading to jumps in the process U on the same interval.

For more details on Spatial Mixed Poisson Processes, on the required technical conditions and on the estimation procedure, refer e.g. to Cerqueti (2014).

4   Conclusions

In this work, two contexts on the stock of water have been presented; they allow to understand the relevance of the human activities and of biological and mechanical factors on the vital cycle of this crucial natural resource.

By one side, we deal with pollution; by the other side, with depletion of water.

Our words suggest that the quantitative analysis is able to foster good practices when pursuing sustainability. In particular, decision makers should implement long-term policies for avoiding the irreversibility pollution of shallow lakes or the depletion of water in a specific region.

The arguments set out call consciences towards greater responsibility on environmental protection, since the damage caused by external events is sometimes irreversible.

References

Anatolyev, S., & Vasnev, A. (2002). Markov chain approximation in bootstrapping autoregressions. *Economics Bulletin, 3*, 1–8.

Ausloos, M., & Kowalski, J. M. (1992). Stochastic models of two-dimensional fracture. *Physical Review B, 45*(22), 12830.

Ausloos, M., & Vandewalle, N. (1996). Growth models with internal competition. *Acta Physica Polonica Series B, 27*(3), 737–746.

Cazcarro, I., López-Morales, C. A., & Duchin, F. (2016). The global economic costs of the need to treat polluted water. *Economic Systems Research, 28*(3), 295–314.

Cerqueti, R. (2014). Exhaustion of resources: A marked temporal process framework. *Stochastic Environmental Research and Risk Assessment, 28*(4), 1023–1033.

Cerqueti, R., Falbo, P., Guastaroba, G., & Pelizzari, C. (2013). A tabu search heuristic procedure in Markov chain bootstrapping. *European Journal of Operational Research, 227*(2), 367–384.

Cerqueti, R., Falbo, P., Guastaroba, G., & Pelizzari, C. (2015). Approximating multivariate Markov chains for bootstrapping through contiguous partitions. *OR Spectrum, 37*(3), 803–841.

Cerqueti, R., Falbo, P., & Pelizzari, C. (2017a). Relevant states and memory in Markov chain bootstrapping and simulation. *European Journal of Operational Research, 256*(1), 163–177.

Cerqueti, R., Falbo, P., Pelizzari, C., Ricca, F., & Scozzari, A. (2017b). A mixed integer linear program to compress transition probability matrices in Markov chain bootstrapping. *Annals of Operations Research, 248*(1–2), 163–187.

Gadomski, A. (1996). Stochastic approach to the evolution of some polycrystalline (bio) polymeric complex systems. *Chemical Physics Letters, 258*(1–2), 6–12.

Gadomski, A. (2003). Multilineal random patterns evolving subdiffusively in square lattice. *Fractals, 11*(supp01), 233–241.

Gadomski, A., & Ausloos, M. (2006). Agglomeration/aggregation and chaotic behaviour in d-dimensional spatio-temporal matter rearrangements. Number-theoretic aspects. In *The logistic map and the route to Chaos* (pp. 275–294). Berlin/Heidelberg: Springer.

Grandell, J. (1997). *Mixed poisson processes* (Vol. 77). Boca Raton: CRC Press.

Huisman, R., & Mahieu, R. (2003). Regime jumps in electricity prices. *Energy Economics, 25*(5), 425–434.

Jones, J. A., & Post, D. A. (2004). Seasonal and successional streamflow response to forest cutting and regrowth in the northwest and eastern United States. *Water Resources Research, 40*(5), 052031–0520319.

Kiseleva, T., & Wagener, F. O. (2010). Bifurcations of optimal vector fields in the shallow lake model. *Journal of Economic Dynamics and Control, 34*(5), 825–843.

Loukas, A., Vasiliades, L., & Dalezios, N. R. (2002). Potential climate change impacts on flood producing mechanisms in southern British Columbia, Canada using the CGCMA1 simulation results. *Journal of Hydrology, 259*(1–4), 163–188.

Łuczka, J., Hänggi, P., & Gadomski, A. (1995). Diffusion of clusters with randomly growing masses. *Physical Review E, 51*(6), 5762.

Martin, E. (2019). Cover crops and water quality. *Environmental Modeling and Assessment, 24*(6), 605–623.

Mitra, T., & Roy, S. (2007). On the possibility of extinction in a class of Markov processes in economics. *Journal of Mathematical Economics, 43*, 842–854.

Norris, J. R. (1998). *Markov Chains* (No. 2). Cambridge: Cambridge University Press.

Perko, L. (2013). *Differential equations and dynamical systems* (Vol. 7). New York: Springer.

Tamerius, J. D., Wise, E. K., Uejio, C. K., McCoy, A. L., & Comrie, A. C. (2006). Climate and human health: Synthesizing environmental complexity and uncertainty. *Stochastic Environmental Research and Risk Assessment, 21*(5), 601–613.

Vandewalle, N., & Ausloos, M. (1996a). The screening of species in a Darwinistic tree-like model of evolution. *Physica D: Nonlinear Phenomena, 90*(3), 262–270.

Vandewalle, N., & Ausloos, M. (1996b). Growth of Cayley and diluted Cayley trees with two kinds of entities. *Journal of Physics A: Mathematical and General, 29*(22), 7089.

Vandewalle, N., & Ausloos, M. (1997). Construction and properties of fractal trees with tunable dimension: The interplay of geometry and physics. *Physical Review E, 55*(1), 94.

Wagener, F. (2013). Shallow lake economics run deep: Nonlinear aspects of an economic-ecological interest conflict. *Computational Management Science, 10*(4), 423–450.

Wang, H., Yang, Z., Saito, Y., Liu, J. P., & Sun, X. (2006). Interannual and seasonal variation of the Huanghe (Yellow River) water discharge over the past 50 years: Connections to impacts from ENSO events and dams. *Global and Planetary Change, 50*(3–4), 212–225.

Yu, H. L., Yang, S. J., Yen, H. J., & Christakos, G. (2011). A spatio-temporal climate-based model of early dengue fever warning in southern Taiwan. *Stochastic Environmental Research and Risk Assessment, 25*(4), 485–494.

Zhang, Q., Xu, C. Y., Tao, H., Jiang, T., & Chen, Y. D. (2010). Climate changes and their impacts on water resources in the arid regions: A case study of the Tarim River basin, China. *Stochastic Environmental Research and Risk Assessment, 24*(3), 349–358.

Chapter 16
Water in Livestock – Biological Role and Global Perspective on Water Demand and Supply Chains

Maria Siwek, Anna Slawinska, and Aleksandra Dunislawska

Abstract This chapter describes issues related to the use of water in agriculture, especially in livestock production, which is the major water consumer in the world. Water is found in every cell of the body, as well as in the intercellular spaces. There is a difference between biological (bulk) water and cellular (nonbulk) water. Biological water is any water surrounding biomolecule that has distinct properties compared to those in the water mass. Cellular water is an ordered molecular structure that is surrounded by the molecules mediating its transfer inside the cells. The amount of water in the individual body depends on the species, gender, age, and body structure. At cellular level, water mediates and modulates intermolecular forces, controls the rate of substrate diffusion and conformational changes. In animal physiology, water influences all bodily functions, such as thermoregulation, fluid balance, and salt concentration. The level of water consumption in livestock varies between animal species. As much as 99% of the water footprint in animal production comes from feed that animals consume, rather than water that they drink. Therefore, water productivity in livestock farming depends to a large extent on selection of the diets and fodder production. There is a vast difference between use of water by different species to produce the same amount of meat or other products (e.g., eggs or milk). Among meat products, beef has the largest water footprint compared to meat obtained from other farm animals. It is related to the high feed efficiency of the beef cattle. Monogastric animals (e.g., pigs and poultry) farmed in intensive, industrial conditions, have lower total water footprint. Since agriculture is the most water-consuming branch of the human activity, the global supply chains are also taken into account when considering food security. The agriculture might be supplied with water in two ways: rainfed or irrigated. Over the last 50 years the global irrigated area has doubled. Irrigated water competes with water intended for human consumption, as well as producing crops and pulses for human feed. Water

M. Siwek (✉) · A. Slawinska (✉) · A. Dunislawska (✉)
Department of Animal Biotechnology and Genetics, Faculty of Animal Breeding and Biology,
UTP University of Science and Technology, Bydgoszcz, Poland
e-mail: siwek@utp.edu.pl; slawinska@utp.edu.pl; Aleksandra.Dunislawska@utp.edu.pl

© The Author(s), under exclusive license to Springer Nature
Switzerland AG 2021
A. Gadomski (ed.), *Water in Biomechanical and Related Systems*,
Biologically-Inspired Systems 17, https://doi.org/10.1007/978-3-030-67227-0_16

is the primary medium which will be affected by the climate change. To conserve the limited resource of freshwater, the agriculture and livestock farming in particular, should be taken into scrutiny.

Keywords Water demand-supply chains · Scales of water distribution · Water in livestock · Water transport through biomembranes · Perspectives on water demand and supply

1   Introduction

Water plays many different roles in living cells and can be considered the matrix of life (Ball 2017). Water is an integral part of biomolecules and contributes to the structure of nucleic acids and proteins. It is necessary to maintain functionality and biological activity of the cell. The value of water in cell life and animal physiology results from both the structural and dynamic features of this complex liquid.

An average person uses 2 liters of water for daily drinking, however the water footprint included in the agricultural products consumed daily per person in EU requires 4815 liters on average (Vanham et al. 2013). Groundwater provides around 50% of drinking water and 43% of the agricultural irrigation. Even though livestock farming is not efficient in using limited natural resources, the global demand on meat increases faster than the actual human population. Based on the current estimates, the increase in human population is 2 billion people by 2050, from the current 7.7 billion people to 9.7 (UN 2019). At the same time, the amount of meat consumed per capita globally will increase by as much as 73% (FAO 2011b). The increase is mediated by the culture of eating meat, which is adopted by the developing countries from the developed ones. Such demand will be satisfied with growing number of intensive farming systems, which process large number of livestock.

In this chapter we present the overview on the relationship between water and farm animals, which are the sources of meat and other products for growing human population. We address this issue at different angles, including using water by individual cells, at animal, and global level. We also include the data on availability of the water for consumption and withdrawal for use in agriculture. The main motivation of this chapter is to increase the awareness and perspective of allocating one of the most important natural resources into production of meat and other products of animal origin.

2   Water Requirements in Living Cells

2.1   Water in Cell Biology

Water is found in every cell of the body, as well as in the intercellular spaces. The amount of water in the body is determined depending on the species, gender, age, or body structure. The water content in mammals decreases with age. In humans, the newborn contains about 75% water, while in the elderly it is reduced to about 50% (Watson et al. 1980). Every part of the body, from brain to skin, needs water for proper functioning. The skin contains 64% water. The brain and heart are made up of 73% water. Lungs and kidneys are about 80% water. Even the bones are aqueous and include 31% water (Mitchell et al. 1945). Body structure is also a significant factor in water proportion, because there is only about 10% water in adipose tissue, while as much as about 80% in the muscles (Grossblatt 2003). Most of the body's water is found in body fluids and blood vessels.

At cellular level, water mediates and modulates intermolecular forces, controls the rate of substrate diffusion and conformational changes. It also participates in the enzyme catalysis (Persson and Halle 2008). It is necessary to understand the differences between biological (bulk) water and cellular (nonbulk) water. The term "biological water" describes any water around a biomolecule that has properties that differ from those in the water mass (Jungwirth 2015). Biological water plays a major role in all life processes, from molecules and cells to the tissues and whole organisms. In general terms, water is recognized as a solvent that acts as a carrier for biological macromolecules. It participates in the movements of the molecules that depend on molecular interactions. But, water also plays an active role in the cell's life cycle. Practically all nucleic acids and proteins are inactive in the absence of water, while hydration determines their function, structural stability and flexibility (Zhong et al. 2011). On the other hand, the term "cellular water" describes a separate layer of nonbulk water that exists around a biomolecule (Persson and Halle 2008). Cellular water is an ordered molecular structure that is surrounded by molecules mediating its transfer inside the cells. The volume of cellular water is influenced by many factors, including cells' age and functionality. It is crucial for the body to reduce cell damage and strengthen cell membranes to retain the correct amount of water.

All cells are surrounded by a semi-permeable cell membrane that separates the internal environment of the cell from the extracellular conditions. It consists of two layers of phospholipids and proteins (Fig. 16.1). The cell membrane forms a fluid two-layer phospholipid molecule to which proteins are bound. Some of the proteins are loosely bound to the surface of the membrane (surface proteins), while others pierce the membrane (transmembrane proteins) or are firmly embedded in it (membrane proteins) (Alberts et al. 2002). The main functions of the membrane proteins include reaction to the specific chemical signals from other cells, and transport of substances across the membrane. Phospholipid molecules have the ability to move within layers, making the cell membrane flexible. A very important property of the

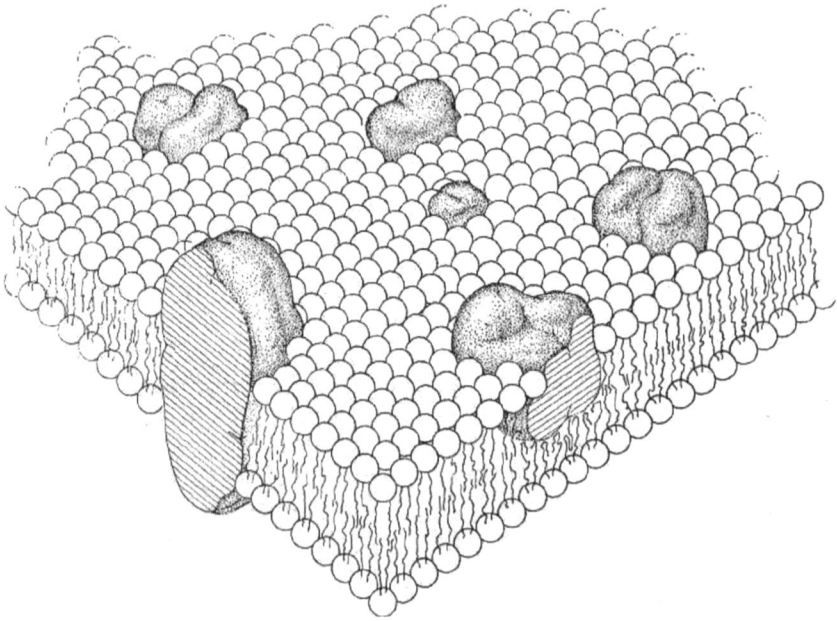

Fig. 16.1 Fluid-mosaic Singer-Nicholson model, showing a simplified scheme of the biological membrane; http://cnx.org/content/m15255/latest/

cell membrane is that it is selectively permeable. The substance is transported through the cytoplasm that fills the cell, between cell elements. Some scientific reports suggest that the cytoplasm is heterogeneous (Ball 2008). Others claim that the cytoplasm is like a gel that maintains cellular integrity (Andralojc 2002). Studies show that water inside the cells acts in significantly different way than the water outside the cell. Extracellular water is unchanged by biomolecules. In contrast, the viscosity of intracellular water is higher, hence the assumptions regarding the gel consistency of the internal environment of the cell may be correct (Andralojc 2002).

Most biological membranes are relatively impermeable to ions and other solutes, while to some extent, are permeable to water. This permeability is determined by the water channel proteins. This is due to osmotic pressure, which causes water to flow through the membranes (Lodish et al. 2000). Water has the ability to travel across the membrane (cell layer or plasma membrane) from a solution with a low concentration of solute to a solution with a high concentration, which is commonly referred to as osmotic flow. Placing cells in a solution in which the concentration of solute is lower than in the cytosol (hypotonic solution) will swell them. In this case, the osmotic flow can also lead to rupture as it occurs in erythrocytes (Goodhead and MacMillan 2017). This phenomenon is called hemolysis. In the case of red blood cells, this has several important practical implications - it is necessary to store them properly (in plasma solution with slightly hypertonic properties) and to properly administer intravenous drugs to a living organism - the drug must be suspended in physiological saline, which has slightly hypertonic properties. Reversely, immersion of the cells in a solution, where the concentration of solutes is higher than in the cytosol (hypertonic solution) leads to cell contraction. Hence, physiologically it is most appropriate to maintain conditions where the concentration of solutes between the cell's cytoplasm and the environment is the same (isotonic conditions).

When considering biological water, it is also necessary to approximate the water-protein interaction. Filling the spaces around and in protein is not the only function of water. A certain amount of water is necessary for the biological activity of proteins. The proteins are generally water-soluble, but some do not dissolve in water, including fibrillar proteins (present in the skin, hair, and tendons), and myosin (a protein found in muscles) (Sutton 2008). Proteins have the ability to bind water molecules. This effect is called hydration. The interaction between water and nucleic acids is much firmer than in proteins due to the strongly ionic nature of nucleic acids. Both, DNA sequence and structure, are highly dependent on interaction with water. This is due to the shrinking and expansion of the DNA helix depending on hydration (Makarov et al. 2002).

2.2 Water Requirement in Farm Animals

In animal physiology, water influences all bodily functions, such as thermoregulation, fluid balance and salt concentration. Water provides protection for the brain, spinal cord, eyeballs, or the fetus, and ensures proper joint mobility. The level of water consumption in livestock varies between animal species. On individual level, it depends on the animal weight, age, reproductive status, and production rate. The amount of the drinking water is strongly linked to the environmental factors, such as ambient temperature, humidity, and the type of feeding. The latter is particularly different between dry feed (10–14% water) and the grazed forages (60–80% water). The total water intake is a sum of the free water intake (water consumed directly by an animal) and water included in-feed (Parker and Brown 2003). The daily requirements of the water intake in livestock has been reviewed elsewhere (Parker and Brown 2003; Ward and McKague 2007).

One of the key aspects of the water consumption by livestock, aside from freshwater availability, is also its quality. Drinking water with the limited amount of toxic agents (e.g., heavy metals, pesticides and herbicides, petroleum, and disinfection products) maintains health of the livestock and safety of the products of animal origin (i.e., lack of toxic residues in meat or milk)(Valente-Campos et al. 2014). Water quality influences also the amount of the water consumed by the livestock. Poultry is the most sensitive among farm animals to water contamination. The poor water quality results in the dehydration of the animals, loss of productivity, and ultimately – death (Ward and McKague 2007).

3 Water Demand for Livestock in a Global Perspective

3.1 Water Efficiency in Livestock – Basic Concepts

Water productivity in livestock farming depends to a large extent on selection of the diets and fodder production. As much as 99% of the water footprint in animal production comes from feed that animals consume, rather than water that they drink (Vellenga et al. 2018). Thus, there are two sides of the story – the drinking water directly consumed by the animals and the water used to produce crops for the animal feed. There is a differentiation between water consumption and water withdrawal. The first one uses freshwater that is removed from a water source and not returned (i.e., consumed). The vast majority of water (85%) is consumed by agriculture, including irrigation and livestock (Rodriguez et al. 2013). Water withdrawal is broken down into thermal electric power (39%), irrigation and livestock (41%), as well as mining (8%). The amount of the consumed water is predicted to raise by 85% by 2035 (Rodriguez et al. 2013). Taking into account that the livestock and irrigation of arable fields uses most of it, one should look into the actual demand of agriculture on freshwater.

The complexity of water use in agriculture is much broader than the water physically included in a product at the moment of its consumption. For example, water content (moisture) in a burger is about 55–60% (the leaner meat the more water). This amounts to 55–60 g of water in 100 g of beef. However, to produce 100 g of beef, one uses 1541.5 l of water on average. The total amount of water needed to produce agricultural commodities is called a virtual water (Allan 2003). It includes water used in all steps of the production chain – for drinking, producing feed, and cleaning the facilities. The virtual water calculated for a given commodity produced in different countries varies, because the estimation accommodates also the local geographical conditions. The concept of the virtual water was devised by John Anthony Allan in 1993, and it linked the water used by agriculture to the climate change, trade, and politics. It was meant to solve the problem of the water scarcity in the countries of Middle East. Trading the virtual water by the countries that were facing water deficit was based on importing commodities, that required large amount of water to produce. In this manner, those countries did not only import commodities, but also the supplies of the virtual water used to produce them.

The concept of the virtual water leads to development of another water-related term: the water footprint. Even though, water footprint is not as widespread as carbon footprint, it became a part of the growing environmental awareness. As of 2020, the occurrence of the climatic changes resulting in severe weather events, including extraordinary heat, floods, droughts, hurricanes, and wildfires, smashed global records. Water stocks became a viable investment strategy, especially in countries, where the water industry is not regulated. Water is a limited resource; even though it covers over 70% of the Earth surfaces, only 2.5% of the total water volume is freshwater, and less than 1% is usable and available for ecosystems and humans (IAEA 2011).

3.2   Water Footprint of Livestock

By definition, the water footprint is a total volume of water required to produce any goods or service. It includes both, the internal and external water used at all steps of the production chain (Chapagain and Hoekstra 2008). The water footprint is broken down into three components: the blue water, the green water, and the grey water. The blue water includes the volume of the freshwater that is consumed from the surface or groundwater; the green water includes the volume of the rainfall stored in soil; and the grey water is the volume of clean freshwater used to dilute water polluted by the process (Chapagain and Hoekstra 2008). In a global perspective, water footprint of humanity amounts to 9087 Gm³/year on average (74% green, 11% blue, 15% grey). Amongst the world water footprint, as much as 92% is attributed to the agriculture (Hoekstra and Mekonnen 2012). Mekonnen and Hoekstra (2012) reported a comprehensive water footprint assessment of farm animals, taking into account blue, green, and grey water footprints and their sources. This report shows that the water footprint of any animal product surpasses the water footprint of plan-based diets with the same nutritional value. Therefore, it is by far more water-efficient for humans to eat plants directly, instead of feeding them to the animals. In relation to saving freshwater, diet is the largest contributor, with vegetarian and vegan diets using the least water (Vanham et al. 2013).

When it comes to the livestock farming, there is a vast difference between efficiency in the use of water by different species to produce the same amount of meat or other products (e.g., eggs or milk) (Fig. 16.2). All animal products have larger

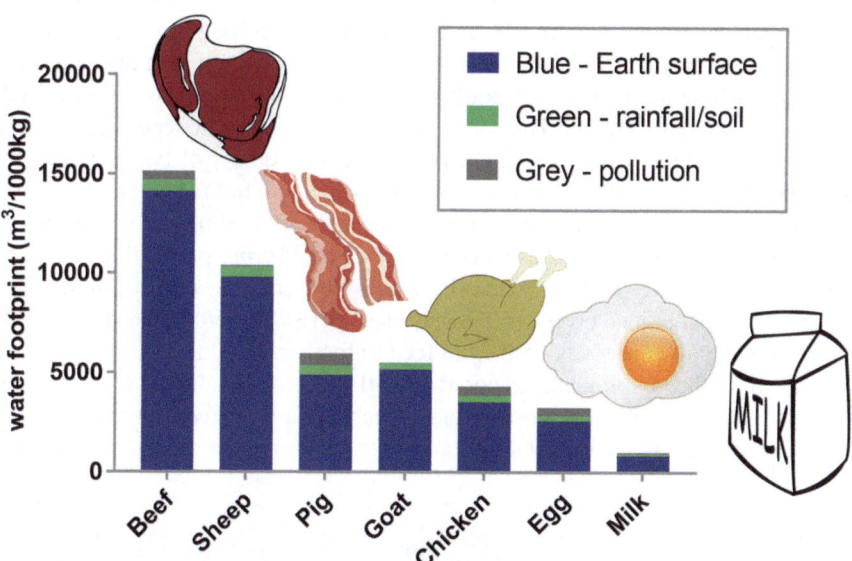

Fig. 16.2 Water footprint of different types of meat and other animal-derived products, divided into blue, green, and grey. (Adapted from Mekonnen and Hoekstra 2012)

water footprint per unit of meat or protein than crops or pulses (Mekonnen and Hoekstra 2012). Among meat products, beef has the largest water footprint compared to the meat obtained from other farm animals. It is related to the high feed efficiency of the beef cattle. Feed efficiency, also known as feed conversion ratio, is defined as amount of feed (dry matter) needed to produce one unit of meat. In intensive (industrial) farming, the feed efficiency amounts to 2.6 kg in chicken, 6.5 kg in pig and 7 kg in beef cattle (Trostle 2010). However, those numbers increase when only edible meat is taken into account as the output. In this case, feed efficiency equals to 4.5 kg for chicken, 7.2 for pig and as much as 20 kg for beef cattle (Smil 2001).

With the growing demand on meat, the livestock farming switched from extensive strategies to the intensive ones (including mixed and industrial farming). The intensive farming allows for better allocation of the resources with improved feed efficiency, reduced land use, and increased reproduction rate (Naylor et al. 2005). One of the sustainable strategies to maintain livestock production under water scarcity pressure is a zero-grazing system implemented by the farmers in Uganda (FAO 2017). Zero-grazing strategy assumes that the cattle, instead of pasture foraging, is kept in the barns and fed drought-tolerant fodder. In this manner, the animals use less arable land, which is a limited resource in face of the global warming. They also spend less energy seeking for food, and are less exposed to infectious diseases. This method also protects the pastures from overgrazing. However, zero-grazing system is justified only in case of genetically superior, high-yield animals.

The growth of the world meat market is based mostly on the intensively farmed monogastric animals, especially poultry and pigs. These animals have much lower lifespan than the cattle; thus a strong genetic selection can be applied to improve growth and feed efficiency traits. With respect to the water footprint, high feed efficiency is obviously beneficial, because it reduces the amount of feed that is needed to produce edible meat. However, poultry and pigs are mostly fed concentrate feed, containing cereal and pulses. The water footprint on concentrate feed is larger than on roughages fed to cattle and small ruminants. Production of concentrate feed competes for land and water with human food. Allocating limited natural resources to feeding farm animals is controversial in the light of feeding the world human population. It is estimated that as much as 36% of world crops production is used as animal feed, and only 12% of its caloric value is consumed by humans in the form of meat and other animal products (Cassidy et al. 2013). Therefore, the allocation of the crops and pulses needs to shift from feed to food, so that the food security will be assured for the growing human population. Based on estimates made by Cassidy et al. (2013), redirecting all calories from concentrate animal feed to human food is sufficient for feeding additional 4 billion people, assuming the same cropland area.

3.3   Mitigation and Adaptation Measures to Water and Heat Stress

The global warming changes the current perspective on freshwater, based on which as much as 56–73% of world area will be under severe water-stress conditions by 2050 (Alcamo et al. 2007; IPCC 2019). Such scenario requires development of the long-term strategies for future livestock production, taking into account three aspects of the climate change, including: increase of temperature, increase of CO_2, and variation of precipitation. Alcamo et al. (2007) predicts that among those three aspects, the increase of the temperature will be the most severe one. Heat is supposed to increase water consumption (by 2–3x), decrease nutrients availability (by lowering the amount of the high-quality feed), decrease feed intake and feed efficiency, reduce reproduction of the farm animals (associated also with decreased eggs and milk production), and influence health and mortality. There are several mitigation and adaptation measures that can be undertaken to adjust livestock farming to the water scarcity (Fig. 16.3). The mitigation strategies are mostly aimed to reduce emissions of the greenhouse gasses associated with farming by decreasing enteric fermentation (by proper nutrition), managing manure storage, and improving grazing land management (Alcamo et al. 2007).

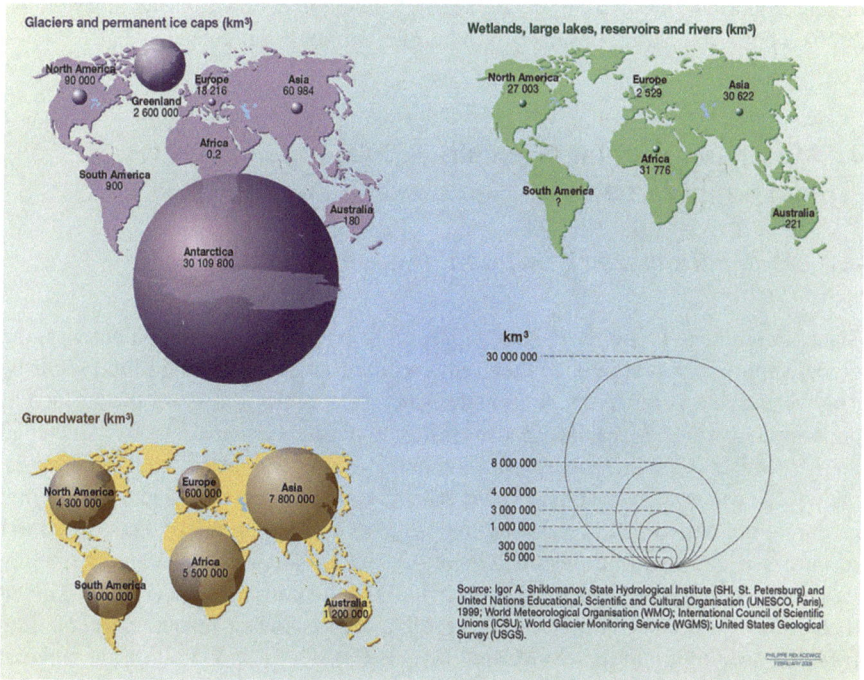

Fig. 16.3 Global freshwater resources – volume by continent (designed by Philippe Rekacewicz, February 2006), https://www.grida.no/resources/5608

The proposed adaptation measures include husbandry practices and genetic selection towards draught-tolerant livestock and feedstuff. According to Scasta et al. (2016), there are several issues to consider while matching the animal to the changing environment, pressured by water and heat stress. Genetic selection of the well-adapted animal of the future should follow the changing environments. For example, the preferred breeds of livestock are those that can handle the thermal regulation and heat stress better than the current highly productive crosses. Considering the most common livestock (i.e., cattle, pigs, and poultry), there are breeds and crossbreeds that are superior in handling heat stress and severe environmental conditions. Cattle is an example of the grazing animal. The selection strategy considers individuals with smaller size and hide color to improve thermoregulation and tolerance to heat (Scasta et al. 2016). One of the selection criterion in the heat-tolerant cattle is lower body temperature during heat that might reduce heat stress effects on production and reproduction parameters. However, such selection objectives will compromise animal productivity (Dikmen et al. 2012). In pigs, heat stress changes the feeding behavior towards lower feed intake, which helps decreasing body temperature through slower metabolism. Processing nutrients increases metabolic heat due to catabolic enzymatic reactions. Poultry has been first domesticated in hot regions of Southeastern Asia, and therefore there are still many native breeds that are tolerant to heat. Genomic studies allowed for pinpointing genes responsible for feeding behavior and growth traits under heat stress in chickens and pigs (Cross et al. 2018; Van Goor et al. 2015; Zhuang et al. 2020). They can be used to select for heat-tolerant poultry breeds.

4 Water Demand for Agriculture, Water Availability, and Supply Chains

4.1 Water Reserves in Different Regions

Since agriculture, is the most water-consuming branch of the human activity, the global supply chains should be taken into account when considering food security. The water cycle is relatively straight-forward. The rainfall flows on the surface or underground water. It reaches sea or returns to the atmosphere. The two universal paths of water cycle include evaporation and consumption by vegetation. The first path of the water cycle is considered a water resource to humans. The natural water resources are called a blue water by hydrologists and agronomists. The total water resources in the world are estimated for 43,750 km³/year. The water distribution varies depending on the geographical region and the climate zone (Fig. 16.3). Out of the four continents, Americas have the largest freshwater resources accounting for 45%, accounting of the world land (FAO 2003). This region has a wide range of climates and variety of hydrological regimes. The Northern America receives 17% of the precipitation. It generates 15.3% of the water resources, what gives 16,000 m³/

year per person. Based on the freshwater resources, the second continent is Asia with 28%, followed by Europe (15%) and Africa (9%) (FAO 2003).

Central America and Caribbean is geographically divided into islands characterized by a humid climate. This region represents 0.6% of the world land. The climate variation in this region generates strong differences in water resources over the seasons and consecutive years. As a region, Central America and Caribbean receives 1.4% of the world's precipitation and generates 1.8% of its water resources. The water resources per person equals in this region 11,900 m³/year. The Southern America represents 13% of the world land. As a geographical region, Southern America is characterized by wide range of climates and variety of hydrological regime. This region receives 26% of the world's precipitation and generates 28% of the water resources. The water resources per person amount to 35,000 m³/year.

Near East region represents 4.7% of the total world land. The general characteristic of the climates in this region is that they are dry. As it should be assumed, the water resources in this part of the world are the lowest. The water resources are less than 1000 m³/year per person in most of the countries in this region. Central Asia covers 3.5% of the total world land. The water resources in this region are rather high and amount to 3320 m³/year per person. Southern and Eastern Asia covers 15% of the total world land. This large geographical region is characterized by a high variety of climates and hydrological regimes. There are large interseasonal variations in the river flows due to monsoon climate. However, overall this region is well endowed with water resources.

The Western and Central Europe represents 3.7% of the world's land area. Due to the great differences in topography and distances from the sea, the distribution of the precipitation in the region varies significantly. The lowest precipitation is detected in the Mediterranean region and equals to 300 mm/year. The highest precipitation of more than 3000 mm/year is noted on the cost of Norwegian Sea and in Balkans area. The water resources in this geographical region are abundant at 2200 km³ in an average year what makes 4270 m³/year per person (FAO 2003). The Eastern Europe covers 13.5% of the world's land. The geographic conditions and well as the climate differs both between and within countries. The water resources reflect this variability with the lowest precipitation of 227 m³ detected in the Republic of Moldova to the highest precipitation equals to 29,000 m³ noted in the Russian Federation. The same relation is for the water resources per persons in this two extremes regions: from less than 1000 m³/year per person in the Republic of Moldova to 190,000 m³/year per person in the Siberian and Far Eastern regions of Russian Federation.

Africa is a large continent which represents 22.4% of the total world land. The water resources equals to 4797 m³/year per person. Africa is a large continent characterized by a great variety of different hydrological conditions. Depending on the geographic location the climate varies from: Mediterranean climate to desert (Northern Africa), arid to tropical climates (Sudano-Sahelian region), humid equatorial climate (Central Africa), from tropical humid or semi – arid (Eastern Africa), from subtropical humid to arid (Southern Africa). Oceania and Pacific covers 6% of the total world land. This region is mostly composed of islands. The water resources vary among countries and islands.

4.2   Water Intended for the Agriculture

Agriculture uses 12% (1.6 billion ha) of the world's land. The remaining part of the global land area is under the forest (28%) or used for grassland and woodland eco-systems (35%) (FAO 2011a). Renewable water resources equals 42,000 km^3 per year. However, more than 60% of all the water withdrawals comes back to local hydrological systems through rivers or groundwater. The remaining 40% is used through evaporation and plant transpiration. The yearly withdrawal of the water resources differs among sectors and continents (Table 16.1). Out of the three major sectors using the water resources, agriculture has the largest water withdrawal. Over the last 50 years the global irrigated area has doubled. Hence, the withdrawals for agriculture has been rising. Considering different geographical regions the lowest water withdrawal is in Europe (29%) and the highest in Africa (86%) (FAO 2011a). Even within the countries water resources are unevenly distributed. In some cases the water amount might be abundant yet the it expensive or not accessible to develop or even located in a distance from the agricultural region.

Agriculture being the sector using the highest amount of water is also the major contributor impacting the water quality (Table 16.1). The major source of water pol-lution are nutrients and pesticides derived from crop and livestock. A separate prob-lem is related to water salinization. This is mostly the case of the countries in Central Asia.

The agriculture might be supplied with water in two ways: rainfed or irrigated. Rainfed agriculture relies on rainfall for water. This type of agriculture is predomi-nant worldwide, and occurs mostly at small farms in poor regions. As much as 80% of the world cultivated areas are rainfed. This type of agriculture produces about 60% of the global crop output. However, less than 30% of rainfall is used by plants for biomass production. The rest of rainfall (70%) evaporates into atmosphere, goes into groundwater or contributes to the river runoff (FAO 2011a).

Table 16.1 Water withdrawal by major water use sectors

Continent	Municipal		Industrial		Agricultural		Total fresh water withdrawal
	km^3/year	%	km^3/year	%	km^3/year	%	km^3/year
Americas	126	16	280	35	385	49	790
Asia	217	9	227	9	2012	82	2451
Europe	61	16	204	55	109	29	374
Africa	21	10	9	4	184	86	215
Oceania	5	17	5	10	19	73	26
World	429	11	723	19	2710	70	3856

Adapted from FAO (2011a)

Table 16.2 Area equipped for irrigation

Continent	Equipped area (million ha)		As % of cultivated land		Of which groundwater irrigation (2006)	
	1961	2006	1961	2006	Area equipped (million ha)	As % of total irrigated land
Americas	22.6	48.9	6.7	12.4	21.6	44.1
Asia	95.6	211.8	19.6	39.1	80.6	38.0
Europe	12.3	22.7	3.6	7.7	7.3	32.4
Africa	7.4	13.6	4.4	5.4	2.5	18.5
Oceania	1.1	4.0	3.2	8.7	0.9	23.9
World	139.0	300.9	10.2	19.7	112.9	37.5

Adapted from The State of the World's Land and Water Resources for Food and Agriculture

Table 16.3 Annual long-term average renewable water resources and irrigation water withdrawal

Continent	Precipitation (mm)	Renewable water resources (km^3)	Water use efficiency ratio (%)	Irrigation water withdrawal (km^3)	Pressure on water resources due to irrigation (%)
Americas	1091	19,238	41	385	2
Asia	827	12,413	45	2012	16
Europe	540	6548	48	109	2
Africa	678	3931	48	184	5
Oceania	586	892	41	19	2
World	809	43,022	44	2710	6

Adapted from The State of the World's Land and Water Resources for Food and Agriculture

The second type of agriculture, the irrigated agriculture uses artificial watering of the land. This type of watering agriculture areas has expanded recently. On one hand, most irrigated farming systems are performing below their potential what makes the room of improvement for water productivity. On the other hand agriculture withdrawals of groundwater are intensifying causing depletion of aquifers. In many countries agriculture irrigation competes with the growing number of domestic and industrial water users. Table 16.2 presents the growing trend of areas equipped for irrigation.

Hence, the same picture is for water withdrawals for irrigation (Table 16.3). It almost doubled in the period from 1961 to 2006, from 1540km^3 to 2710 km^3. Most of the irrigated farms are located on the areas previously watered solely by the rainfall. However, part of the irrigation takes place on arid land and deserts. Out of the 219 Mha irrigated in the developing countries, 40 Mha are on arid land and deserts.

The water for irrigation comes from rivers, lakes and aquifers. Major part (62%) of the irrigated area is supplied from surface water, the remaining part (38%) from the groundwater. Wastewater or desalinated water are a minor source of irrigation water (FAO 2011a).

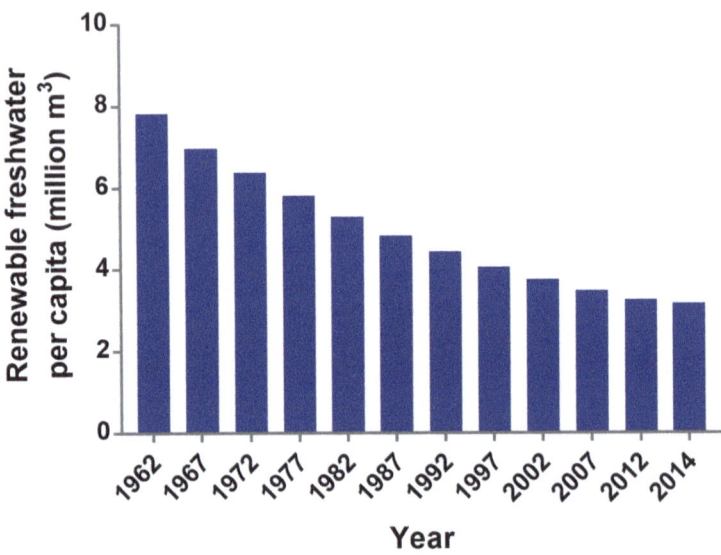

Fig. 16.4 Renewable internal freshwater resources per capita. (Adapted from OECD 2017)

4.3 Long-Term Perspective Against Climate Change

Water is the primary medium which will be affected by the climate change. The experts says that "Climate change is to large extend water change". Water availability is less predictable in many geographical regions. High temperatures and extreme weather conditions will have an impact on: rainfall, snowmelt, river flows and groundwater (OECD 2017).

On the global scale, due to the climate change, the world is facing: regional water scarcity, increasing sea level, and more frequent flooding events. The annual number of disasters caused by weather-related natural hazards doubled between 1908s and 2010. Droughts were particularly severe in southern Brazil, southwestern United States, northern Chile, southern Africa and Thailand in 2015. Number of storms increased by 200% in the East North Atlantic between December 2014 and March 2015. Extreme rain events in French Cevennes are three times more likely now comparing to 2014.

All of the above will have a severe impact on the water resources and availability of the internal freshwater resources per capita (Fig. 16.4). Water crisis has been identified as a single greatest impact on economies worldwide in the medium term. The highest risks have been attributed to: water availability for more than 40% of the population; risks of flood for 20% of population; groundwater depletion in many regions; decrease in the quality of surface water. The newest analysis predict that the water cycle will accelerate more frequent episodes of strong precipitation in high latitude. There is expected a significant reduction of surface water and increased seasonal variation of the precipitation in Southern Europe. More intense rainfalls during the winter season cause negative recharge of groundwater. Many countries, such as Australia or SEA, are expected to cope with rising sea level.

The availability of water resources in not the only problem in the near future. The other issue related to water is its quality. The main reasons of water quality deterioration are human activity and climate factors. The most severe causes of declining the water quality are: eutrophication and salinization. Eutrophication is related to nitrogen and phosphorus dischargement. It affects the most: Sub-Saharan Africa, India and Southeast Asia. Salinization is caused by raising sea level and saline intrusion in costal aquifers. The countries affected by this process are for example Japan, Mexico and The Netherlands, and many more.

From the perspective of the agriculture, all its activities will be affected by the climate change. Changes in precipitation pattern will influence rainfed agriculture. Irrigated agriculture will need to compete for water supply with other sectors. Water quality (due to for example salinization) will be an issue with regard to water availability. Extreme climate events such as droughts and floods will affect crop production and livestock production (OECD 2017).

References

Alberts, B., Johnson, A., Lewis, J., et al. (2002). Molecular biology of the cell. *Membrane proteins* (4th ed.). New York: Garland Science. Available from https://www.ncbi.nlm.nih.gov/books/NBK26878/

Alcamo, J., Flörke, M., & Märker, M. (2007). Future long-term changes in global water resources driven by socio-economic and climatic changes. *Hydrological Sciences Journal, 52*, 247–275.

Allan, J. A. (2003). Virtual water-the water, food, and trade nexus. Useful concept or misleading metaphor? *Water International, 28*, 106–113.

Andralojc, J. (2002). *Pollack, GH cells, gels and the engines of life. (A new, unifying approach to cell function)* (1st ed.). Oxford: Oxford University Press.

Ball, P. (2008). Water as an active constituent in cell biology. *Chemical Reviews, 108*, 74–108.

Ball, P. (2017). Water is an active matrix of life for cell and molecular biology. *Proceedings of the National Academy of Sciences, 114*, 13327–13335.

Cassidy, E. S., West, P. C., Gerber, J. S., & Foley, J. A. (2013). Redefining agricultural yields: From tonnes to people nourished per hectare. *Environmental Research Letters, 8*, 034015.

Chapagain, A. K., & Hoekstra, A. (2008). *Globalization of water: Sharing the planet's freshwater resources*. Oxford: Blackwell Publishing. 220p.

Cross, A. J., Keel, B. N., Brown-Brandl, T. M., Cassady, J. P., & Rohrer, G. A. (2018). Genome-wide association of changes in swine feeding behaviour due to heat stress. *Genetics Selection Evolution, 50*, 11.

Dikmen, S., Cole, J., Null, D., & Hansen, P. (2012). Heritability of rectal temperature and genetic correlations with production and reproduction traits in dairy cattle. *Journal of Dairy Science, 95*, 3401–3405.

FAO. (2003). *Review of world water resources by country*. Rome: Food and Agriculture Organization.

FAO. (2011a). *The state of the world's land and water resources for food and agriculture (SOLAW) – Managing systems at risk*. Rome/London: Food and Agriculture Organization of the United Nations/Earthscan.

FAO. (2011b). *World livestock 2011: Livestock in Food Security*. Rome: FAO.

330 M. Siwek et al.

FAO. (2017). *Improved cattle breeds zero grazing with drought tolerant fodder in Uganda*, http://www.fao.org/3/CA2565EN/ca2565en.pdf

Goodhead, L. K., & MacMillan, F. M. (2017). Measuring osmosis and hemolysis of red blood cells. *Advances in Physiology Education, 41*, 298–305.

Grossblatt, N. (2003). *Nutrient requirements of nonhuman primates*. Washington, DC: National Academies Press.

Hoekstra, A. Y., & Mekonnen, M. M. (2012). The water footprint of humanity. *Proceedings of the National Academy of Sciences, 109*, 3232–3237.

IAEA. (2011). *All about water*, IAEA Bulletin (p. 17).

IPCC. (2019). An IPCC special report on climate change, desertification, land degradation, sustainable land management, food security, and greenhouse gas fluxes in terrestrial ecosystems. In Shukla, P.R., Skea, J., Calvo Buendia, E., Masson-Delmotte, V., Pörtner, H.-O., Roberts, D.C., Zhai, P., Slade, R., Connors, S., van Diemen, R., M. Ferrat, R., Haughey, E., Luz, S., Neogi, S., Pathak, M., Petzold, J., Portugal Pereira, J., Vyas, P., Huntley, E., Kissick, K., Belkacemi, M., Malley, J. (Eds.).

Jungwirth, P. (2015). Biological water or rather water in biology? *Journal of Physical Chemistry Letters, 6*(13), 2449–51. https://doi.org/10.1021/acs.jpclett.5b01143. PMID: 26266717. ACS Publications.

Lodish, H., Berk, A., Zipursky, S. L., Matsudaira, P., Baltimore, D., & Darnell, J. (2000). *Osmosis, water channels, and the regulation of cell volume, molecular cell biology* (4th ed.). New York: WH Freeman.

Makarov, V., Pettitt, B. M., & Feig, M. (2002). Solvation and hydration of proteins and nucleic acids: A theoretical view of simulation and experiment. *Accounts of Chemical Research, 35*, 376–384.

Mekonnen, M. M., & Hoekstra, A. Y. (2012). A global assessment of the water footprint of farm animal products. *Ecosystems, 15*, 401–415.

Mitchell, H., Hamilton, T., Steggerda, F., & Bean, H. (1945). The chemical composition of the adult human body and its bearing on the biochemistry of growth. *The Journal of Biological Chemistry, 158*, 625–637.

Naylor, R., Steinfeld, H., Falcon, W., Galloway, J., Smil, V., Bradford, E., Alder, J., & Mooney, H. (2005). Losing the links between livestock and land. *Science, 310*, 1621–1622.

OECD. (2017). *Water risk hotspots for agriculture*. Paris: OECD Publishing.

Parker, D. B., & Brown, M. S. (2003). Water consumption for livestock and poultry production. In B. A. Stewart & T. A. Howell (Eds.), *Encyclopedia of water science* (1st ed.). New York: Marcel Dekker.

Persson, E., & Halle, B. (2008). Cell water dynamics on multiple time scales. *Proceedings of the National Academy of Sciences, 105*, 6266–6271.

Rodriguez, D. J., Delgado, A., DeLaquil, P., & Sohns, A. (2013). *Thirsty energy*. Washington, DC: World Bank.

Scasta, J. D., Lalman, D. L., & Henderson, L. (2016). Drought mitigation for grazing operations: Matching the animal to the environment. *Rangelands, 38*, 204–210.

Smil, V. (2001). *Feeding the world: A challenge for the twenty-first century*. Cambridge, MA: MIT press.

Sutton, R. (2008). *Chemistry for the life sciences*. Boca Raton: CRC Press.

Trostle, R. (2010). *Global agricultural supply and demand: Factors contributing to the recent increase in food commodity prices* (Rev. ed.). Diane Publishing.

UN. (2019). *World population prospects 2019: Highlights*. New York: United Nations Department for Economic and Social Affairs.

Valente-Campos, S., Nascimento, E. d. S., & Umbuzeiro, G. d. A. (2014). Water quality criteria for livestock watering: A comparison among different regulations. Acta Scientiarum. *Animal Sciences, 36*, 01–10.

Van Goor, A., Bolek, K. J., Ashwell, C. M., Persia, M. E., Rothschild, M. F., Schmidt, C. J., & Lamont, S. J. (2015). Identification of quantitative trait loci for body temperature, body weight, breast yield, and digestibility in an advanced intercross line of chickens under heat stress. *Genetics Selection Evolution, 47*, 96.

Vanham, D., Mekonnen, M., & Hoekstra, A. Y. (2013). The water footprint of the EU for different diets. *Ecological Indicators, 32*, 1–8.

Vellenga, L., Qualitz, G., & Drastig, K. (2018). Farm water productivity in conventional and organic farming: Case studies of cow-calf farming Systems in North Germany. *Water, 10*, 1294.

Ward, D., & McKague, K. (2007). *Water requirements of livestock*. Edmonton: Alberta Agriculture, Food and Rural Development.

Watson, P. E., Watson, I. D., & Batt, R. D. (1980). Total body water volumes for adult males and females estimated from simple anthropometric measurements. *The American Journal of Clinical Nutrition, 33*, 27–39.

Zhong, D., Pal, S. K., & Zewail, A. H. (2011). Biological water: A critique. *Chemical Physics Letters, 503*, 1–11.

Zhuang, Z.-X., Chen, S.-E., Chen, C.-F., Lin, E.-C., & Huang, S.-Y. (2020). Genomic regions and pathways associated with thermotolerance in layer-type strain Taiwan indigenous chickens. *Journal of Thermal Biology, 88*, 102486.

Index